Chemical Recovery in the Alkaline Pulping Processes

A Project of the Alkaline Pulping Committee of the Pulp Manufacture Division

Edited by Gerald Hough

TAPPI PRESS

International Standard Book Number 0-89852-046-0
Library of Congress Catalog Card Number 85–050956

TAPPI
Technology Park/Atlanta
P.O. Box 105113, Atlanta, GA 30348, USA

Printed in the United States of America

Preface

A monograph on Chemical Recovery in Alkaline Pulping was published by TAPPI in 1968. It was a valuable guide used extensively in the industry and filled a definite need. The impact of environmental control and the escalation of energy costs during the 1970s has brought about significant advances in chemical recovery. The Alkaline Pulping Committee recognized the need to update the monograph in the late 1970s and a committee was established to take on the assignment.

Joe Land of International Paper initially chaired the committee attended by Dave Wetherhorn, Art Boniface, Venki Venkatesh, Tom Grace and Jerry Hough. Joe Land later had to resign due to pressures of his new job and a new mill startup. Venki Venkatesh picked up the evaporator section after Dave Wetherhorn's death. The economic situation in the early 1980s put extreme pressure on people, in that reductions of staff increased responsibility and workload of others. Several volunteers who had offered to work on the document had to back out. We appreciate their intent and can sympathize with their situation.

Four major operations in chemical recovery have been developed as individual chapters. These are the evaporation and concentration of black liquor, the chemical recovery boiler, the preparation of cooking liquor or recausticizing department, and lime reburning. We have tried to keep this treatise on a practical vein and have not dealt with some of the rather exotic and experimental developments in the field. Some in the academic field may feel this was an oversight; however, our effort has been to make this a practical document which can serve as a reference and guide for the engineers and other technical people concerned primarily with operations.

This book on chemical recovery is the result of a lot of effort by a lot of people. Undoubtedly there are many people in the industry who could have contributed even more to this document. However, our problem was to find people who could and would take the time to develop a particular section. All of the major equipment suppliers have contributed by reviewing and editing their particular section.

We sincerely hope this volume will fill the need the Committee recognized, and be as useful as the monograph it replaces.

Gerald W. Hough

Acknowledgements

In addition to the authors who contributed so much of their time to the contents and worth of this book, the committee would also like to express appreciation to the many people who have been consulted for information and advice. We would also like to thank the secretaries and typists who have spent many thankless hours in preparation of the final documents which went to TAPPI.

We wish to acknowledge the reviewers who read and critiqued various chapters. To the best of our ability we have recorded the reviewers and wish to thank the following:

C. B. Almond	Goslin-Birmingham
P. A. Angevine	Dorr-Oliver Co.
R. J. Albert	Charles T. Main
J. T. Badyrka	Unitech
G. Dahlinger	Combustion Engineering
Jay Dockendorf	HPD, Inc.
J. Dickenson	Babcock and Wilcox Co.
T. M. Fosberg	R. C. Co.
W. J. Frederick	Oregon State University
John Karoly	HPD, Inc.
Tom Montgomery	Babcock and Wilcox Co.
Henry Moser	Babcock and Wilcox Co.
R. V. McCamish	MacMillan Bloedel
Wharton Nelson	Combustion Engineering
Wells Nutt	Union Camp
John Rauscher	Rosenblad Corp.
Axel Rosenblad	Rosenblad Corp.
K. V. Sarkanen	University of Washington

R. R. Shafer	Fuller Co.
D. Smith	Whiting Corp.
R. L. Springer	Retired (Champion Int'l)
Stan Suda	Babcock and Wilcox Co.

P. H. Glatfelter Co., Spring Grove, PA
Crestbrook Forest Industries Ltd., Cranbrook, BC

Omissions to these acknowledgements will be inevitable; however, we want each person who contributed in any way to the success of this book to accept the thanks of the subcommittee and of the Alkaline Pulping Committee.

We thank the suppliers for providing the majority of the illustrations used in the text:

Allis Chalmers Corp.
Babcock and Wilcox Co.
Combustion Engineering, Inc.
Combustion Engineering, Inc., Raymond Div.
Dorr-Oliver, Inc.
Fuller Company
The Ducon Company
Goslin Birmingham, Inc.
Lamb-Cargate Industries, Ltd.
Silver Eagle Manufacturing Co.
Rosenblad Corp.
HPD, Inc.
Unitech
Kamyr, Inc.

For the original Chemical Recovery Monograph Subcommittee.

Jerry Hough, Chairman
Joe Land
Tom Grace
Art Boniface
Venki Venkatesh
Dave Wetherhorn (deceased)

Contributors

Arthur Boniface, Regional Sales Manager, Dorr-Oliver, Inc., Atlanta, Georgia

Conrad F. Cornell, Marketing Manager, Goslin-Birmingham, Inc., Birmingham, Alabama

Thomas M. Grace, Professor and Senior Research Associate, Institute of Paper Chemistry, Appleton, Wisconsin

Robert P. Green, Senior Associate, Central Research, Mead Corporation, Chillicothe, Ohio

Gerald W. Hough, President, Hough Consulting Services, Portland, Oregon (formerly with Sandwell International)

Douglas J. Kramm, General Manager of Pyroprocessing, Fuller Company, Bethlehem, Pennsylvania (now retired)

Narendra K. Mehra, Senior Staff Engineer, Corporate Engineering, Champion International, Hamilton, Ohio (formerly with Dorr-Oliver, Inc., Stamford, Connecticut)

Xuan N. Nguyen, Research Specialist, Central Research, Mead Corporation, Chillicothe, Ohio (now with Champion International, Pensacola, Florida)

Wallace L. Schultz, Manager, Chemical Sales, Allis Chalmers Corporation, West Allis, Wisconsin (now retired)

Venki Venkatesh, Consulting Engineer, Facilities Analysis, Mead Corporation, Dayton, Ohio

Contents

List of Figures

Chapter 3

Chapter 6

List of Tables

Chapter 5

Chapter 6

1
Introduction

Arthur Boniface

Early developments

Soda pulping was pioneered by Hugh Burgess and Charles Watt during the mid-19th century in England. The industrial revolution had greatly increased the need for paper, which, in turn, had outstripped the supply of rags utilized in paper manufacturing. When the British government rejected the soda process, Burgess came to the United States, and, in 1866, opened the Manayunk Pulp Works in Manayunk, Pennsylvania. The production rate was 15 tons per day (*1*).

C. F. Dahl of Danzig, Germany, is generally credited with the development of the kraft pulping process. It was discovered that the addition of sodium sulfide to sodium hydroxide in a soda cook accelerated the delignification process, and that stronger pulps were achieved with the resulting shorter cooks. Kraft in both the German and the Swedish languages means strong. Dahl was issued a patent on kraft pulping in 1879. Because of the failure to complete the necessary financing, Dahl went to Sweden, and in 1891, the first kraft pulp and paper were manufactured at the Munksjo mill in Jonkoping, Sweden (*2*). Dahl substituted sodium sulfate for sodium carbonate to replace lost alkali, as sodium sulfate was readily available from England as a byproduct of their chemical industry (*3*). The use of sodium sulfate as makeup chemical gave this particular type of pulp the name sulfate pulp.

In addition to Dahl's work, A. K. Eaton of Brooklyn, N.Y., was issued a U.S. patent entitled "Improvement in the Manufacture of Paper-Pulp" in 1871 (*4*). The object of this invention was to reduce the cost of pulping by substituting sodium sulfide for caustic soda. The economic advantage of sodium sulfide over caustic soda was due to the availability and lower cost of sodium sulfate from which sodium sulfide could be formed.

The first kraft pulp in North America was produced at the Brompton Paper Company in East Angus, Quebec, in 1907. Shortly thereafter, in 1909, the first American kraft pulp was produced at the Roanoke Rapids Paper Manufacturing Company in North Carolina, and at Pensacola, Florida, in 1911. Production in 1909 at Roanoke Rapids was 12 tons per day. The mill is now a part of Champion International Company, and production at this mill is now 1100 tons per day.

Rapid growth of kraft pulping

Kraft pulping grew rapidly in the 1930s, due to the development of the modern recovery furnace, which gave kraft pulping an economic edge over sulfite pulping. The kraft process was adaptable to practically all types of wood, including pines, where sulfite is generally unsuitable. The availability of pine and its rapid growth in the Southeastern U.S. led to the extensive development of kraft pulp and papermaking in this area in the past few decades. The adaptability of the kraft process to other fiber sources, including sawdust, partially decayed wood, and sawmill waste products has spurred additional growth.

In 1965, the total kraft pulp production in North America was 26 million tons, sulfite was seven million tons, and soda less than one million tons. By 1980, kraft production was 48.2 million tons, a growth of 85%. During the same period, sulfite pulp production decreased from seven million tons per year to just over four million tons. For the U.S. only, kraft pulp production has increased from 22.5 million tons in 1965, to 38.6 million tons in 1980, for a growth rate of 71%. During the same period, total U.S. pulp production increased by 83%, from 31 million tons to 60 million tons (*5*).

Pollution control

Unfortunately, the sulfur component of the kraft pulping liquor contributes to the release of odorous compounds, total reduced sulfur (TRS) compounds, such as hydrogen sulfide, methyl mercaptan, and dimethyl mercaptan. These can be discharged from the digester relief and blow, pulp washer hood vent, evaporator noncondensible gas relief, recovery furnace, lime kiln stack, and tank vents.

Since the middle 1960s, the parameters connected with environmental control and energy have greatly influenced the chemical recovery plant design, equipment developments, and the method of operation of the various components of the recovery system. Efforts to comply with environmental regulations have resulted in several new developments in the field of chemical recovery.

The controlled odor recovery boiler, whereby the direct contact evaporator is eliminated and the proper black liquor solids are developed in the evaporator plant, is now the industry standard.

A highly efficient electrostatic precipitator with a collection efficiency above 99%, is now required to control particulate emissions from the recovery boiler.

All noncondensible gases from the digester and evaporator areas have to be incinerated to destroy the odorous TRS compounds. Also, in many modern mills, all black liquor tank vents are collected and incinerated. The burning of noncondensible gases is normally accomplished in the lime kiln.

A highly efficient scrubber is now required to control particulate emissions from the calciner. There are emission limits for TRS from the lime kiln, which require efficient washing of the lime mud, and close control of the calcining operations. Lime mud filters have been increased in size to facilitate the oxidation of certain soluble sulfide compounds and hence reduce the TRS emissions from the lime reburner.

The newest recovery furnaces are producing less dregs per ton of pulp than the older designs, but the dregs produced are more difficult to settle. As a result, the unit type green liquor clarifier with storage above has replaced the tray type clarifier. In many mills, dregs are no longer sewered, due to their tendency to upset primary clarifier overflows and increase the suspended solids in the effluent. In addition, inert material in the dregs restricts the burning of effluent sludge. Therefore, the filtration washing and thickening of dregs, rather than the sedimentation approach, is now common to allow the trucking of dregs to landfill.

Chemical recovery equipment, including recovery furnaces, calciners, and causticizing equipment are less frequently operated in an overloaded state. Instead, bottleneck areas are relieved with additional equipment in an effort to operate as efficiently and as economically as possible. Increased liquor and pulp storage is being provided to allow more flexibility between departments. Many causticizing plants now feature a general service clarifier, which may be utilized to hold or store green liquor, white liquor, lime mud, or dregs that are being emptied from the respective clarifier tanks. This inventory of material then is recycled, rather than being dumped to the sewer.

The recycle and reuse of various effluent streams gained momentum in the 1970s. Through the reuse of effluent streams, energy is conserved, effluent requirements are reduced, and fiber and chemicals are retained in the mill. Also, fresh water consumption is appreciably reduced. However, the reuse of white water filtrates or condensates is not without some penalty. Increased corrosion can occur, requiring the use of higher grades of stainless steel and other corrosion resistant materials. Maintenance costs can increase, and fouling of heat exchanger surfaces can occur. Also, there will be an increased tendency for foaming to occur and, therefore, the increased usage of defoamers.

Energy

The rapidly rising cost of energy in the 1970s spurred many equipment developments and changed many operating practices. These include:

- greater use of insulation
- increased recovery of low level heat
- extensive instrumentation and use of the computer to better control and manage the pulping and chemical recovery processes
- improved pulp washing to remove and recover more chemicals and dissolved solids
- installation of digester blow heat evaporators
- use of vapor recompression evaporation
- concentration of black liquor to higher solids prior to firing in the recovery furnace
- improved efficiency of the recovery boiler and elimination of the direct contact evaporator with installation of extended economizers to increase steam output
- use of larger lime mud filters to improve lime mud discharge cake dryness
- satellite coolers on the rotary lime kilns
- acceptance of the fluid bed calciner to better fuel economy
- a greater amount of recycling to reduce fresh water addition and to conserve energy

The pulp and paper industry has taken a serious look at energy requirements, and, as a result, energy productivity increased appreciably during the 1970s. The industry has increased the use of such energy sources as spent cooking liquor, hogged wood, bark, and hydroelectricity. According to the American Paper Institute, the industry has, for the years between 1972 and 1981, reduced fossil fuel and purchased energy by 30.7% per ton of product. This is equvalent to approximately 22 million barrels of oil per year (*6*).

Energy conservation projects continue to receive high priority in the early 1980s.

Process control

Instrumentation has become more reliable and more accurate during the past decade. Electronic instrumentation with electrical transmission of measurements and control signals have obsoleted pneumatic controls as used in the 1960s.

Computer process simulation is being used extensively to design and develop the material and energy balances for the chemical recovery processes. Dynamic simulation is being developed to assist in the design and, in turn, is being used for operator training. Programmable logic controllers are replacing conventional push button electrical controls for startup and shutdown of equipment.

Centralized control is now the standard in new mills and is being implemented in mill modernization and expansion projects. It is now accepted that the entire chemical recovery complex can be controlled from a single control area. Many mills have the power and recovery boilers and evaporators controlled from one control station, and the recausticizing and lime mud reburning controlled from a common control room.

The computer is being used extensively to assimilate, digest and display process information through the use of cathode ray tubes (CRT's). Dedicated computers are being used to control the individual processes in the chemical recovery complex. The trend is to transmit information on data highways to management computers. This will provide management with information and allow improved control of the various processes.

Future

New developments are continually being made in the pulping field to increase pulp yield, reduce chemicals, reduce and conserve energy, and modify the pulping and chemical recovery process to make them more energy efficient and to reduce environmental impacts. The use of oxygen in bleaching is now accepted in Sweden, and is gaining momentum in the U.S. The use of oxygen to continue the delignification process allows solids, which would normally be discharged to the sewer, to be returned to the recovery system. Thus additional solids are recovered for conversion to thermal energy in the recovery boiler. The caustic soda used in the oxidation process can upset the sodium balance in the cycle, and therefore oxidized white liquor is used as the source of caustic.

The chemical recovery system is a major reason for the high capitalization of the pulp and paper industry. During the past decade, the size of mills has continually increased in an effort to reduce fixed operating costs by increased volume of sales. Efforts are continually being made to improve the thermal efficiency of the chemical recovery plant.

Considerable research has been carried out to modify or completely replace the conventional kraft recovery system in an attempt to improve thermal efficiency and reduce the capital cost.

St. Regis experimented with a hydropyrolysis system in the 1970s as an alternate to the conventional Tomlinson kraft recovery boiler (7). In this process black liquor is heated under pressure in the presence of water and in the absence of oxygen. The organics are converted to a carbonaceous char and the sulfur present is reduced to sulfide. Improved energy recovery was claimed as well as the elimination of the potential smelt/water explosions. However, St. Regis dropped plans to develop a full scale demonstration unit in 1980 (*8*).

A new process called the "direct alkali recovery system" (DARS) is being piloted in Australia. The basic process was invented by Toyo Pulp Co. in Japan (*9*) and Australian Paper Manufacturers Ltd. has a licensing agreement to develop the technology. In this process the spent liquor from the soda based pulping process is evaporated and burned with ferric oxide in a fluidized bed reactor to form sodium ferrite, $Na_2Fe_2O_4$ (*10*). The product is dissolved in water and the sodium hydroxide and ferric oxide are then separated on a filter with the ferric oxide being reused.

A new kraft recovery process (NSP) was reported by the Swedish Forest Products Research Lab in 1980 (*11*). The NSP is based on four process steps: 1) pyrolysis of the black liquor, 2) gasification of "char" and formation of Na_2CO_3, 3) absorption of released H_2S, and 4) final combustion with steam generation. At this point in time, basic engineering and design criteria are being developed and experimental models are being studied.

The Zimpro wet oxidation process is a process for soda pulping chemical recovery, in which weak spent liquor from the pulp washing plant is introduced into the reactor under high pressure, 3000 psig, and air is used to oxidize the sodium to sodium carbonate (*12*). The sodium carbonate solution can be causticized to sodium hydroxide in the conventional causticizing plant. There is only one unit operating on soda chemical pulping recovery in Australia and one on semichemical pulping in the U.S. The high operating pressure and relatively sophisticated equipment appears to be limiting the application of this process in the pulp industry.

In the future, we will see a much broader use of instrumentation and computer control in an effort to maximize efficiency and reduce energy costs. There will also be a broader use of fuel sources for calcining, such as coal or wood. Design of new mills will be more conservative and equipment will be operated at high efficiency levels, with fewer people.

Literature cited

1. Clark, James d'A., *Pulp Technology/and Treatments for Paper.*
2. *Pulp & Paper Manufacture* "Vol. I—The Pulping of Wood". 1950 First Edition pp. 364, 365.
3. G. S. Witham, Sr., *Modern Pulp and Paper Making.*

4. U.S. Patent No. 119,224, *Improvement in the Manufacture of Paper—Pulp*, Asahel K. Eaton, Sept. 26, 1871.
5. 1982 *Post's Pulp and Paper Directory*, pp. 539-542.
6. *Patterns of Fuel and Energy Consumption in the U.S. Pulp and Paper Industry 1971-1981*, by Thomas J. Grant and Ronald J. Slinn, A.P.I., June 1982.
7. *Pulp & Paper*, September 1978, pg. 29.
8. *Pulp & Paper*, February 1980, pg. 29.
9. Nagano, R., Miyao, S., and Niimi, N.: *US Patent 4000 266 (Dec. 28, 1976).*
10. G. H. Covey, *Development of the Direct Alkali Recovery System and Potential Applications*, Int'l Conf. on Recovery of Pulping Chemicals, 1981.
11. H. Magnusson, and B. Warnqvist. *The NSP Project: An Alternative to the Conventional Recovery Furnace*, CEP-7 - Feb. 1980, 54-56.
12. Morgan, J.E., *Associated Pulp & Paper Mills Ltd.*, TAPPI Non-Wood Fiber Conference, Oct. 1973.

2
Principles of Chemical Recovery

Gerald W. Hough

Objectives of chemical recovery

The main objectives of the chemical recovery process are 1) to recover the chemicals used in the cooking of wood and other fibrous raw materials used for the production of pulp, and 2) to recover and beneficially use the thermal energy from combustion of the organics. The black liquor, containing the materials dissolved from the wood and the spent chemicals, is washed from the pulp, concentrated in the evaporator plant and burned in the recovery furnace. The recovery boiler converts the combustible materials extracted from the wood into useable steam energy and reduces sulfur compounds to sodium sulfide and sodium organic compounds to sodium carbonate; the recovered chemicals are then discharged from the furnace bottom as a molten smelt. The smelt is dissolved in a weak wash solution from the recausticizing department to become green liquor. The green liquor is reacted with lime in the recausticizing plant to form white liquor, the name for the fresh cooking liquor. Lime mud formed in the causticizing reaction is separated, washed, thickened, and converted to lime in the lime reburning plant.

There are losses in the cycle which require the addition of makeup chemicals. The makeup chemical can be salt cake (sodium sulfate) which is added to the black liquor at the recovery boiler; soda ash (sodium carbonate) which is added to the green liquor at the slaker; caustic soda (sodium hydroxide) which is added directly to the white liquor; or spent chemical solutions containing any of the above chemicals, as well as sodium sulfide.

The objective of the pulp washing department is to recover the maximum amount of black liquor with a minimum amount of fresh water dilution. Modern pulp washing facilities normally recover at least 98% of the chemicals applied in the digester. This is accomplished with a minimum of wash water such that only 2 to 3 tons of water per ton of pulp is added to the black liquor.

General characteristics of black liquor

Black liquor washed and extracted from the pulp generally contains 14% to 17% dissolved solids composed of about 1/3 inorganic chemicals which were in the white liquor added to the digester and 2/3 organic chemicals extracted from the wood. The black liquor has to be concentrated to above 60% solids before it will burn without supplemental fuel. As the solids are increased by evaporation of water in the evaporator plant the black liquor becomes more viscous and its boiling point rises. At 65% solids the specific gravity will be 1.34 and the boiling point elevation will be in the order of 15°C.

The calorific value of black liquor ranges from 5800 to 6600 Btu/lb of dry solids (13,490 to 15,350 kJ/kg). The burning of black liquor in a modern recovery boiler will produce between 12,000 and 14,000 pounds of steam per ton of pulp (5400-6400 kg/Mt). The entire pulping process is thus self sufficient in thermal energy and additional energy is required only to dry the pulp or paper.

Evaporation of black liquor

The evaporation or concentration of black liquor is normally carried out in multiple effect evaporators (MEE) using low pressure steam (50 psig) from the turbine extraction. In multiple effect evaporators, steam is used in the first effect and boiling vapor from each effect is condensed in the next effect with the vapor from the last effect condensed by a water condenser. The pressure decreases progressively to about 2.2 psi absolute (15 kPa) in the condenser.

The number of effects is limited by the viscous, fouling nature of the black liquor and the boiling point rise as the solids are increased. Kraft black liquor evaporators are generally limited to seven effects with a maximum expected economy of 5.75 pounds of water evaporated per pound of steam. Generally, the temperature is limited to about 280°F (138°C) in the first effect to minimize the fouling effect of the black liquor.

Prior to the 1970s it was standard practice to concentrate the black liquor to about 50% solids in the MEE and complete the concentration to above 60% in a direct contact evaporator (DCE) using the flue gases from the recovery boiler. As the cost of energy increased in the 1970s it became more economical to complete the concentration of the black liquor in the evaporator plant and use the energy in the flue gases to produce steam by extending the economizer section of the boiler. Stringent odor control requirements also encouraged the need to eliminate the DCE.

Evaporators raising solids concentration to 65% and higher have been characterized as concentrators. The industry trend is to operate the concentrator integral with the evaporator with liquor circulation being required above about 55% solids.

The increasing viscosity and elevation of the boiling point as the liquor solids increase, complicates the concentration of the liquor. The fouling nature of the black liquor requires the cleaning of evaporator surfaces with water or weak black liquor boils at periodic intervals. At extended intervals, acid cleaning or mechanical cleaning may be required.

The chemical recovery boiler

The chemical recovery boiler is a steam generator using black liquor as its fuel. The black liquor is burned at about 65% solids and the boiler has a recovered energy efficiency of about 65%. In recent years as the cost of energy has escalated, the operating pressure has increased to allow maximum electrical power generation by passing the steam through a turbine generator. Process steam used in the pulp and paper mill is extracted from the turbine at about 160 and 60 psig (1100 and 415 kPa). Modern recovery boilers are generating steam at 1250 psig/900°F (8620 kPa/490°C) and up.

The kraft recovery boiler is also a chemical reactor and as the organic materials are oxidized by combustion to produce energy, the sulfur compounds as well as the sodium sulfate are reduced to sodium sulfide. The smelt produced is comprised mainly of sodium carbonate and sodium sulfide. In a soda boiler, only combustion occurs and sodium carbonate smelt is produced.

Recovery boilers have become larger and larger in recent years with the largest to date started up at the ITT-Rayonier plant at Jessup, Georgia, in 1982. This unit is designed to burn 4.9 million pounds (2700 metric tons) of black liquor solids per day and to produce 810,000 lb/h (370,000 kg/h) of steam at 1325 psig/900°F (9740 kPa/490°C). The unit was reported to cost 110 million dollars.

Up until the early 1960s the recovery boiler was the major source of air pollution in the pulp mill. Due to its odorous gas emission, the kraft pulp industry suffered publicly. Air emission regulations decreed the control of both particulates and total reduced sulfides (TRS) and today the modern recovery boiler emits less than 5 parts per million of TRS, equivalent to 0.15 pounds per ton of pulp (0.075 kg/Mt), and 2.3 pounds of particulates per ton of pulp (1.2 kg/Mt).

Oxidation of the sulfides in the black liquor to thiosulfates prior to introduction into the DCE minimizes the stripping of sulfides from the black liquor by the CO_2 in the flue gas. Oxidation of black liquor thus became the accepted principle for TRS emission control from the recovery boiler until about the mid 1970s.

The accepted practice today is to concentrate black liquor to firing solids in the evaporator plant and use an extended economizer to maximize the thermal efficiency of the recovery boiler. Thus, the DCE is not required and the need for black liquor oxidation is eliminated.

The TRS emissions from the recovery furnace are controlled by proper control of the firing conditions and air distribution. The particulate emissions are controlled by control of the firing conditions and the use of high efficiency (99+%) electrostatic precipitators to remove particulates from the boiler flue gas.

Conversion of smelt to cooking liquor

Conversion of smelt to cooking liquor is accomplished in the recausticizing plant. Molten smelt from the bottom of the recovery furnace is dissolved in water to become green liquor, so called due to the greenish color from iron impurities. Insoluble elements called dregs are removed by settling, and the clarified green liquor is reacted with lime in the slaker. The reaction involves the conversion of sodium carbonate in the green liquor to sodium hydroxide and is called causticizing. The lime mud or calcium carbonate resulting from the reaction is separated from the liquor by settling or filtering and the clarified solution, called white liquor, is the cooking liquor. The white liquor solution has an amber color and is highly alkaline with a pH of about 13.

Lime mud separated from the white liquor is washed by dilution in settling tanks called clarifiers, then by displacement on filters. The lime mud is removed from the clarifier at 40% to 45% solids and thickened to about 70% solids before entering the calciner, which converts the lime mud back to lime.

The green liquor solution from the smelt dissolving tank contains sodium carbonate and sodium sulfide, with sodium hydroxide and minor amounts of sodium sulfate and miscellaneous elements extracted from the wood or introduced with the wash water. The total concentration of chemicals expressed as Na_2O is approximately 120 g/L (7.5 lb/cu ft). In the causticizing reaction the sodium carbonate reacts with lime to form sodium hydroxide and calcium carbonate. Sodium hydroxide and sodium sulfide are the active ingredients in the white liquor with a normal concentration approximately 100 g/L (6.2 lb/cu ft) as Na_2O. The causticizing reaction continues until about 85% complete, so the white liquor contains about 15% sodium carbonate as Na_2O. The white liquor solution is clarified to less than 100 parts per million of undissolved solids to eliminate the possibility of inert impurities contaminating the pulp.

The sodium sulfide concentration is established in the dissolving tank and remains unchanged through the causticizing reaction, unless makeup sodium sulfide is added.

Lime reburning

Lime reburning is performed in a rotary kiln or in a fluidized bed calciner. The lime mud or calcium carbonate is calcined to lime or calcium oxide by the removal of carbon dioxide (CO_2) with heat. Calcination of lime commences at about 1500°F (300°C). In a rotary kiln, the maximum temperature is about 2100°F (1150°C), whereas in a fluidized bed calciner the maximum temperature is about 1600°F (870°C). Heat of dissociation is 1385 Btu per pound of CaO (770 kcal/kg). The total energy requirement ranges from 7 to 12 million Btu per ton of lime (8000 to 14,000 mJ/Mt) depending upon the moisture content of the lime mud entering the kiln and the length of the kiln.

Before the early 1970s, relatively short kilns with length to diameter ratios below 20:1 were common, and no real effort was made to reduce the energy requirement. In the early 1970s the cost of fuel oil escalated by a factor of 10 in just a few years and ways to reduce the energy requirements for calcining the lime mud became necessary. A greater effort was then made to reduce the moisture content of the lime mud by installing and operating efficient lime mud filters. Long kilns with length to diameter ratios over 25:1 and hot lime coolers became common. These are devices through which the hot lime and secondary air pass countercurrently, the hot lime transferring part of its energy to preheat the incoming air. The most common type of cooler is tubes mounted around the hot end of the kiln, and the incoming air is heated by the hot lime.

The fluidized bed reactor is an energy efficient calciner and has gained acceptance and is now competitive with the standard rotary kiln.

Currently, the use of wood waste and coal as fuel for the lime kiln is being given serious consideration. Introduction of impurities into the chemical recovery cycle can result from the direct firing of wood or coal. This can be avoided by use of an external gasifier which produces low Btu fuel for combustion in the kiln.

Makeup lime is fresh lime which is fed directly to the slaker or crushed limestone which is calcined in the lime kiln. Although limestone is usually cheaper, most mills provide for fresh lime makeup in the event the lime kiln is out of service.

Until the 1970s the only emission control from the kiln was particulates which were recovered from the flue gas by water scrubbers. As the pulp mill was tightened up, it became evident that the kiln could be a major source of TRS emissions. Control of TRS emissions is generally achieved by adequate washing of the lime mud and proper firing of the lime kiln with an excess of air to oxidize residual TRS in the mud to sulfur dioxide (SO_2). Recently, several mills have installed electrostatic precipitators for particulate recovery from the exhaust gas in place of wet scrubbers.

3

Evaporation and Concentration of Black Liquor

V. Venkatesh
X. N. Nguyen

Introduction

Black liquor is the spent pulping liquor that is collected from brown stock washing after alkaline pulping. The designation of "black liquor," is based on its appearnace. Black liquor contains complex organic/inorganic components in an alkaline aqueous medium. The chemical composition of black liquor is a function of the pulpwood species and cooking conditions. The equilibrium properties of black liquor are dependent mainly on the concentration, inorganic content, and temperature. Variations also occur because of the different alkaline pulping proçess used besides kraft, such as soda-anthraquinone or soda-oxygen pulping.

Black liquor concentration is generally measured by drying a liquid sample under specified conditions and is expressed as percent solids. The term weak black liquor is used to describe the lower concentration (usually 14–20% dissolved solids) liquor from brown stock washing and strong black liquor refers to high concentration liquor (usually 50% or more dissolved solids). Typically, in evaporation plants black liquor is concentrated to about 50% solids in multiple effect evaporators. In older recovery boilers, this black liquor is subject to oxidation to stabilize reduced sulfur compounds for odor control prior to evaporation in direct contact evaporators using recovery boiler flue gases. In most modern boilers, direct contact evaporators have been replaced by concentrators in which the liquor is raised to 62%–68% solids. The current trend is towards the high solids concentration of black liquor to 70% solids or higher. Condensate segregation based on pollutant levels, followed by steam stripping integral with evaporator operations, is practiced in newer systems for cost effective effluent control.

The basic process requirements of black liquor evaporation are as follows:

1. Make efficient use of the energy required for evaporation, which typically consumes up to 25% of the total energy needs of a bleached grade kraft pulp mill.
2. Efficiently separate the water vapor formed during evaporation from the black liquor, which has a relatively high chemical and energy value. The vapors are condensed and used in the mill with or without treatment for odor and biochemical oxygen demand (BOD) control.
3. Ensure proper separation of tall oil soap (if present), which, besides its economic value, can also cause foaming and scaling problems.
4. Concentrate black liquor to solids levels (62% or more) acceptable for firing in the recovery furnace.

General properties of black liquor

Physical properties

The properties of black liquor that are of general interest relate to its processing characteristics or to heat transfer. These are listed below:

1. Density and specific gravity.
2. Relative volatility—boiling point elevation/vapor pressure.
3. Specific heat.
4. Viscosity.
5. Thermal conductivity.
6. Surface tension.
7. Chemical composition.

Density and specific gravity The density of spent liquor is customarily reported as specific gravity referred to the density of water at a reference temperature, commonly chosen to be 60°F (15.5°C). The specific gravity is measured with a hydrometer which is often calibrated in degrees Baumé. The Baumé gravity compares densities at 60°F (15.5°C).

Hydrometer readings taken at any other temperature should be corrected for temperature (based on a chart prepared for each liquor). The definition of the Baumé gravity scale for liquids heavier than water is as follows:

$$\text{Be}' = 145 - \frac{145}{\text{specific gravity}}$$

where, specific gravity is at 60°F (15.5°C)

The density characteristics of different spent liquors are very similar (*1*). The specific gravity concentration isotherms exhibit slight curvatures as shown in Figure 3.1. The deviation from a linear relationship becomes more and more pronounced as concentration increases. The higher the inorganic content, the steeper is the initial slope of these isotherms.

The dependence of density on temperature may be taken to be linear in the range from 60°F (15.5°C) to the boiling point. The temperature effect

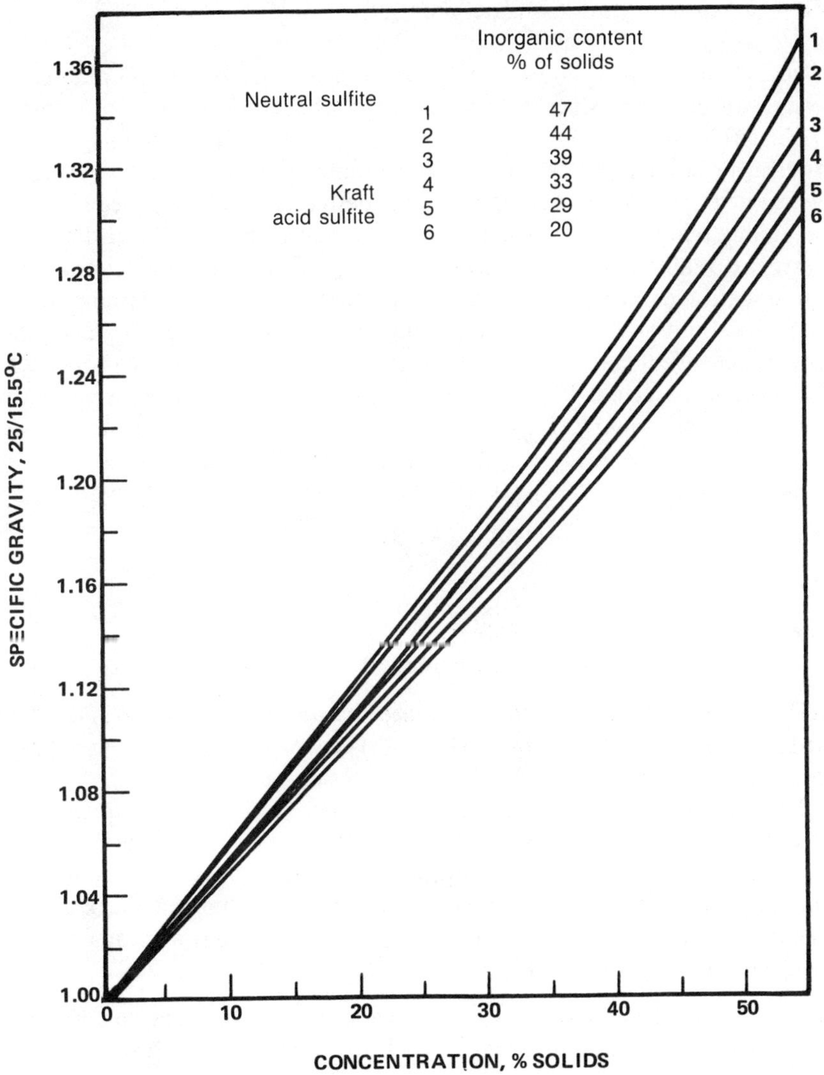

Fig. 3.1 Specific gravity of spent liquors.

increases slightly as concentration increases. Regestad(*2*) has presented an equation relating the density of black liquor to temperature and concentration.

$$\text{Density, kg/m}^3 = 1007 - 0.495(\text{T}) + 6.0(\text{S})$$

where,

T = temperature of liquor, °C
S = concentration (solids content), %

Relative volatility Black liquor vapor pressure is a significant parameter in evaporator operation since it determines pressure-temperature relationships. The volatility of black liquor may be measured in terms of vapor pressure lowering or boiling point elevation. The boiling point rise (BPR) is generally determined by direct measurement and is defined as the difference between the boiling temperature of black liquor and that of pure water at the same pressure (usually 1 atm., 101.4 kPa). Vapor pressure data, besides enabling determination of BPR, can provide important information pertaining to the solubility limits of certain black liquor components (*3, 4*) and the physical/chemical nature of black liquor such as thermodynamic properties, hygroscopic characteristics, etc. The Institute of Paper Chemistry is presently investigating alternative methods to develop accurate black liquor vapor pressure data particularly at solids levels above 50% (*3, 4*).

Hultin (*5*) has shown that the BPR can be expressed as follows:

$$\text{BPR} = \text{K (S)/(1} - \text{S)}$$

where,

BPR = boiling point rise, °C
S = black liquor solids concentration (as a weight fraction)

The constant, K, depends on the concentration of all dissolved species in the black liquor and according to Hultin (*5*) generally varies between 6.5 and 7.5.

BPR data for general use is commonly correlated against black liquor solids content as shown in Figure 3.2 (*1*). Observe that the BPR is smaller at lower inorganic content. Frederick, et al. (*3, 4*) have also shown that beyond the solubility limit of sodium carbonate and sodium sulfate in black liquor (usually about 50% solids), the organic components exert a greater role in BPR changes as shown in Figure 3.3 (*4*).

Specific heat The specific heat of black liquor may be determined calorimetrically. The data of Harvin and Brown (*6*) for a kraft black liquor

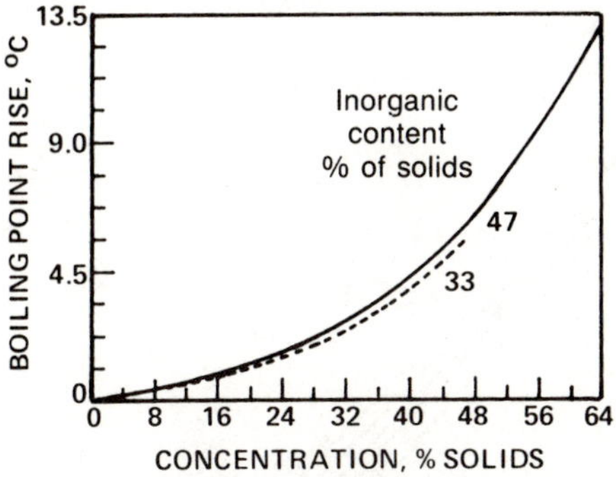

Fig. 3.2 Boiling point rise of spent liquors at atmospheric pressure.

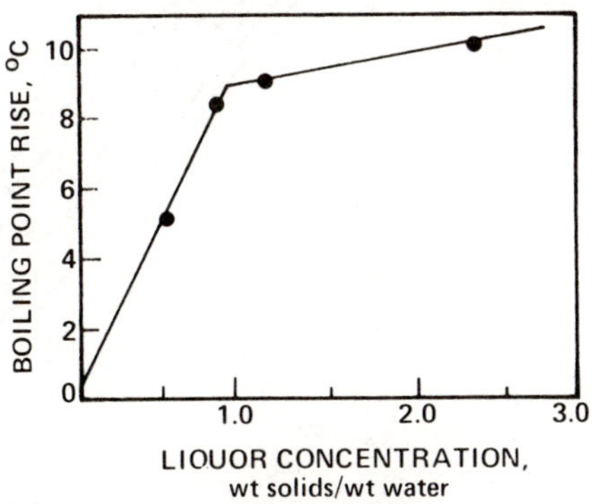

Fig. 3.3 Boiling point rise curve.

with an inorganic content at 29% of the solids are shown in Figure 3.4. Their results were also represented by the equation:

$$C_p = 1.246 + 14.4 \times 10^{-3}T - (0.34342 - 11.52 \times 10^{-4}T)S$$

where,

C_p = the specific heat of the liquor at T°C, Kcal/(Kg)(°C)
S = black liquor solids concentration (as a weight fraction)

The previous data of Kobe and Sorenson (7) were less precise. Their results averaged over the range 77° to 200°F (25° to 95°C) correlated as follows:

$$C_p = 0.98 - 0.52(S)$$

which gives lower values than Harvin's at the mean temperature of 150°F (65°C). The difference could be in the different inorganic contents of the two liquors. Regestad (2) has also proposed a similar correlation:

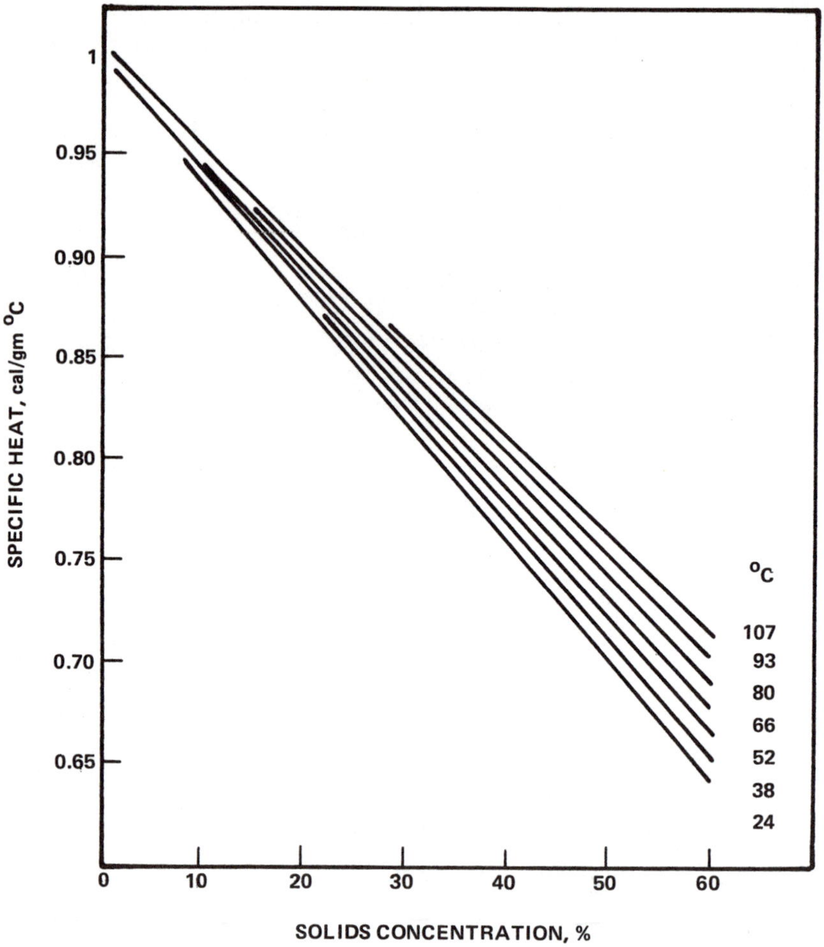

Fig. 3.4 Specific heat of a kraft black liquor.

$$C_p = (1.0 - 0.0054S)(.4187)$$

In the absence of actual data, the following approximation is often used:

$$C_p = 1.0 - (1 - C_{p,s})S$$

where $C_{p,s}$ is the specific heat of the black liquor solids. Its value is assumed to be in the range of 0.3–0.5. At $C_{p,s} = 0.5$, the estimated black liquor values correspond closely with Harvin's at 200°F (95°C). The reciprocal of specific gravity is sometimes used as specific heat in rough calculations.

Viscosity The viscosity of black liquor as a function of temperature and total solids concentration has been reported by several workers using Brookfield

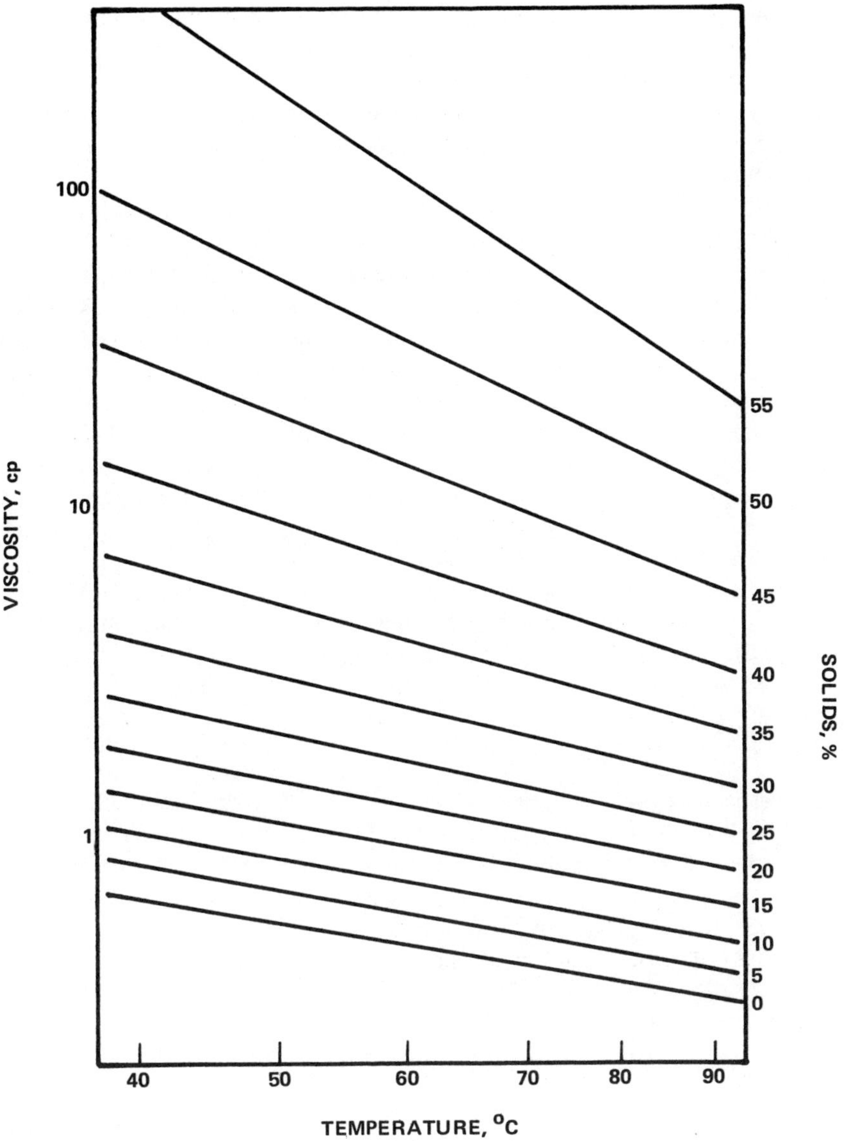

Fig. 3.5 Viscosity of a kraft black liquor.

type viscometers (*5, 8–12*). These data are in general agreement in spite of differences in liquor composition. Figure 3.5 is a reproduction of Harvin's (*10*) data. The original data were smoothed by the Othmer method (*13*), using water as the reference substance.

When the viscosity of a solution is plotted on logarithmic scales against that of water at the same temperature, a straight line results. From the smoothed data a viscosity-temperature diagram may be constructed with concentration as a parameter. The resulting plot usually shows slight curvatures at lower temperatures, which may be straightened by a modified temperature scale (Figure 3.5)

Published viscosity data are generally restricted to a total solids concentration limit of 50–55% and to temperatures below 212°F (100°C). Kraft black liquors are generally Newtonian up to solids level of 50%, which is the operating limit in most multiple effect evaporation systems. As black liquor concentration to over 65% solids is being increasingly practiced, higher solids viscosity data are now required for design and operational purposes. Concentrated liquors tend to be thixotropic, with the apparent viscosity decreasing with an increasing rate of shear. Several workers (*14–17*) have described the appropriate viscometry and published data on the rheological properties of black liquor at high solids (60 + %) concentration.

Frequently, the viscosity and other rheological characteristics of black liquor are altered by the addition of other process streams such as chlorine dioxide generator spent acid, tall oil plant brine, or semichemical spent liquors. As most of these streams are acidic, their addition results in reducing pH and precipitates lignin. The oxidation of black liquor is also known to change black liquor viscosity. Strong black liquor (50% solids or higher) oxidation practiced for odor control is reported to have caused a ten-fold increase in viscosity (*18*).

Thermal conductivity A complete set of data for the thermal conductivities of a sulfate black liquor were reported by Harvin (*10*). His results are shown in Figure 3.6 as interpolated from the original data. The slope of the conductivity-concentration isotherms varies linearly with temperature in accordance with the following correlation:

$$m = 8.233 \times 10^{-4} - 2.515 \times 10^{-6}\ T$$

where,

m = thermal conductivity, cal/(s) (cm^2) (°C/cm)

These linear isotherms terminate at the conductivities of water.

Surface tension Very little published information is available on the surface tension of kraft black liquor. Surface tension is important when considering black liquor drop formation during spraying or nucleate boiling and in characterizing surface active agents present in black liquor. Figure 3.7 gives surface tension data published by Beckwith, et al. (*24*).

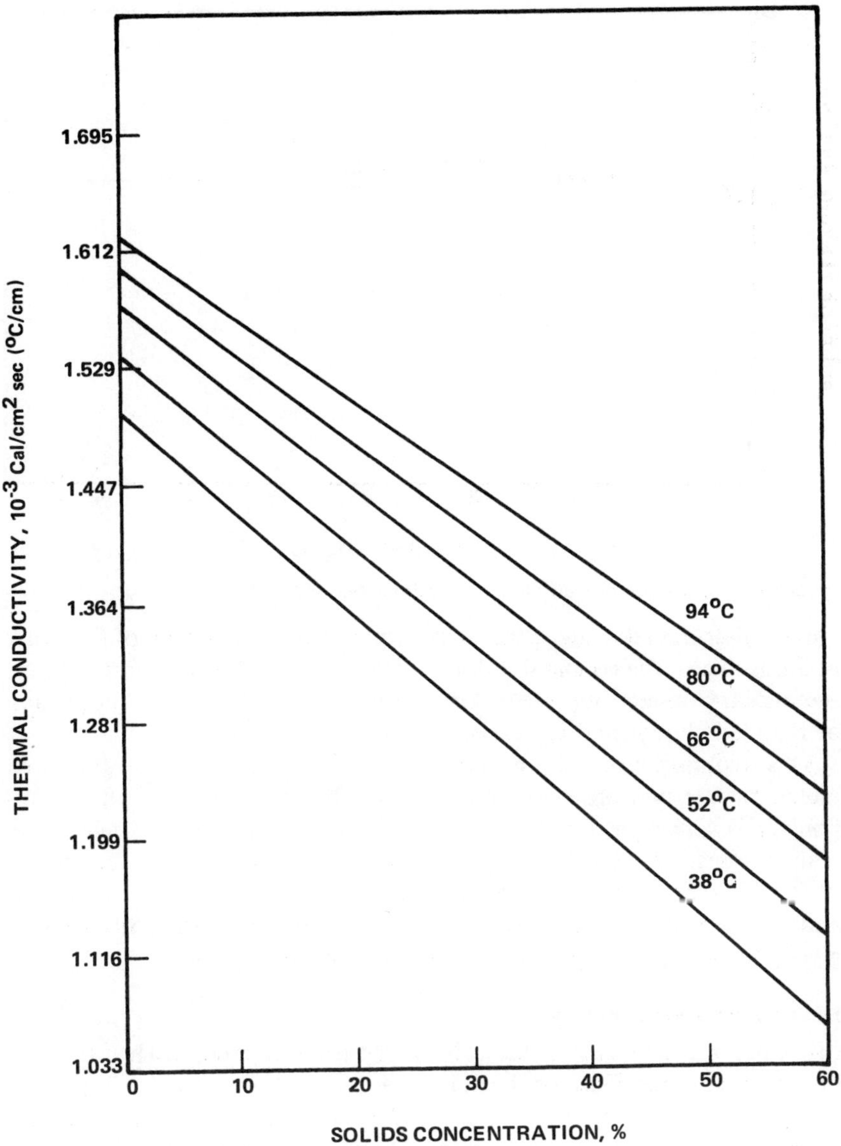

Fig. 3.6 Thermal conductivity of a kraft black liquor.

Composition of black liquor The chemical composition of black liquor is important in multiple effect evaporation when considering scaling potential, soap skimming, and loss of volatile compounds. To prevent the formation of organic scales, a residual alkali content of 1.1% (as Na_20) in the dry black

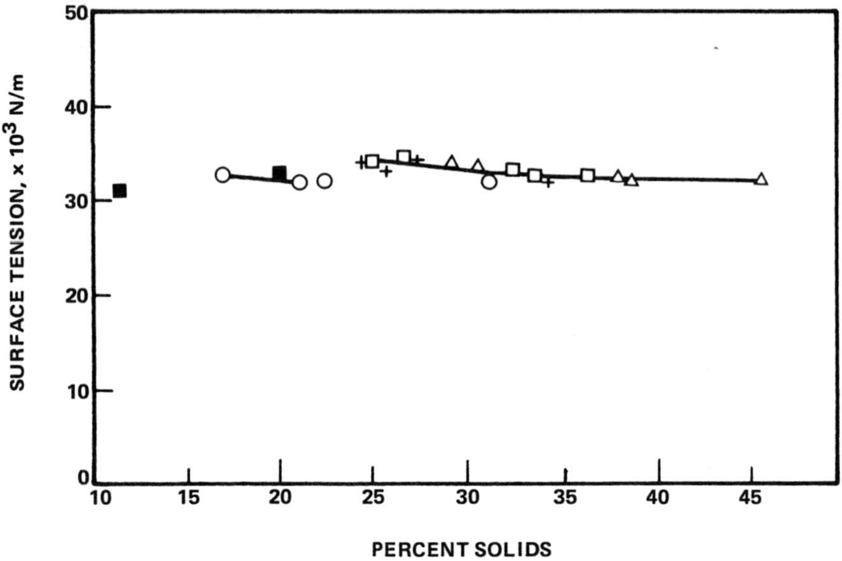

Fig. 3.7 Surface tension of kraft black liquors from several Southern mills.

liquor solids has been suggested. Also, when the solids content of the black liquor is raised, the colloidal stability of the fatty acid and resin acid soaps is reduced. Consequently, these components separate and rise to the surface of the black liquor and are easily skimmed.

This skimming is usually accomplished at 25–28% solids concentration in one of the intermediate evaporation stages. The heating value of these components is very high, but their removal is essential in view of their severe scaling effect. Removal of tall oil components is required only when processing softwood black liquors. Malodorous organic sulfur compounds and some organic acids are also released during evaporation. Table 3.1 (*25*) gives chemical analyses of the components of black liquor.

Heat transfer fundamentals

Black liquor evaporation is basically a heat transfer process. Heat transfer in a spent liquor evaporator is generally accomplished by condensing steam or vapor on one side of a tubular or flat plate heating surface. With saturated steam or vapor, the temperature is fixed by the pressure in the shell or vapor chamber of the heater or heat exchanger, and this temperature remains essentially constant over all of the heating surface in the effect or body. As the liquor absorbs heat from the condensing steam, its temperature rises until the boiling point is reached. Clearly black liquor evaporation involves sensible heat transfer to heat the flowing liquor to raise its temperature to the boiling point, followed by latent heat transfer at boiling to vapor. The

Table 3.1. Chemical components in black liquors (expressed as percentage of dry solids)

Wood	*Pine*	*Pine*	*Spruce*
Dry solids in sample, %	17	58	
Lignin	28.9	30.7	41
Hemicellulose and sugars	1.14	0.11	
Extractives	6.69	2.53	
Saccharinic acids			28
Acetic acid	3.52	2.08	5
Formic acid	4.48	2.70	3
Other organic acids	5.5	2.22	
Methanol			1
Unknown organic compounds	19.0	29.5	
Inorganic salts	18.6	18.5	
Organically combined Na	10.1	10.3	
Unknown inorganic compounds	2.08	1.35	
Sulphur, S			3
Sodium, Na			16
TOTAL	100.0	100.0	100

mechanism by which boiling takes place depends upon the particular hydraulics employed to move the liquor. As discussed later, three methods have been employed for moving liquor past heat transfer surfaces to bring it to boiling: rising liquor film, falling liquor film, and forced liquor circulation.

Several workers (*19, 20, 26, 27*) have proposed detailed mechanisms for heat transfer in tubular type black liquor evaporators. A detailed discussion of the theoretical aspects of heat transfer during black liquor evaporation is beyond the scope of this chapter. The flow of heat (i.e., the evaporation rate) from the heating media (steam or flash vapor) through a metal wall to the black liquor is controlled by three factors; namely, the area of the heating surface (A), the total working temperature drop (T_d), and the overall heat transfer coefficient (U).

The rate of heat transfer (Q), can then be expressed as:

$$Q = UAT_d$$

The heat transfer coefficient, U, is a measure of the overall resistance to the flow of heat from the heating medium to the liquor.

The above equation can then be expressed as follows:

$$Q = \frac{AT_d}{R}$$

where, R = overall resistance to the flow of heat.

The resistance encountered to the flow of heat comprises several individual resistances listed below.

R_C = resistance of the steam-condensate film
R_W = resistance of the metal wall (heating surface)
R_S = resistance due to any scale deposited on the metal surface
R_L = resistance of the liquor side film.

These resistances are shown schematically in Figure 3.8 and are combined to give the overall resistance to the flow of heat, R:

$$\text{thus } Q = \frac{AT_d}{R_C + R_W + R_S + R_L}$$

For clean heating surfaces, the resistance of the liquor side film, R_L, is usually the largest contributor. The condensate film resistance, R_C, is also significant. The resistance of the metal wall, R_W, is generally small and can be neglected. The liquor film resistance, R_L, varies greatly depending upon the system thermodynamics as well as the liquor flow regime. For a stagnant film, the resistance is high, but if the liquor flow is turbulent, the thermal resistance can be lowered by a factor of ten to one. Scale formation on the liquor side impedes heat flow and the rough scale surface also catches fibers and other residues in the black liquor. At high black liquor concentration levels (45% or more), the scale resistance (R_S) can exceed all the other individual resistance terms combined.

It is the inverse of the total of all these resistances that is normally expressed as the overall heat transfer coefficient, U. Table 3.2 gives typical U values for black liquor evaporators from various published sources. Falling film evaporators generally achieve higher U values.

The temperature difference (T_d) is the driving force causing heat flow in the evaporator. Theoretically, the total temperature drop would be the difference between the temperature of steam used in the first effect and condensing vapor in the condenser and is usually 125°–145°F (50°–60°C). The available temperature drop is, of course, considerably reduced by liquor characteristics, such as the boiling point rise (BPR). Falling film evaporators

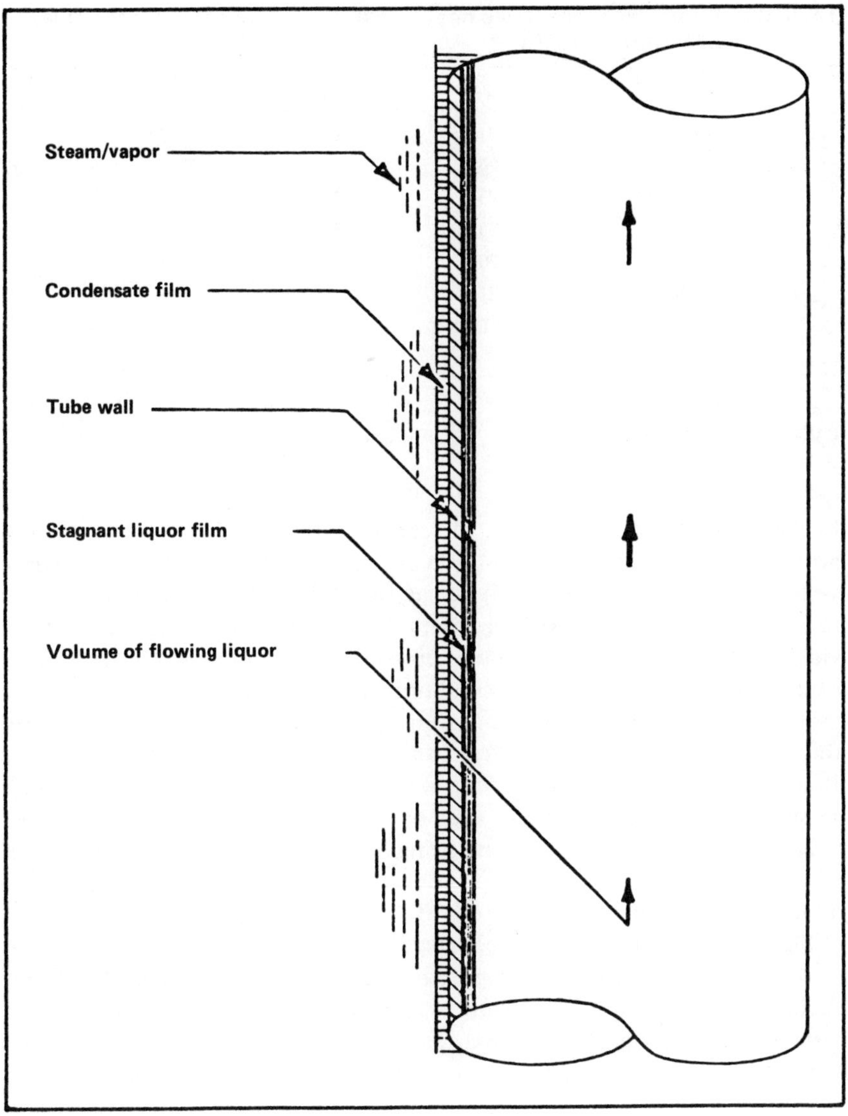

Fig. 3.8 Heat transfer resistances across an evaporator surface (courtesy of Goslin-Birmingham, Inc.).

can be operated at relatively lower temperature drops. For a given evaporator equipment and configuration, the temperature difference is often the only factor that can be manipulated to alter evaporation rates.

Table 3.2. Summary of available heat transfer coefficient (U) (data values in BTU/(hr) (ft^2) (°F)

Source	(21)	(22)	(23)
Effect 1A	214	96.2	250-300
1B	248		250-300
2	257	195	300-350
3	293	297	300-350
4	260	203	250-300
5	211	178	200-250
6	186	---	150-175
Concentrator	---	---	150-200

Evaporation equipment

Introduction

In the pulp and paper industry, the concentration of spent black liquor is normally carried out with a multiple effect evaporation system. In such a system, steam is supplied mostly to the first effect at the highest operating pressure, and the vapor evolved from a high-pressure effect is used to provide heat to the next effect at a lower pressure. The vapor from the last effect, usually operated under vacuum conditions, is evacuated and condensed. The principle of multiple effect evaporation allows the use of the latent heat of the evaporated water in one effect to be used again in a subsequent effect. In a well-operated sextuple effect evaporation plant, for example, one ton of steam evaporates about five tons of water. Where energy costs justify, vapor compression evaporators are being used either as pre-evaporators prior to the conventional multiple effects or to concentrate weak black liquor to 50% solids or higher.

Black liquor evaporators can be categorized according to the characteristics of the liquor movement or heating section design as shown in Table 3.3. These designs have evolved from attempts to improve evaporator performance in terms of steam usage, scaling, foaming, and operational simplicity. Equipment most often encountered in pulp mill evaporation plants are rising film type long tube vertical (LTV) evaporators, which generally deliver up to 50% solids. Falling film plate and tube type evaporators are also in use and final discharge solids of 65–70% are achieved. A variety of special evaporator equipment employing the forced circulation concept have been used to deliver black liquor solids up to 73%. Evaporator equipment used in place of direct contact evaporation in kraft recovery boilers for better odor control are commonly called concentrators, and are operated at solids levels above 50%.

Table 3.3. Summary of evaporator types

Film evaporators	Rising film
	  Long tube vertical (LTV) type
	Falling film
	  Long tube vertical type
	  Plate type
	  Horizontal tube
	  Spray film
	Combination
	Rising/falling film
	  PFR type (P-pre-heat, F-falling film, R-rising film)

Natural circulation evaporators	Short tube vertical type

Special evaporators using forced circulation	LTV with recirculation
	Forced circulation type
	  Integrally mounted vertical tube section in single pass
	  Externally mounted tube section in single pass or two passes:
	    horizontal tube
	    inclined tube
	    vertical tube
	Evaporator/crystallizer

Rising film long tube vertical (LTV) evaporator

The rising film long tube vertical (LTV) evaporator has been commonly used in most kraft pulp mill multiple effect evaporation systems. The number of effects in an evaporator train is normally established based on steam economy and capital cost considerations. Five to seven effect LTV systems are typical. The rising film LTV evaporator consists of a vapor head, a steam chest, and a main liquor box as shown in Figure 3.9. All vapor heads have a deflector directly over the heating element to break foam and impart an initial downward direction to the liquor. The vapor head is provided with a vapor outlet, liquor outlet, manhole, and peephole. It is generally fabricated of welded steel. However, the lower portion of the head is made of stainless steel in effects where high solid concentration and high temperature can be corrosive to mild steel.

The steam chest is fabricated of steel and the trend today is to use stainless steel tubes. Tubes are normally about two in. in diameter and 24 to 28 ft long. The heating element has a vapor inlet, condensate outlet, flash condensate inlet, vent, and gauge connections. The liquor from the main liquor

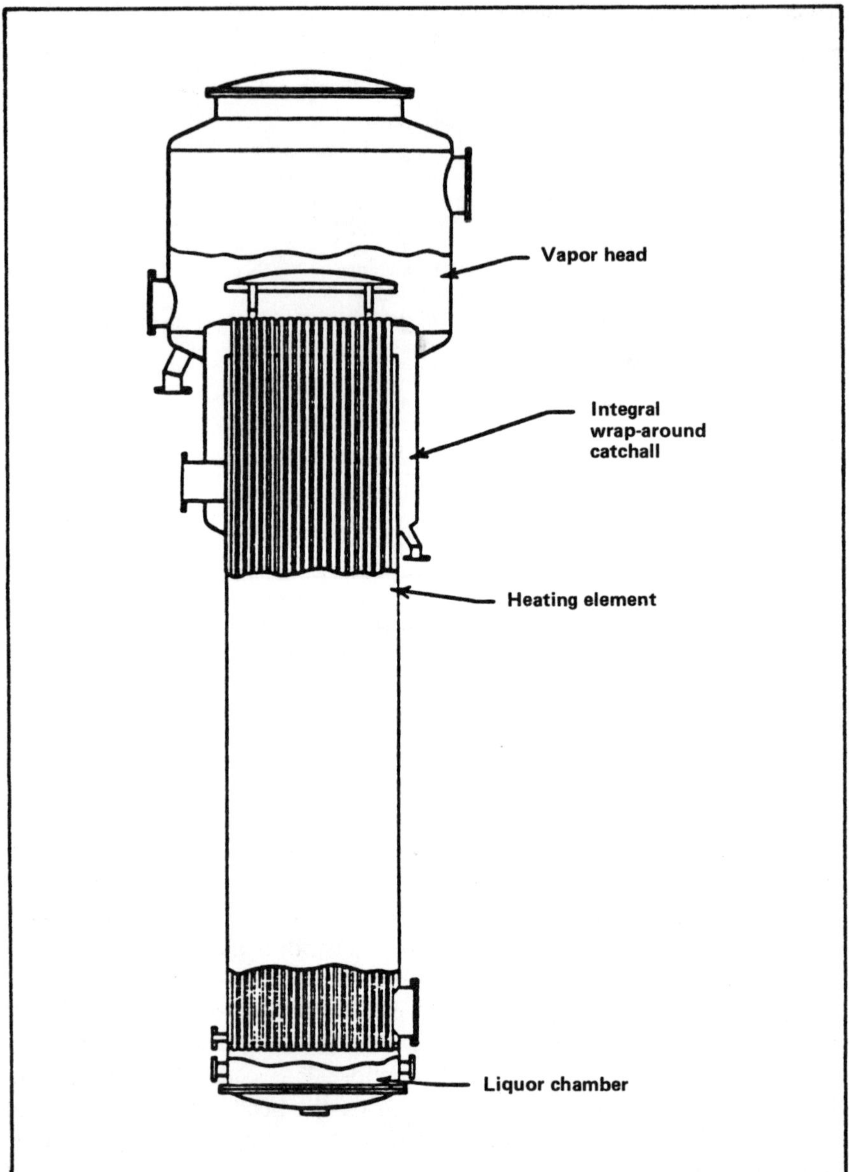

Fig. 3.9 LTV evaporator.

box rises under pressure through the tubes and is heated by the surrounding steam. At the same time, the static head of the liquor above the tube decreases, and thus, at some point along the tube, the liquor starts to boil. The point of initiation of boiling is the point of maximum temperature. The region below the boiling point is known as the heating or nonboiling section. The velocity of the black liquor in the nonboiling section is very small. A Reynolds number of about 800 is typical. Depending upon the rate of bubble formation, the liquor steam mixture is pushed through the boiling section at a velocity higher than that in the nonboiling section.

The liquor is pumped to the main liquor box at the bottom of the LTV evaporator and rises through the tubes. Since the cross section area of the liquor inlet pipe is much smaller than that of the tubes, the liquor velocity into the main liquor box is much larger than the liquor rising velocity. As a result of the velocity differential, flow eddies are formed and channeling of the black liquor could occur. The extent of this channeling depends on the position of the liquor inlet, the relative velocities of the liquor flow at inlet and exit, and the interior design of the main liquor box.

Each effect of a black liquor evaporator system is provided with an entrainment separator to remove entrained liquor from the vapor and, therefore, reduce the chemical loss. Centrifugal type catchalls mounted inside or external to the vapor head, or in the upper portion of the evaporator heating element, were used in early designs. Mesh pads have also been used as entrainment separators to achieve low chemical losses, but frequent plugging of the mesh is a major problem with this type of device. Impingement mist eliminators, shown in Figure 3.10, for vertical flow or horizontal flow have largely replaced catchalls, and separation efficiencies of 99.9% for droplets larger than 30 microns have been achieved. Properly designed systems are nonclogging and achieve high performance at low pressure drops.

Preheater and/or afterheater sections are incorporated into the LTV heating element. Preheaters or postheaters comprise a small portion of the total volume of the tubes in any one effect. The incoming black liquor is isolated from the evaporating liquid in that effect and is heated under forced circulation (see Figure 3.11). Since the cross section area is much smaller for the preheater or postheater than for the evaporation region, the flow velocity and, thus, the heat transfer coefficient through the preheater (postheater) is considerably higher than that in the nonboiling tube section.

The use of preheaters or postheaters, therefore, reduces the nonboiling section length in that effect (in the next effect forward for postheaters) and thereby provides a higher overall heat transfer coefficient and, thus, additional capacity. Postheaters also improve steam economy by providing lower grade heat from the following effect for sensible heat.

Falling-film (FF) evaporators

As its name suggests, in a falling film evaporator the liquor film progresses downward on the heating surface (Figure 3.12). The falling film concept is

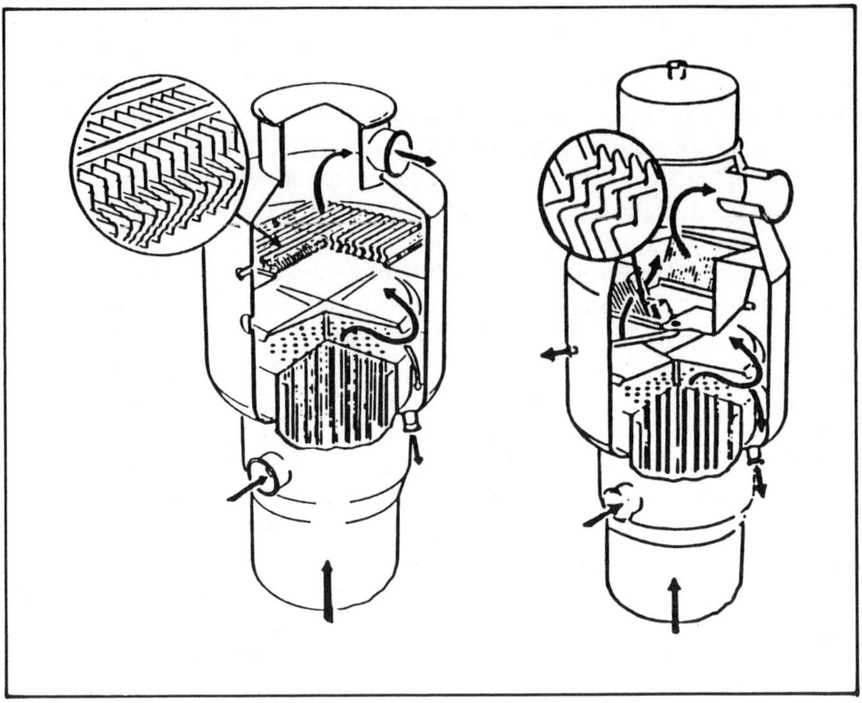

Fig. 3.10 Impingement mist eliminator.

employed in many new installations. The following evaporator types are used for black liquor service:

1. Vertical tube type (liquor inside of tubes);
2. Vertical plate type (liquor outside of plates).

Falling film evaporators comprise a liquor distribution system, heating element, vapor body (optional), and separator system. It may also be provided on a once-through basis or with liquor circulation depending on the ratio of liquor to evaporation. In the circulation type FF units, the circulation is achieved by a pump which transfers the liquor from the bottom to the top of the evaporator. Since the circulation rates here are only about 10% of those required for forced circulation evaporators, pump size and power requirements are lower.

Falling film type evaporators may be used where a low temperature differential (T_d) exists. The relatively higher heat transfer coefficients of FF evaporators make them an attractive alternative for low T_d applications. Efficient liquor distribution over the heat transfer surfaces in falling film

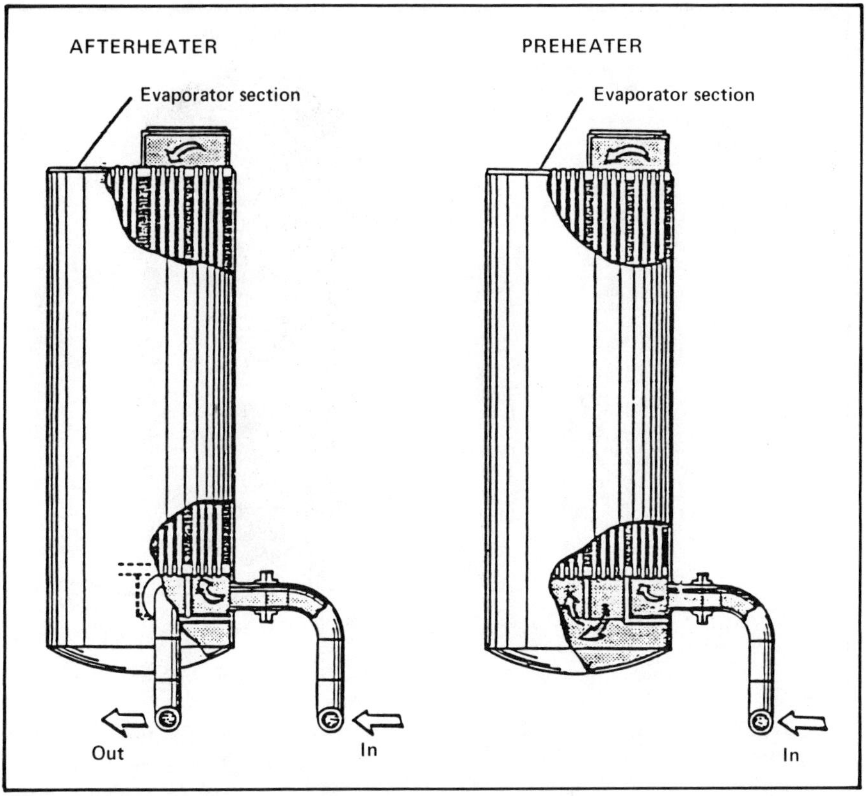

Fig. 3.11 LTV with built-in heater sections (courtesy of Unitech Corp.).

evaporators is necessary to enable stable opertion at low T_d values (*28*). These features make possible the design of multiple-effect systems with seven or more effects for evaporation to 50% solids or higher. Several seven effect FF plate type evaporators raising liquor concentration from 20% to 68% solids are in operation. In rising-film evaporators, unstable operation results at the lower T_d values associated with more than six effects and their end concentrations rarely exceed 50–52%. Unlike LTV units, FF evaporators may be turned down to operate at low feed and evaporation rates without fouling the heat transfer surface.

Plate type FF evaporators may be turned to 25% or more of full load. In mill operations with a kraft-semichemical liquor, it was found that tube fouling was minimal and the regular boilout cycles often required for rising-film evaporators was not required (*28*). The reduced fouling tendency of FF units compared to rising film LTV evaporators has enabled the application

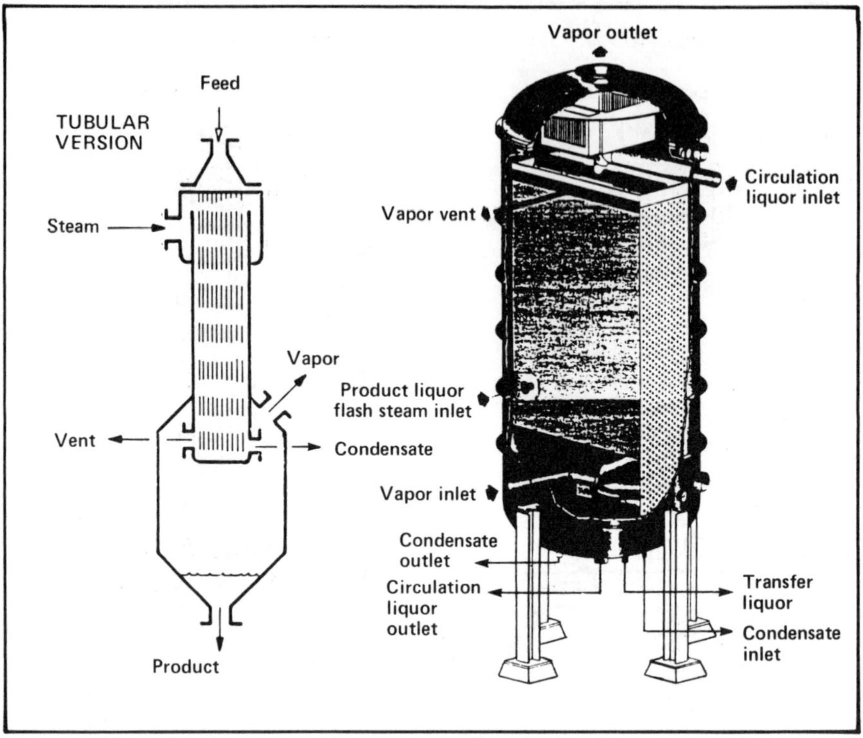

Fig. 3.12 Tubular and plate type falling film evaporators (courtesy of Rosenblad Corp.).

of plate type FF units to 65–70% solids service. Black liquor carryover with condensates is minimal, with plate type units reporting values as low as 1 ppm Na_2O.

Forced circulation (FC) evaporators

The use of FC units is usually limited to the concentration of black liquor above 50% solids after lower concentrations have been achieved in LTV evaporators. FC evaporators have been applied to high solid, viscous, corrosive, and scaling liquors. The forced circulation evaporator shown in Figure 3.13 is very similar to the LTV. In general, however, it has a smaller diameter 1.25–2 in. (3.17–5.05 cm) and shorter tubes 14–24 ft (4.267–7.315 m). Smaller tubes, however, increase the risk for pluggage and the high tubular velocities could cause increased vibrations. The vapor head is higher vertically, and there is a large recirculating line from the vapor head to the suction of the main circulating pump. The main circulating pump is a high-capacity, low-head pump which operates at low speed. The circulating pump maintains

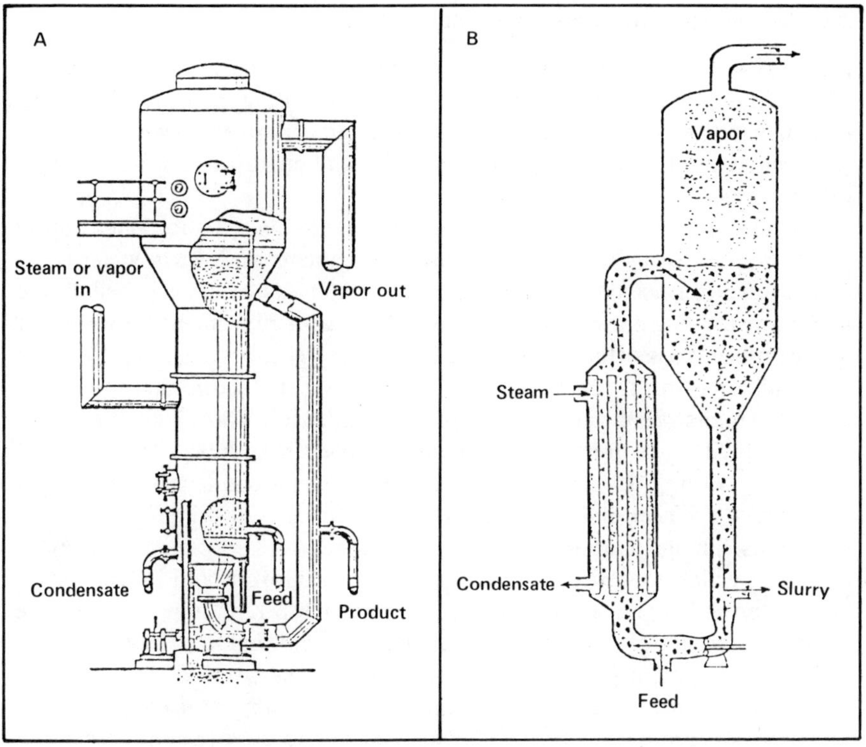

Fig. 3.13 Forced circulation evaporator, (A) internally mounted tube section, and (B) externally mounted tube section.

a steady volume of liquor to each heat exchanger tube. Liquor velocities are in the 6–12 ft/s (1.829–3.66 m/s) range. The high liquor velocity promotes higher overall heat transfer rates compared to LTV units in similar service. This type of evaporator typically allows boiling to occur in the tubes. When the liquor exhibits high scaling characteristics, a submergence weir can be used to partially suppress boiling in the tubes. When boiling is suppressed in the tubes due to the liquor head maintained by the weir over the top tube sheet, scale that may otherwise be formed due to localized concentration in the boiling region may be prevented. The heater can be oriented either vertically, horizontally, or inclined in one-pass arrangement. Two-pass arrangement is common with the horizontal tube units. Another type of forced circulation evaporator in which boiling of liquor in the tubes is suppressed is the evaporator/crystallizer. The liquor is heated in an indirect heat exchanger and subsequently flashed in a flash tank. Scaling is said to be negligible and evaporator/crystallizers have been used to raise solids to 73%.

Auxiliary equipment in the evaporation plant

Condensers Direct contact and surface type heat exchangers have been generally used in pulp mill evaporation plants are shown in Figure 3.14. Direct contact or barometric condensers—either countercurrent or cocurrent (jet)—were widely used, since they were inexpensive and operate at low temperature differences.

Surface condensers are now used in most kraft black liquor evaporation plants because of pollution abatement requirements and warm water reuse or conservation needs. A surface condenser consists of a steel shell closed at the top and bottom with tube sheets. Multipass arrangement for the water side or tube side is often considered. Stainless steel tubes are normally used to avoid corrosion from the vapor. The cost of a surface condenser is considerably higher than that of a barometric condenser. This higher cost is accentuated when the unit is supplied with stainless steel tubes. At many mill locations, cooling towers are used for recirculating condenser water.

Recently, surface condensers have been used in series as primary and secondary units. The primary unit is designed to condense approximately 85% of the vapor which contains only small amounts of odorous and BOD producing compounds. The secondary unit condenses the remaining 15% of the vapor containing major quantities of these objectionable compounds.

This arrangement permits separation and concentration of foul condensates and in turn would reduce the condensate stripping steam requirements if that was being considered. Other approaches to condensate segregation for steam stripping are discussed later.

Evaporative condensers shown in Figure 3.14 which combine the functions of a surface condenser and cooling tower have also been used. Heat from the circulating cooling water is transferred to the air which is drawn in by a fan. Lower capital costs and minimal space requirements have been claimed, compared to surface condenser/cooling tower systems.

Another trend in recent years has been the use of liquid ring vacuum pumps in place of steam jet ejectors. Vacuum pump operating costs are approximately 12% of the cost of operating an ejector (*29*) as shown in Table 3.4, but capital costs are higher.

Pumps Generally, pumps used with black liquor evaporators are of the centrifugal type. They are designed to handle foamy liquor and are capable of withstanding varying suction and discharge head conditions. All parts in contact with the liquor and combined condensates are usually made of stainless steel or noncorrosive materials. Submerged gland pumps, where the gland is submerged in a bath of water, have been used in the vacuum effects to

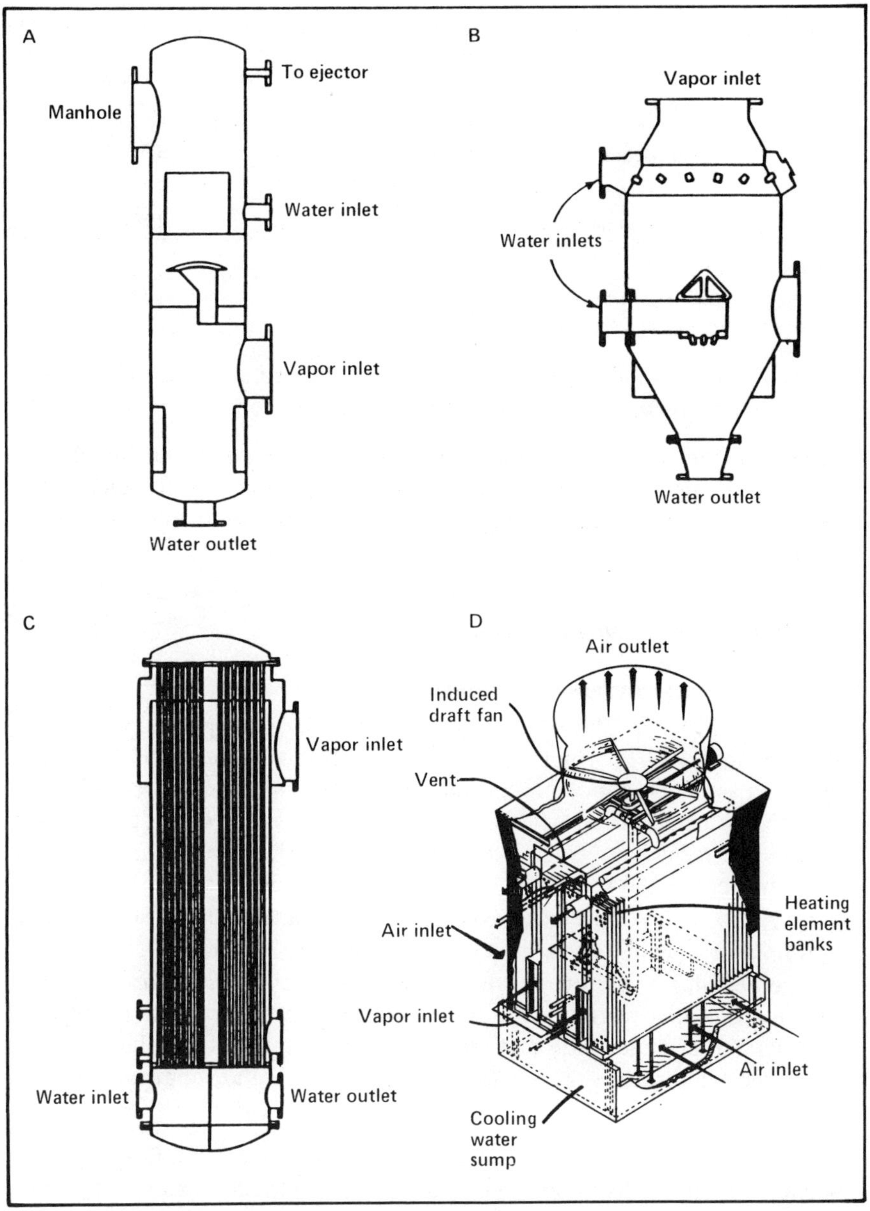

Fig. 3.14 Various types of condensers in evaporator plant. (A) Countercurrent condenser, (B) jet condenser, (C) surface condenser, and (D) evaporative surface condenser (courtesy of Rosenblad Corp.).

Table 3.4. Energy requirements for steam ejectors and vacuum pumps (29)

Equipment	*Energy used*	*Annual cost** *\$/year*
Steam ejector	450 kg steam/ 100 kg/gas	19,000
Vacuum pump	16.5 kw/100 kg gas	2,300

*8400 h/year.
\$11.03/1000 kg - steam cost.
\$300/kg year - power cost.

eliminate the possibility of drawing air into the pump if the gland leaks. Mechanical seals are also coming into increasing use.

Flash tanks Concentrated black liquor is flashed to atmospheric temperature with a 1–2% gain in solids concentration. Flash vapor usually goes to the third effect vapor head. Condensate flash tanks recover steam for reuse in evaporation.

Multiple effect evaporators

Evaluation and analysis of performance

As indicated earlier, black liquor concentration is accomplished in a multiple effect evaporator which comprises several units (effects) that are connected in series by vapor piping, and in which the water vapor boiled off from the liquor in one effect acts as heating steam in the steam chest of the following effect. The principle behind multiple effect operation is to conserve the heat in the vapor by causing it to be condensed in another effect operating at a lower pressure and temperature, and thereby raising black liquor solids concentration in stages and reducing live steam usage.

Steam economy and evaporator capacity (i.e., evaporation rate) are the twin performance variables of major concern in evaporator operations. Another performance factor of major concern in evaporator plants is scaling of the evaporator tubes which also causes a loss in evaporator productivity and steam economy. The following is a discussion of these aspects.

Steam economy Steam economy is the amount of water evaporated per unit of live steam used. As indicated earlier, increasing the number of stages (effects) raises the steam economy. Besides the number of effects, the steam economy is also influenced by other factors such as evaporator type, the characteristics of the feed/discharged liquors, steam, and condensates; liquor/vapor flow sequence, and condensate flashing, as discussed below:

1. The temperature of the feed liquor significantly influences the steam economy. As the feed liquor temperature increases, the steam economy improves.
2. The operating steam pressure in the first effect of the evaporator affects the steam economy. The use of high steam pressures decreases the economy. This is caused partly by the higher temperature levels in the evaporator and partly by the lower latent heat of the steam at high pressures. This means that when an evaporator becomes scaled and the steam pressure is increased to maintain evaporator capacity, the steam economy will be decreased.
3. The total solids concentration of the evaporator feed and discharged liquors influences the steam economy. If the feed solids concentration is raised, the steam economy will decrease. The same effect will occur if the discharged liquor concentration is lowered. This is caused by the fact that more liquor is pumped through the evaporator in relation to the evaporation rate. The additional liquor must be heated to the boiling point in the evaporator effects and less heat is available to cause evaporation.
4. Most black liquor evaporators are equipped with discharged liquor flash and condensate flash systems. The number of stages of liquor and condensate flash, as well as the temperatures of the discharged liquor and condensate, affect the steam economy. Flashing the steam condensate is, however, not desirable as it reduces overall mill energy conservation by requiring additional boiler feed water heating.
5. The liquor flow sequence within the evaporator system can markedly influence the steam economy. In past years, it was common to introduce the feed liquor into the high vacuum effects and pump the liquor in a straight backward flow to the first or steam effect. Many modern evaporators in operation use variations of the older flow sequence. Feed liquor is introduced into the intermediate effects with part of the liquor pumped in backward flow and part in forward flow to the high vacuum effects, where it is preheated and pumped with the remainder of the liquor in backward flow. The steam economy with arrangements of this type is higher than with straight backward liquor flow. This possibility is due to the use of integral liquor pre(post)heaters in the evaporator bodies. Many liquor flow sequences

are possible with integral liquor pre(post)heaters, and the best system will depend on all of the process conditions for a specific situation.

6. The vapor flow sequence could have a direct impact on the steam economy. Falling film multiple effect evaporators which have lower required temperature differential and specific liquor loading may enable designs with larger number of effects than rising film LTV units. As the number of effects increases, the steam economy generally improves.

The preceding discussion emphasizes the point that it is not possible to determine the steam economy of a given evaporator by simple relationships. Steam economy for a sextuple-effect system can typically range from 4.7 to 5.0. Owing to the many factors which influence steam economy, there are no shortcut methods of predicting results. The differences in practical results are sufficiently important to justify a complete heat balance in order to determine steam economy.

Capacity The evaporator capacity (i.e., the heat transfer rating) is determined by the heating surface area, available temperature drop, and overall heat transfer coefficients. The capacitites of two evaporator systems with equal total heating surfaces would not necessarily be the same. To compare the capacities of evaporators with a different number of effects, the average heating surface per effect should be considered. For example, a sextuple-effect evaporator with the same amount of total heating surface as a quintuple-effect evaporator would have a lower evaporation capacity. This is because the capacity of a multiple effect evaporator is also directly proportional to the total temperature difference, which is the sum of the temperature drops for the various effects. Because of the boiling point rise of the liquor being concentrated, as more effects are added the total effective temperature difference is decreased. A quintuple-effect operating at same process conditions will have more capacity than the sextuple-effect system.

The temperature distribution across a multiple effect evaporator depends principally on the individual heat transfer coefficients in the various effects. This is because the heat transferred in each effect is equal or very nearly equal. Therefore, the temperature drop for each effect is inversely proportional to the heat transfer coefficient for that effect. Until recently, there was no completely adequate method of calculating the heat transfer coefficients for black liquor evaporators from basic principles. Heat transfer coefficients were estimated based on design experience. The heat transfer coefficient computation method proposed by Gudmundson (*19, 20*), though reliable, does not lend itself to easy calculations.

Heat balance calculations

It is clear from the above discussions that heat and material balance calculations for the multiple effect evaporator system will serve as a useful tool to evaluate the design of new systems or to analyze the performance of existing operations. These calculations use either design or mill operating data to determine the following:

1. Overall heat-transfer coefficients;
2. Steam economy;
3. Evaporation rate (of water) in each effect;
4. Solids concentration in each effect.

Coats (*30*), Wetherhorn (*31*), and Clay (*32*) have described methods for evaporator heat balance calculations using the data available from the operator's log sheet. An example of the calculation based on typical black liquor operating data listed in Table 3.5 for a six-body sextuple-effect LTV evaporator system, shown in Figure 3.15, is presented in Table 3.6. Note that the heat transfer coefficients of the individual effects increase as the liquor temperature increases from the sixth effect up to the second effect and compare favorably with the data in Table 3.2.

The higher liquor temperatures lower the liquor viscosity and increase the heat transfer coefficients. Although the viscosity rises as the solids concentration is increased, the higher temperature level reduces the effect of increased liquor viscosity. In the first effect, the high liquor temperature may not be sufficient to offset the high liquor viscosity and increased scaling tendency, and could result in lower heat transfer rates.

The dry and saturated steam used in the first effect of black liquor evaporators is generally between 239 and 480 kPa (20 and 55 psig). It is frequently found that steam pressures above 515 kPa (60 psig), though providing greater temperature difference, cause accelerated scaling of the evaporator tubes in the first effect due to the high temperature of the liquor.

The vacuum in the last effect of an LTV black liquor evaporator train is usually between 61–68.5 cm/Hg (24–27 in./Hg). In most cases, it is uneconomical to operate at vacua below 61 cm/Hg (24 in./Hg), because the low total available temperature drop significantly decreases the evaporation capacity of a given amount of heating surface. At vacua above 68.5 cm/Hg (27 in./Hg), the increased temperature drop is ineffective in regard to capacity, as the lower liquor temperatures reduce the heat transfer coefficients in the high vacuum effects. The required quantity of condenser water increases rapidly at high vacua. Falling film type evaporators permit operation at higher vacuum of 68.5–71 cm/Hg (27–28 in./Hg), where cooling water is available,

Table 3.5. Operating conditions in a sextuple-effect evaporator

	Flash Tank	IA	IB	II	III	IV	V	VI
Steam Pressure								
psi (kpa)		31.5 (217.2)	31.5 (217.2)	9.4 (64.8)	1.2 (8.3)			
Vacuum, in. Hg (kpa)						8.3 (28.1)	16.2 (54.9)	22.0 (74.5)
Steam temperature, °F (°C)		276 (135.5)	276 (135.5)	238 (114.5)	216 (102)	196 (91)	175 (79.4)	152 (66.6)
Working temp. drop, °F (°C)		24 (13.3)	26 (14.4)	13 (7.2)	13 (7.2)	15 (8.3)	17 (9.4)	24 (13.3)
Liquor temp., °F (°C)	232 (111.1)	252 (122.2)	250 (121.2)	225 (107.2)	203 (95)	181 (82.8)	158 (70)	128 (53.3)
Boiling pt. rise, °F (°C)	14 (7.7)	13 (7.2)	11 (6.1)	7 (3.9)	5 (2.8)	4 (2.2)	3 (1.6)	3 (1.6)
Saturated vapor temp., °F (°C)	218 (103.3)	239 (115)	239 (115)	218 (103.3)	198 (92.2)	177 (80.5)	155 (68.3)	125 (51.7)
Vapor pressure								
psi (kpa)		9.8 (67.6)	9.8 (67.6)	1.8 (12.4)				
Vacuum, in. Hg (kpa)					7.4 (25.1)	15.6 (52.8)	21.4 (72.5)	26.0 (88)
Latent heat of vapor, Btu/lb (Kcal Kg)		953 (529.4)	953 (529.4)	967 (537.2)	979 (543.8)	992 (551.1)	1005 (558.3)	1023 (568.3)
Pressure drop								
psi (kpa)		0.4 (2.8)	0.4 (2.8)	0.6 (4.1)				
Vacuum, in. Hg (kpa)					0.9 (3.1)	0.6 (2.)	0.6 (2.)	
Vapor pressure								
Next effect, psi (kpa)		9.4 (64.8)	9.4 (64.8)	1.2 (8.3)				
Vacuum, in. Hg (kpa)					8.3 (28.1)	16.2 (54.9)	22.0 (74.5)	
Feed, lb/h (Kg/h)	90,605 (41,098)	109,775 (49,794)	135,825 (61,610)	178,815 (81,110)	219,585 (99,604)	251,995 (114,305)	166,520 (75,533)	166,520 (75,533)
Discharge, lb/h (Kg/h)	89,200 (40,461)	90,605 (41,098)	109,775 (49,794)	135,825 (81,610)	178,875 (81,137)	219,585 (99,603)	130,475 (59,184)	121,520 (55,121)
Solids in, %	51.1 (51.1)	42.2 (42.2)	34.2 (34.2)	26.0 (26.0)	21.2 (21.2)	18.4 (18.4)	13.9 (13.9)	13.9 (13.9)
Solids out, %	52.0 (52.0)	51.1 (51.1)	42.2 (42.2)	34.6 (34.6)	26.0 (26.0)	21.2 (21.2)	17.8 (17.8)	19.1 (19.1)
Heating surface, ft^3 (m^2)		4,400 (409)	4,400 (409)	8,800 (818)	8,800 (818)	8,800 (818)	8,800 (818)	8,800 (818)
Heating transfer coefficient, $Btu/h/ft^2/°F$ (Cal/cm^2 s °C)		174 (0.0236)	242 (0.033)	392 (0.053)	386 (0.052)	316 (0.043)	240 (0.033)	190 (0.026)

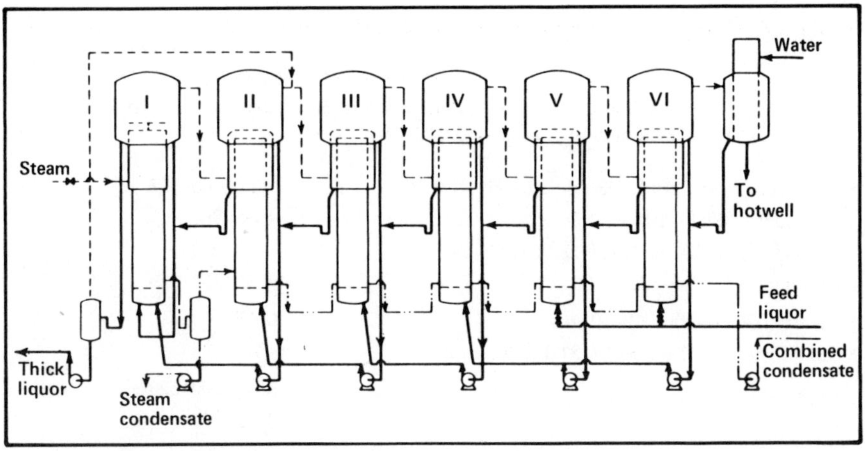

Fig. 3.15 Sextuple-effect evaporator.

since heat transfer coefficients are higher than for rising film LTV systems. Vapor velocities are relatively high under high vacuum operation and the vapor heads and piping of the letter effects must be large to prevent excessive entrainment losses. The most suitable operating vacuum for a given evaporator system will depend on many factors, including water temperature, water availability and pumping costs, materials of construction, and type of condenser used.

The liquor flow sequence in a black liquor evaporator should be designed to give the highest possible capacity per square foot of heating surface consistent with the high steam economy. The number of effects and their area and steam/liquor flow arrangement are important factors in determing energy usage and process economics of a particular multiple effect evaporator configuration. Itahara and Stiel (*33*) have proposed an optimization algorithm to determine the optimum number of effects with minimum heating area to minimize evaporation costs. Their study disclosed that the usual practice of equal area effects is not necessarily consistent with a minimum cost design, particularly when more than five effects are considered.

North American evaporation plants generally employ a (5–6–4–3–2–1) liquor flow sequence and the weak liquor is fed to the last (5th or 6th) effect. This has been characterized as "backward" feeding. In this method, liquor and steam flow in opposite directions. In Scandinavia, most plants operate in what is known as the "intermediate" feeding sequence, with weak black liquor fed to the third effect. Edwards and Baldus (*34*) have shown that the liquor flow sequence (5–6–4–3–2–1) achieves slightly higher capacity compared to the standard Scandinavian configuration (3–4–5–1A–1B–2). The steam economy for the Scandinavian system was, however, higher at 5.2, as compared to 4.9 for the North American case. Bolmstedt and Gudmund-

Table 3.6. Operating results in a sextuple-effect evaporator

Feed	151,067 Kg/h at 13.43% solids	
Discharge	40,461 Kg/h at 52% solids	
Evaporation	110,605 Kg/h	

Flash Liquor from 1A (41098) (122.2 – 111.1) 0.75	342,140/537	637
Steam to 1A (9299) (515) (0.97)	4,644,850	
Heating Feed to 1A (49794) (122.2 – 121.1) 0.79	+ 43,590	
Evaporation 1A	4,601,231/529	8,698
Steam to 1B (13925) (515) (0.97)	6,955,643	
Heating Feed to 1B (61610) (121.1 – 107.2) 0.83	– 710,053	
Evaporation 1B	6,245,590/529	11,816
From 1A	+ 4,601,231	
Steam Condensate Flash (9299 + 13925) (135.5 – 115)	11,324,053	
Heating Feed to II (81138) (107.2 – 95) 0.845	– 837,046	
Evaporation II	10,487,007/537	19,527
Condensate Flash (8695 + 11816 + 903) (115 – 102.2)	+ 273,640	
Heat to III	11,103,075	
Heating Feed to III (99604) (95 – 82.8) 0.87	– 1,061,300	
Evaporation III	10,041,775/544	18,466
Condensate Flash (21415 + 19527 + 637) (102.2 – 91.1)	+ 461,862	
Heat to IV	10,503,638	
Heat Feed from V (59184) (82.8 – 70) 0.9	– 679,060	
Heat Feed from VI (55122) (82.8 – 53.3) 0.91	– 1,472,766	
Evaporation IV	8,351,810/551	15,155
Condensate Flash (41529 + 18466) (91.1 – 79.4)	+ 700,478	
Heat to V	9,052,288	
Heat Flash to V (75533) (71.1 – 70) 0.93	+ 78,111	
Evaporation V	9,130,398/558	16,350
Condensate Flash (599830 + 15155) (79.4 – 66.7)	+ 960,763	
Heat to VI	10,091,162	
Feed Flash to VI (75533) (71.1 – 53.3)	+ 1,252,545	
	11,343,705/568	19,958
Total Evaporation		110.605
Steam Economy		4.8

son (*35*) have also shown detailed heat balance calculations to evaluate alternative vapor and liquor flow sequences for the same set of evaporators. The changes in the evaporator capacity and steam economy are, of course, related to the changes in heat transfer coefficients and liquor boiling point rise that occur for different vapor/liquor flow schemes. It is evident that computer-aided heat balance calculations are a valuable tool to quickly screen evaporator alternatives to identify an optimum system.

Computer-aided heat balance calculations

In recent years, several digital computer programs designed to analyze the operation of existing evaporator plants, as well as to evaluate alternative evaporator designs and equipment arrangements have been developed (*19, 20, 34, 35–38*). The INDUNS computer program (*35*) uses the unit cell concept to represent an evaporator effect. Each modular unit cell consists of a liquor flash tank, mixing vessel, evaporator body, condensate level regulation tank, condensate flash tank, and heat exchanger (*19, 35, 36*). INDUNS estimates the overall heat transfer coefficients of each effect in any existing or proposed evaporator configuration based on the following parameters: heat flux, boiling point, liquor feed rate, liquor feed temperature, total solid content, surface tension, and foaming characteristics of the liquor.

Another modular computer system, GEMS (Generalized Energy and Material Simulation), developed at the University of Idaho (*39*) and used widely in the pulp and paper industry, has also been applied to multiple effect evaporator calculations (*34*). The GEMS approach is similar to INDUNS, and each evaporator effect is represented by a set of process models describing mixing, heat transfer, flashing, etc. Other computer programs such as GEMCS and EVAP are also described in the literature (*37, 38*).

EVAP (*38*) developed at the University of Western Ontario can be used to determine equipment parameters and operating constraints for any evaporator configuration or used to troubleshoot existing operations based on mill measurements. It may be stressed that the overall kraft mill energy balance will often determine evaporator design and selection. For instance, it may be economical in some mills to operate or install a relatively lower steam economy evaporator system to balance low pressure steam usage against power demand.

Evaporator scaling

Scaling in evaporators is a major cause of reduced evaporator productivity in many mills. Severe scaling may cause black liquor flow interruptions. This is a serious bottleneck to pulp production, especially when the mill is dependent on a single train of evaporators and weak liquor storage capacity is limited.

Black liquor evaporators are subject to scaling or fouling on both the inside and outside of the heating surface. The vapor side of the heating surface can become scaled by sulfur-containing corrosion products. The high-density, high temperature evaporator effects are usually likely to become scaled on the liquor side. Vapor-side scaling is most common in the low temperature effects. In general, vapor-side scaling is prevalent only when carbon steel is used, as is the case in older evaporators. As scales are solids, they originate from materials which are insoluble or have low solubilities in the weak or strong black liquor. Table 3.7 characterizes these scale components. Insoluble materials such as dirt, fiber, or pipe linings are not only part of scales, but could also plug the tubes or restrict flow. Other materials such as tall oil soap, lignin, calcium carbonate, or sodium salts generally coat the tube surface. There are two major types of scales from inorganic salts present in black liquor:

1. Water soluble scale: sodium sulfate-sodium carbonate double salt (Burkeite), which generally precipitates at higher black liquor solids (above 50%).
2. Water insoluble calcium scale: calcium carbonate precipitate at all solids levels and temperature above 250°F (120° C).

As these are the main types of scales encountered in kraft pulp mill evaporators, various studies have been conducted to elucidate the mechanism of scaling and its prevention (*40–46*). Salient results disclosed by the Institute of Paper Chemistry from an industry-wide study of evaporator scaling (*40–43*) are summarized below:

1. Calcium carbonate scaling was the major contributor to evaporator scaling and is the main cause of loss in evaporator capacity. High calcium content in the weak black liquor is responsible for these scales.
2. Water soluble (sodium sulfate-sodium carbonate) scales, while prevalent, are less of a problem with respect to sustaining evaporator capacity, as they are relatively easy to remove.

Table 3.7. Scale materials in black liquor

A) Materials insoluble in weak (15% solids) liquor
   1) Fiber
   2) Sand, dirt, etc.
   3) Miscellaneous (pieces of pipe, gasket materials, etc.)

B) Materials insoluble in stronger black liquor
   1) Tall oil soap (at 25 to 30% solids)
   2) Lignin (at pH below 11, precipitation)
   3) Calcium scale ("Water Insoluble")
      Forms at 250°F or higher from calcium-organic complexes.
   4) Water soluble scale (sodium carbonate-sodium sulfate)
      Double salt forms when critical solubility limit is reached (usually 50% solids or higher).
   5) Silica scales
      Difficult to remove.

3. Insoluble calcium carbonate scales can also form as rapidly as the soluble sodium sulfate-carbonate scales and thus affect short-term and long-term evaporator performance. The widely held belief that calcium scaling was longer term and short-term scaling was due to water soluble inorganic components is not quite correct.
4. Main source of calcium is from the wood supply. Calcium scaling is more in mills which pulp hardwoods, particularly barky whole-tree chips.
5. Calcium carbonate scaling rate depends strongly on temperature. Scaling rates do not appear to be strongly dependent on liquor solids content since scale distributions were found to be reasonably uniform in the boiling region.

Both soluble and insoluble scale deposition is complicated by the presence of organics in the black liquor. It was postulated that such organic components as lignin or hydroxy acids, which contain adjacent hydroxyl groups, form a chelate with calcium which greatly increased its solubility in black liquor. These organic components may be divided into two groups with respect to their impact on $CaCO_3$ scaling. The first group includes those whose ability to complex with calcium depends on pH in the normal pH range or black liquor or whose stability constants decrease with increasing temperature. This group will contribute to $CaCO_3$ scaling by releasing ionic calcium as the temperature increases. The second group consists of those whose complexing ability with calcium does not change significantly with pH and whose stability constants are temperature-independent or increase with temperature. They do not contribute to $CaCO_3$ scaling.

Consequently, as long as the first group of compounds and sufficient calcium are present, $CaCO_3$ scaling will form. Because of the relationship between pH and the release of complexed calcium ion, caustic addition to weak black liquor is sometimes practiced to control $CaCO_3$ scaling.

It may be noted that scaling caused by the precipitation of Na_2SO_4 from black liquor is not significant, since modern recovery boilers operate with high reduction efficiencies. The use of improved furnace linings and low silica makeup lime and saltcake have also greatly reduced the silica scaling problems in evaporator tubes. The practice of soap skimming in mills pulping high resin content woods such as southern pine has reduced the scaling caused by precipitation of less soluble resins in the high solids evaporator effects. Several kraft pulp mills practice cross recovery of neutral sulfite semichemical (NSSC) pulping liquors in the kraft black liquor evaporators. These problems become more pronounced as the ratio of NSSC liquor solids to kraft liquor solids increases. Evaporator operating problems encountered from the mixing of NSSC liquor with weak black liquor are:

1. Evaporator capacity is decreased due to increased inorganic scale formation on the liquor side of the tubes. More frequent water boilouts and chemical cleaning of the heating surface are required to maintain evaporator capacity.
2. More noncondensible gases are released from the mixed liquor than from kraft liquor. Proper venting of the evaporator steam chests is necessary to prevent a loss of capacity.
3. Corrosion of mild steel parts is accelerated. This applies especially to the lower section of the heating element shells and the condensate piping.
4. Organic residues are found in the evaporator system, especially on the tubes, lower liquor chambers, and liquor piping. The organic material often includes a considerable amount of fine fibers. In such cases, filtration of the NSSC liquor prior to mixing reduces the fine fibers.

Increasing the pH of the mixed liquor generally helps prevent scale formation. NSSC liquors can be separately concentrated in multiple-effect, forced-circulation evaporators. When the ratio of NSSC solids to kraft solids is high, a separate evaporator train for each liquor is recommended.

The formation of scales on evaporator surfaces causes a decrease in heat transfer rate and a consequent loss of capacity. If significant scaling occurs in any effect, its heat transfer coefficient would noticeably decrease, and the temperature drop would increase for the same heat transfer rate. Since the total temperature drop is fixed within narrow limits by the available pressure and the flow and temperature of condenser water for an evaporator operating at the maximum available steam pressure, a reduction in steam flow would occur, causing a loss in evaporator capacity.

Periodic analyses of the evaporator system heat transfer and temperature drop data are thus useful to maintain good performance since the source of malfunctions can be quickly identified. Besides scaling control, improper venting of noncondensible gases from the evaporator steam chest and the use of superheated steam can also cause a significant loss of capacity. Employing superheated steam results in a lower overall heat transfer rate, as a significant portion of the heating surface of the first effect is required for de-superheating with a consequent reduction in evaporator capacity. The high temperatures of superheated steam can also cause scale formation inside the tubes, resulting in a loss of capacity.

Heit (*47*) has suggested a method of monitoring the changes in concentration of inorganic elements such as calcium, magnesium, iron, aluminum, and silicon in the ash of the weak and strong black liquor to detect evaporator fouling. Evaporator cleaning schedules can be planned based on the particular scaling pattern that evolves in a mill.

Tables 3.8 and 3.9 suggest strategies for the prevention and control of water soluble and insoluble calcium scales. Water and weak liquor boilout are also routinely practiced to keep fouling at tolerable levels. Chemical cleaning with inhibited muriatic or sulfamic acid is recommended when scaling is severe as a longer-term solution. Hydroblasting is resorted to when tubes are plugged. The following cleaning procedures are suggested:

On-stream boilout This is practiced as part of the usual startup and shutdown procedure and should also be done routinely as required with water or weak liquor to clean the tubes on the liquor side. In this procedure, the water valve on the liquor feed line is kept open after opening the valves on the liquor transfer pumps. Water is allowed to flush the evaporator tubes and is then allowed to boil by starting the steam supply. After the feed water, steam, and effect temperatures stabilize, the evaporator is switched to liquor. Thermal shocks during the water boilout greatly aid in short term descaling. Similarly, at shutdown, a water flush and boilout may be in order.

Regular boilout This procedure is recommended for longer-term applications (say every 4–6 months). The cleaning can be done on both the vapor and tube side, and initially involves leaching the scales with white liquor or caustic at high temperature for 6–24 hours. This is followed by a water boilout. The scale materials, usually fibers and sodium sulfate-sodium carbonate (soluble scale), are flushed out at the bottom.

Table 3.8. Prevention/control of water-soluble scales

1) Liquor composition control
   * Improve recovery furnace reduction—reduce sodium sulfate.
   * Raise causticizing efficiency—reduce sodium carbonate.
   * Add ClO_2 generator spent liquor to strong black liquor.
   * Maintain at least 6 gm/liter residual alkali in weak liquor to avoid lignin precipitation.
   * Install fiber filter or keep fiber level low (fiber reinforces scales).

2) Evaporator operation control
   * Run final product solids below critical solids (below 50%).
   * Weak liquor boilouts to remove scales.
   * Sustain stable operation.

3) Evaporator modifications
   * Raise liquor velocity by recirculation in first and/or second effect.
   * De-superheat steam, if necessary.

Table 3.9. Prevention/control of insoluble calcium scales

1) Evaporator operation
   * Low steam temperatures to minimize scaling rate.
   * Liquor recirculation pumps to prevent tube pluggage.
   * Calcium deactivation by heating (expensive).

2) Cleaning
   * Steam shocks during water boilout.
   * Periodic dumping of flow boxes to remove calcium scale flakes.
   * Periodic washing of high-temperature effects with acid or chelant solution.
   * Hydroblast if necessary to clean plugged tubes.

3) Other
   * Calcium scales are not removed by chemical additives (polyacrylates, polyphosphates, etc.) or prevented by alkali addition.

Acid cleaning This is resorted to when short-term and regular boilouts cease to be effective. This is primarily a result of severe calcium carbonate (and in some cases silica scales) scaling on the liquor side and inside the tubes. Black liquor also gets baked onto the tube surfaces with the inorganic calcium. This is best removed by treatment with a 5% inhibited muriatic acid solution or with sulfamic acid.

Interaction with pulp mill operations

In the preceding discussions, the multiple effect evaporator plant has been considered as a single unit operation in the chemical and heat recovery process of the alkaline pulp mill. However, the input and output streams in this system, such as the feed black liquor, steam, condensates, vapors, and concentrated liquor all have an important impact on the operation of the entire pulp mill. Critical aspects of the interaction of the evaporation plant with the rest of the pulping and recovery processes are examined in the ensuing sections.

Brown stock washing efficiency

Spent cooking chemicals are separated and displaced from the brown stock in washers. The wash water used increases the dilution of the black liquor, obviously imposing an additional load on the evaporation system. The need to balance high brown stock washing efficiency against steam costs involved in evaporating the added wash water has been addressed in several papers (*48-51*). An important process variable determining washing efficiency is the washer dilution factor. The salt cake lost with the washed pulp decreases as the dilution factor is increased.

Dilution factor may be defined as the pounds (tons) of water added in the washing operation to the black liquor per pound (ton) of air dried pulp. This is the difference between the water applied as wash water in the last washer and that portion of the wash water that is discharged with the pulp from the last washing stage. The salt cake loss can be generally reduced by increasing the dilution factor, reducing the washer loading, or by installing a greater number of washing stages.

Figures 3.16 and 3.17 (*51*) show a typical relationship among dilution factor, salt cake loss, and evaporation costs. For each mill there is an optimum (economic) dilution factor which generally varies between 2.5 and 4.0 for existing three-stage washers. In recent years, rising energy costs and environmental constraints have justified the addition of a fourth or fifth washer stage to minimize salt cake loss to less than 20 lb/ton (10 kg/metric ton) pulp at dilution factors below 2.5. Lowering the washer dilution factor will, of course, raise weak black liquor solids and reduce steam use for evaporation needs. The weak black liquor solids concentration delivered to the evaporator plant in most mills is usually 13–16%. Multi-stage belt-type displacement

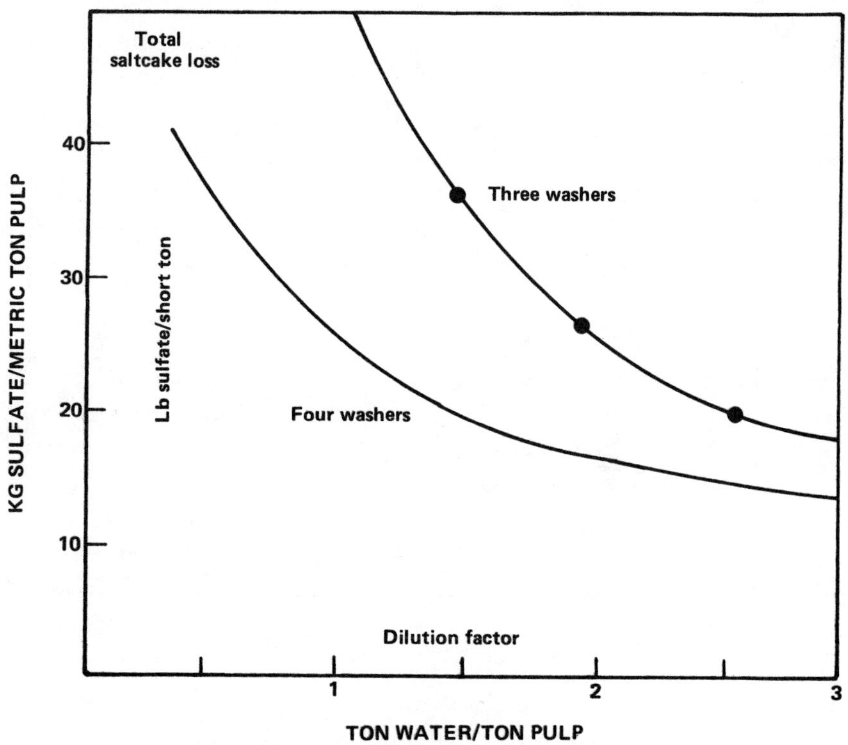

Fig. 3.16 Salt cake loss and dilution factor relationship.

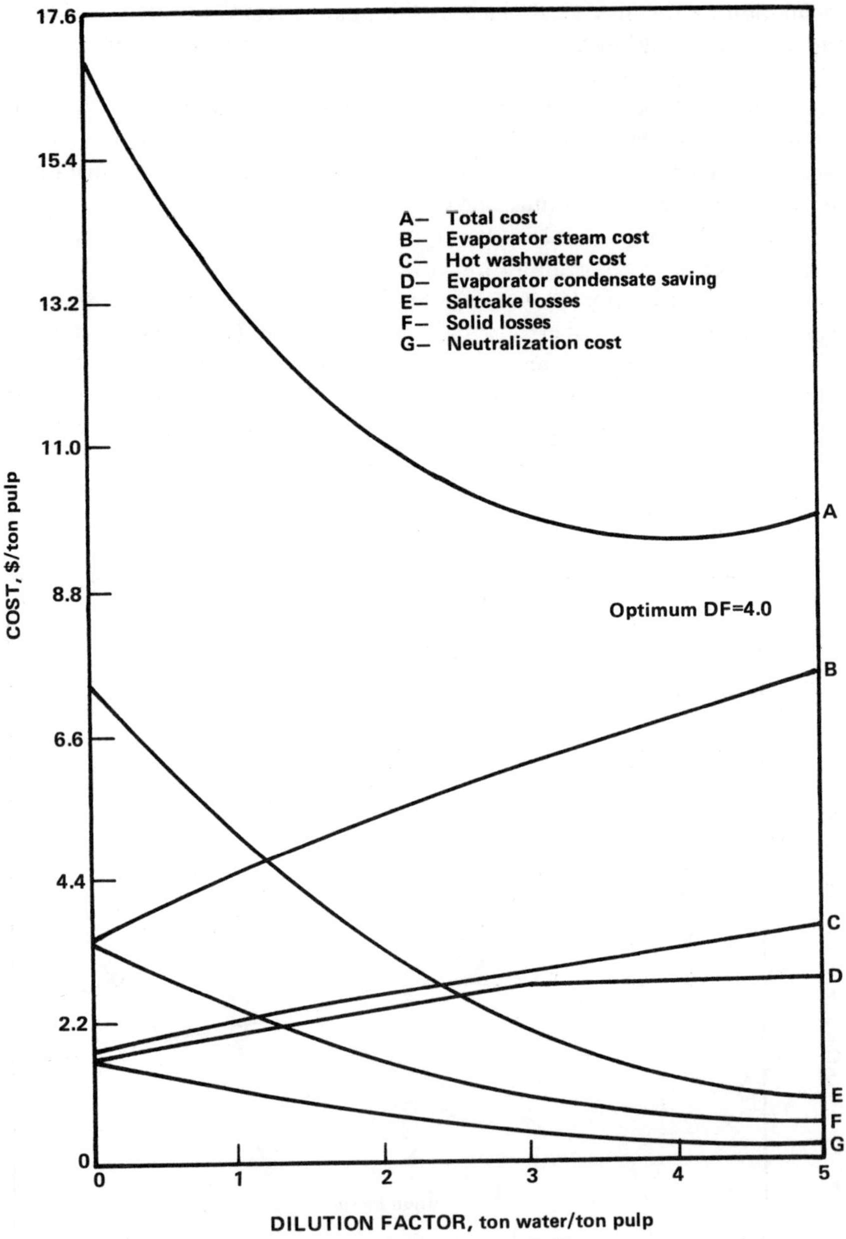

Fig. 3.17 Washing variable cost vs. dilution factor.

washers are reportedly operated at dilution factors of 1.5 or less to deliver weak black liquor at solids concentration in excess of 18% at salt cake loss levels below 25 lb/ton (12.5kg/metric ton) pulp (*52*).

Strong black liquor concentration

Before the 1950s, it was generally considered disadvantageous to burn black liquor of solid concentration higher than 55% because of higher dust loads in the flue gas and fouling of the superheater and the boiler tube bank. It was later established that recovery boiler operation with about 60–65% solid liquor proved to be more stable and steam generation was better (*53*). The incremental increase in the firing solids concentration, from 55% to 65%, also alleviated boiler problems regarding corrosion and odor. At lower solid concentration, the recovery boiler often operated with black hearth spots which increased the rate of corrosion on the tube and the emission of H_2S. Higher black liquor concentration and better air and liquor distribution appeared to help reduce these problems substantially.

Multiple effect LTV evaporators generally concentrate black liquor to 49–52% solids. Their operation is limited by the highly viscous nature of the back liquor at solids levels above 50%. Scaling also becomes severe in LTV evaporators. Final evaporation to firing solids concentration of 62–65% is carried out by direct contact evaporators or in concentrators.

In direct contact evaporation, the hot flue gas exiting the furnace at about 500°F (260°C) is cooled to 300°F (149°C) by direct contact with the liquor from the last effect. The liquor is concentrated from about 50% solids to 63% or more. The flue gas normally contains sodium carbonate and sodium sulfate fume, as well as sulfur dioxide and sulfur trioxide. Electrostatic precipitators installed after the direct contact evaporators recover the sodium compounds. The flue gas temperature from the direct contact evaporator must be kept at or below the design value in order that the precipitator may be operated at or below design flue gas velocities, thereby maintaining design collection efficiency. Also, as the sulfur-containing gases may cause severe corrosion problems to the precipitator, the exit gas temperature from the direct contact evaporator must be maintained above the dew point of sulfur dioxide at about 300°F (149°C).

The use of concentrators to further concentrate the black liquor after multiple effect evaporators has become standard in recent years. This reduces odor problems associated with direct contact evaporation and improves boiler efficiency. Concentrators are merely evaporators operated at solids levels above 50% without direct contact of the black liquor with the flue gas. A major portion of the malodorous sulfur compounds leaving the recovery furnace stack are produced during direct contact evaporation. Thoen, et al. (*55*) established that acidic flue gases (primarily $C0_2$) react with sodium sulfide (Na_2S) and sodium methyl sulfide (CH_3SNa) to release hydrogen sulfide (H_2S) and mercaptan (CH_3SH). These results also indicated that black liquor oxidation may reduce 18-fold the emission of H_2S in direct contact evaporation, from about 72 to 4ppm (v/v). Black liquor oxidation prior to direct contact evaporation is now required to meet stringent odor control standards in older recovery furnaces.

In recovery boilers without direct contact evaporators, final cooling of the recovery boiler flue gas may be done in the economizer, which is used for heating boiler feedwater and indirect heating of air (*56*). For such a recovery system with a concentrator and a larger economizer, the return on additional investment requirements realized from better heat economy may be as high as 45% (*54*). The high cost of energy was a major incentive for Scandinavian mills to install concentrators in the 1960s. Economizer clogging problems which have often been encountered in earlier operations have been practically eliminated with improved design and modern steel tube economizers. Kallenfors (*53*) and Vegeby (*54*) have discussed the use of concentrators in parallel on the steam side with the multiple effect evaporator to achieve final solids concentration of about 60–65%. Figure 3.18 shows a typical arrangement of a five-effect LTV evaporator with two concentrators employed commonly in Scandinavian mills. Because of the viscous nature of the black liquor, it is difficult to operate LTV units at solids levels exceeding 50%. Consequently, the early concentrator units comprised two bodies with forced circulation. One of these is concentrating the black liquor from the second effect from 55% to 64–65%. The other body is simultaneously concentrating the intermediate black liquor (about 35%) flow from the fifth effect. A five-effect falling film plate type evaporator system has been described (*57*) to raise solids to 65–70%. The first effect has three bodies, two of which operate as concentrators while the third body is washed with weak black liquor. The retrofitting of an evaporator/crystallizer type concentrator to a sextuple LTV evaporator system to raise black liquor solids to 69–73% has also been described (*58, 59*). Concentrator operation is generally integrated with the multiple effect evaporators in newer systems to improve overall steam economy.

Three concentrator design concepts are in common use for kraft black liquor service:

1. Combination falling/rising film forced circulation evaporator (Figure 3.19);
2. Falling film evaporator (tube or plate type) (Figure 3.12); and
3. Evaporator/crystallizer (Figure 3.20).

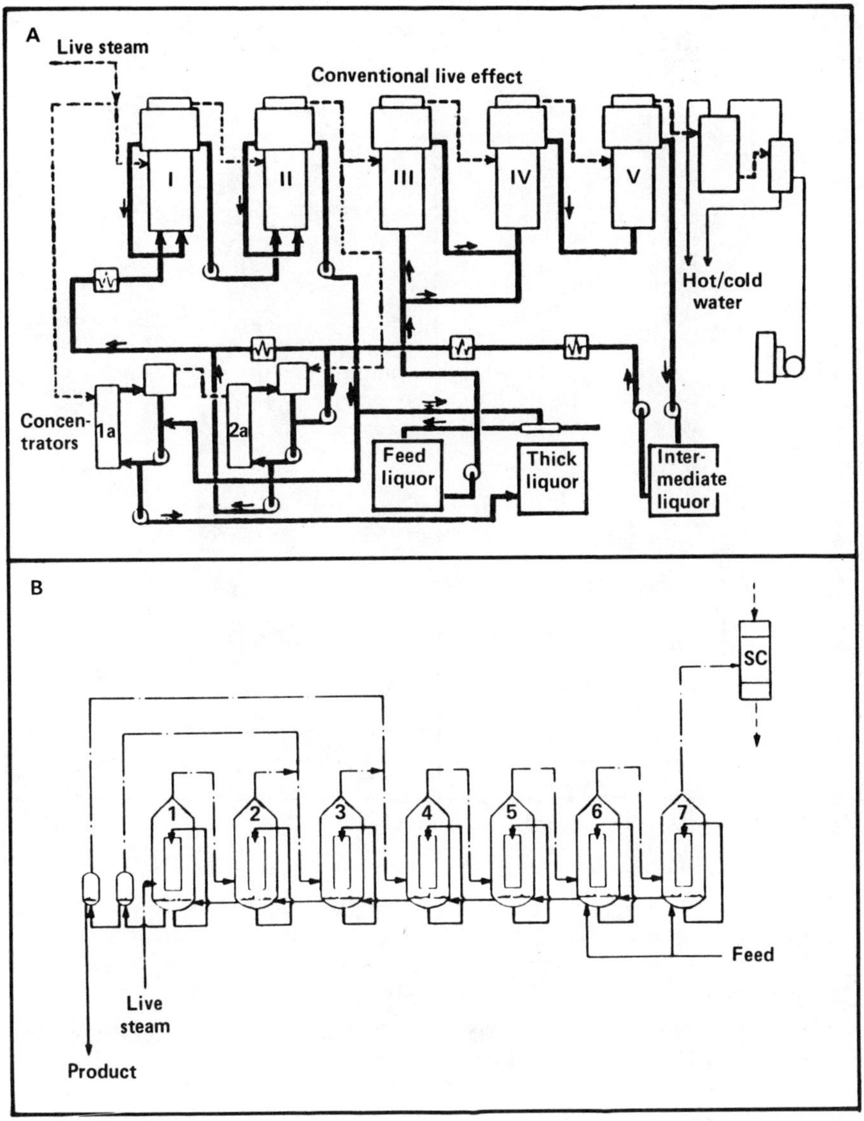

Fig. 3.18 (A) Five-effect evaporators with concentrators and (B) free flow falling film seven-effect evaporator for full end concentration (courtesy of Rosenblad Corp.).

Perhaps the most widely used concentrator in North America for raising solids to 65% employs the combination rising/falling film concept shown in Figure 3.19. These have been characterized as PFR* concentrators and

*PFR is a registered trademark of the Unitech Division, Econdyne Corporation.

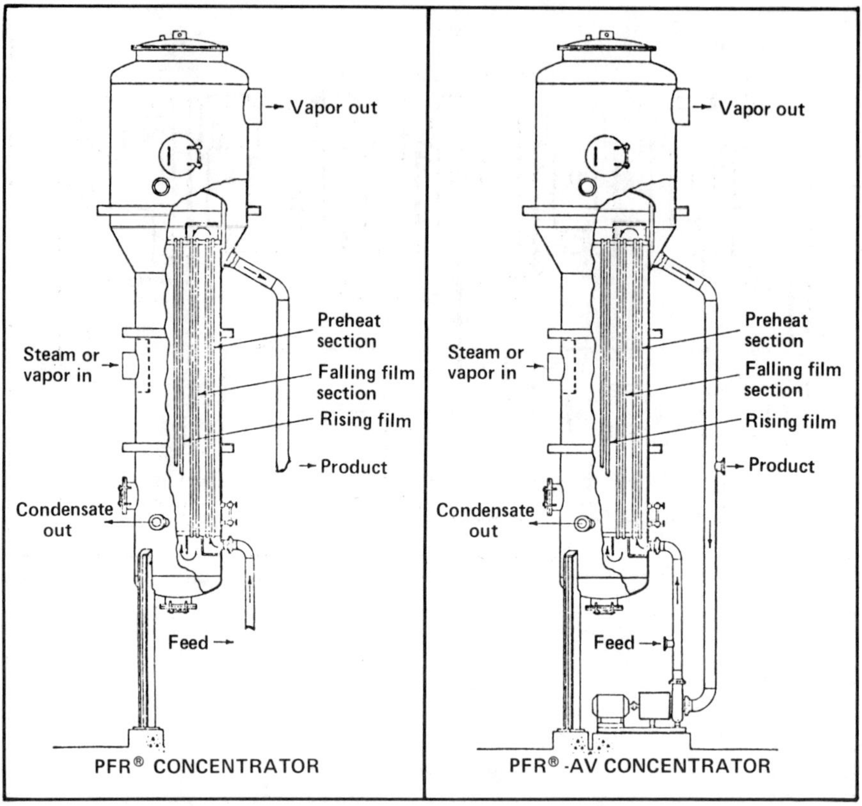

Fig. 3.19 PFR concentrator (courtesy of Unitech Corp.).

are in use in many older installations. (The acronym is derived from three different liquor passes: P—preheats; F—falling film section; and R—rising film section.) High liquor velocities are maintained by using a large circulation rate to minimize the highly viscous and scaling nature of the black liquor at solids levels above 55%. The high circulation rate, while increasing pumping energy requirements, supports film boiling in the tubes and minimizes formation of local dry spots. For liquors with a severe scaling tendency, a circulation pump is provided (Figure 3.19), and this is known as the augmented volume (AV) model. A variation of the design shown in Figure 3.19 includes two circulation loops to increase heat transfer coefficients and minimize scaling. In general, proper selection of the area ratios for the preheat, falling, and rising film sections and the circulation pump capacity are critical for optimum concentrator design. Plate type falling film evaporators and evaporator/crystallizers predominate in new installations.

Falling film concentrators (*28, 57, 60*) are being used when discharge solids over 65% are desired. Plate type falling film concentrators are operated to final solids of 67–70% and operation to 75% or more solids is apparently possible (*60*).

Fouling problems of plate type falling film units are said to be less than in the tubular PFR type units. As already discussed, these units have high heat transfer coefficients even at low T_d and have good turndown capacity. Proper liquor distribution on the heating surface is critical for high performance operation. It has been suggested that a greater part of the circulation pump power goes to the boundary layer, and thus promotes effective film boiling at relatively low power consumption.

Crystallizer type concentrators employing the forced circulation concept (*58, 59, 61*) are also widely used for concentrating black liquor to 67–73% with higher concentration to 75% or more possible. Figure 3.20 shows a typical configuration with two horizontal heaters. The concentrator opertion is based on the controlled precipitation and crystallization of inorganic salts in the black liquor when their solubility limits are exceeded. A sludge-type black liquor is produced with the small inorganic crystals apparently acting as sites for crystal growth in preference to the tube walls. The sludge is kept in circulation at high velocity which reduces the operating viscosity by shearing action (*59, 61*) with a consequent enhancement of the heat transfer

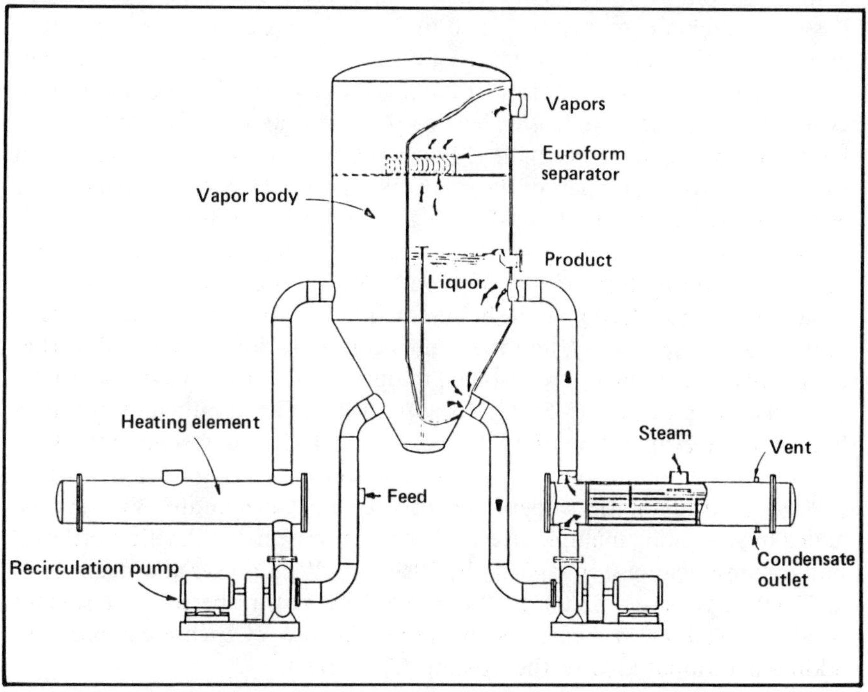

Fig. 3.20 Forced circulation high solids concentrator (courtesy of HPD, Inc.).

coefficient. Evaporation itself occurs by flashing in the vapor body. While circulation pump power requirements could be high, the energy usage is minimized by improving the heat transfer rate and by taking advantage of the reduction in high solids operating viscosity by shear action.

New evaporation technology

Rising energy costs have contributed to the development of evaporator systems such as mechanical vapor compression evaporation, blow heat evaporation, and flash evaporation to reduce the cost of concentrating spent black liquor. High solids concentration of black liquor to 75–80% solids as a means to improve boiler thermal efficiency is also gaining favor.

Mechanical vapor compression evaporation

Mechanical vapor compression evaporation (VCE) systems for the pre-evaporation of kraft weak black liquor (*62, 63*) and concentration of black liquor over 50% have been described (*60*). VCE can be considered for kraft black liquor in areas such as the Pacific Northwest or Quebec where the cost of electric power is inexpensive compared to steam.

In vapor compression evaporation, the vapor produced by the boiling liquor passes through a compressor, which increases its pressure and temperature and returns it to the heating element of the evaporator. Figure 3.21 illustrates the principle of VCE operation. Liquor is circulated over the falling film evaporator surface. Condensate leaving the unit is used to preheat the feed, hence improving overall economy. Foam generation can be significant with weak liquor; therefore, part of the product liquor is recycled to keep the solids concentration above the foam-forming solids concentration.

VCE systems generally use single-stage centrifugal compressors with a pressure ratio (discharge/inlet) of about 1.5 to 1, producing a temperature of about 20°F (11°C) above the boiling point of water. Because of the sharp boiling point rise of black liquor as solids concentration is increased, earlier commercial VCE units were installed to operate on liquors having a boiling point rise of up to 10°F (5.55°C). Its application in kraft pulp mills was thus limited to pre-evaporation of weak black liquor, followed by a multiple effect evaporator to obtain solids concentration appropriate for boiler firing.

Table 3.10 summarizes operating costs for a pre-evaporator VCE system added to an existing multiple effect evaporation system at a Pacific Northwest mill to boost evaporator capacity by 30% (*64*). With power and steam costs at \$0.005/kW·h and \$2.00/1000 lb (\$4.5/1000 kg), respectively, a savings of about \$900,000 per year was projected for the VCE case compared to adding additional area to the existing LTV effects.

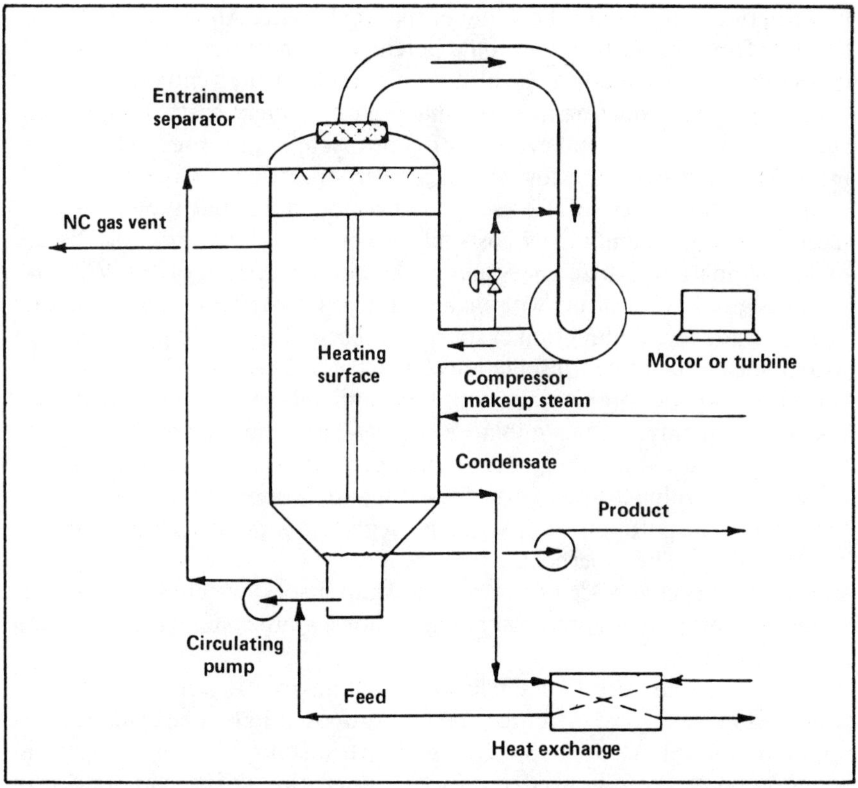

Fig. 3.21 Vapor compression evaporation.

Table 3.10. Operating costs for a six-effect multiple-effect evaporator vs a mechanical vapor recompression evaporator

Evaporator type	*Six-effect multiple-effect evaporator*	*Mechanical vapor recompression evaporator*
Steam required, Kg/h	26,300	0
Power required, kW	200	3,000
Cooling water required, liter/min	11,355	378
Annual operating cost (8,400 h/yr), $	1,032,000	140,000

Also, the VCE pre-evaporator facilitated foul condensate segregation, as 50% of the total condensate BOD was concentrated in the pre-evaporator condensate stream. The volume of this foul condensate may be even further reduced by arranging for co-current flow of steam and condensate in the pre-evaporator to avoid subcooling of the condensate. Allowing a 10% venting rate from the VCE system could potentially concentrate more than 60% of the BOD and malodorous sulfur compounds in the vent vapors.

A substantial reduction in foul condensate treatment costs is anticipated since only 3% of the total evaporator condensate requires secondary treatment. Multi-stage compressors or compressors operating in series can be used to increase the differential pressure and temperature, but the consequent increase of capital and power costs have to be carefully evaluated. The use of low temperature differences across the boiling surface of the VCE implies design of evaporators with high heat transfer coefficients and minimal fouling tendency. Falling film evaporators, both of tube and plate type, are used in the design of new installations (*28, 60, 65*). Good entrainment separation characteristics of the evaporator body and a short, well-insulated vapor line are often employed to minimize droplet carryover which is harmful to the compressor blades. Some degree of superheat at the vapor inlet is maintained by returning a small part of the discharge vapor to the suction inlet. Since this is wasteful on compressor horsepower, superheat should be kept at a minimum. The superheat, of course, protects the impeller from any entrained droplets. The use of centrifugal compressors, which mostly replace the earlier rotary type compressors, have allowed larger system capacity and lower capital cost.

A VCE system with two single-stage compressors to concentrate black liquor from 15% to 65% solids has been proposed (*60*). The evaporator is split into multiple sections that operate in parallel on the steam side and in series on the liquor side, so that only a small portion of the evaporator will be exposed to the most concentrated liquor with its low heat transfer coefficient and high boiling point rise. Steam for this section of the evaporator has two-stage compression which provides ample temperature differential for good evaporation performance. The specific power consumption is estimated at 29 kW·h per 1000 kg of evaporation.

Several advantages claimed for VCE systems are listed below:

1. High evaporator availability because of low boilout or water wash requirements.
2. Surface condenser is not required due to low or no cooling water requirements. (A small amount of cooling may be required to condense the vapor associated with noncondensible gases). If a mill has excess warm water, elimination of cooling water needs represents a net savings in energy as shown in Table 3.11 (*29*).
3. High turndown rate.

Table 3.11. Energy use in VCE system (*29*)

Evaporation system	*Energy consumption Kcal/Kg water evaporation*
Multiple effect LTV	
(5 effects)	140*
(6 effects)	110*
Mechanical vapor compression	28

*Assumes heat loss to surface condenser cooling water is not reused for other process requirements.

4. VCE used as a pre-evaporator is not connected to the regular multiple effect LTV evaporators on the vapor side and is, thus, flexible in its capacity and operating rates.
5. Power to run the compressor can come from onsite or purchased electricity.

It may be mentioned that VCE systems could involve higher capital investment than conventional evaporators because of the added compressor costs and the lower available temperature difference which dictates greater heating surface areas. This added cost may be recovered in savings from operating expenses in certain situations. The application of VCE systems in various alternative black liquor evaporator schemes has been evaluated using the GEMS computer program (*66*). As steam costs have been escalated more rapidly than electric power in many areas, VCE has become an attractive option at several locations.

Blow-heat evaporation

Pulp mill batch digesters blow pulp into a blow tank to reduce pressure from 100 psig (790 kPa) to atmospheric pressure. The steam released is usually condensed in a barometric condenser mounted on top of a large hot water accumulating tank. This condenser produces condensate at about 205°–210°F (96°–99°C). In blow-heat evaporation, this hot water is used as the heating medium in a two or three effect unit to partially evaporate weak black liquor. The blow-heat evaporator was developed and patented by Jacoby (*67*) in the early 1960s. It has gone through a gradual evolution, and current configurations accomplish much more than initially foreseen (*64, 68*). A three-effect blow heat system using falling film evaporators is schematically shown in Figure 3.22 (*68*). Since the falling film evaporator does not require a large temperature differential for proper operation, it permits the operation of a triple effect system in place of the earlier double effect systems. It allows

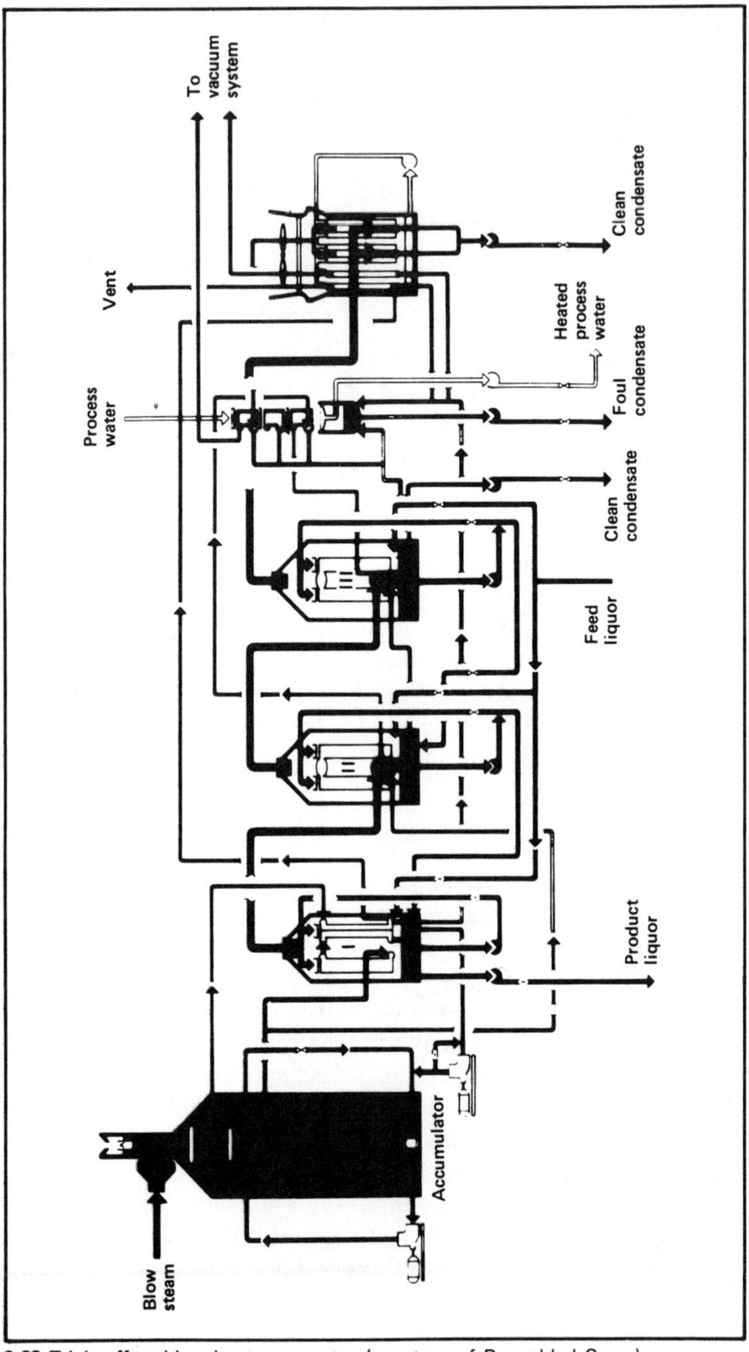

Fig. 3.22 Triple-effect blow heat evaporator (courtesy of Rosenblad Corp.).

segregation of evaporator and accumulator condensates. This is accomplished by making the first effect of the blow heat evaporator into the secondary condenser for the accumulating system. With this arrangement, volatiles are segregated into a separate stream of condensates, and accumulator condensates remain low in BOD and odor. By passing the weak liquor through the blow heat evaporator which does about 35% of the total evaporation, this liquor is essentially steam stripped and will generate low-odor, low-BOD condensates in the subsequent evaporation.

The blow heat system permits heating of boiler feedwater or mill process water to 190°F (90°C). Recently, a three-effect free flow falling film system was put into operation at a southern mill. The unit is capable of 470,000 lb/h (213,000 kg/h) of evaporation while heating 1000 GPM (3800 liter/h) of boiler feedwater from 80° to 190°F (26° to 88°C). Cooling water requirements are minimal, as an evaporative condenser (Figure 3.14D) has been incorporated.

Flash steam evaporation

Evaporation systems utilizing flash steam from continuous cooking systems have been developed (*69, 70, 71*) to better utilize the higher heat content of the extraction stage black liquor. This liquor, usually at about 320°F (160°C), is flashed in two stages. Steam from the first flash stage is used for pre-steaming wood chips and second flash stage steam is used to produce hot water. The two systems shown in Figure 3.23 utilize flash steam to concentrate weak black liquor in a pre-evaporation step prior to the multiple effect evaporation plant.

Figure 3.23A shows a system in a southern mill using three falling film evaporator effects. Steam from the second flash tank and pre-steaming vessel exhaust are used in one or two subsequent stages before generating hot water (*71*). Steam consumption from the main evaporation plant is reduced by 20–30%. As contaminated condensates are segregated, cleaner condensates are produced in the evaporation plant.

The Lockman Multiflash evaporator (Figure 3.23B) has been used for kraft liquor pre-evaporation up to 18% solids concentration (*70*). In contrast to conventional multiple-effect systems where the liquor first passes through the heat exchanger section followed by flashing, in this evaporator, the liquor first passes through a series of heat exchangers and then through the multiple flash chambers in series. Circulating liquor is pumped from the bottom flash chamber upwards through the liquor heat exchangers in series which are heated by flash vapors. Final heating by live steam takes place in the primary heat exchangers on top of the column. The hot liquor from the primary heat exchangers passes downwards through the flash chambers in series, flashing across the orifices. Thus, the pressure and temperature are successively reduced to a temperature of about 131°F (55°C) in the bottom of the flash chamber. The heat exchangers in the two lower flash chambers operate as surface condensers. The combined vapor condensates and non-condensibles are drawn off from the lowest heat exchanger to a separation vessel where noncondensible gas is drawn to a vacuum pump.

In the Lockman pre-evaporation plant, the extraction liquor from the digester is generally expanded to 212–248°F (100–120°C) in two consecutive flash tanks. The vapor from the first flash tank is used in the pre-steaming vessel, and the vapor from the second flash tank is combined with the relief steam from the pre-steaming vessel in the primary heat exchangers of the evaporator column. Alternatively, vapors from both flash tanks may be used in the primary heat exchangers (and the steaming vessel is fed with flash steam from the Lockman column).

High solids concentration

The prior section stated that the concentration of black liquor has been normally limited to about 65–68% solids because of the high boiling point elevation, high viscosity, and increased fouling of the heating surface. The Institute of Paper Chemistry and the University of Maine have research programs to develop data on the properties of high solids black liquor. Current trend, however, has been towards developing concentrator technology to raise solids to well over 70% and to operate these concentrators integral with multiple effect evaporators to increase overall steam economy. High solid concentration is attractive from the energy conservation viewpoint, because over 40%

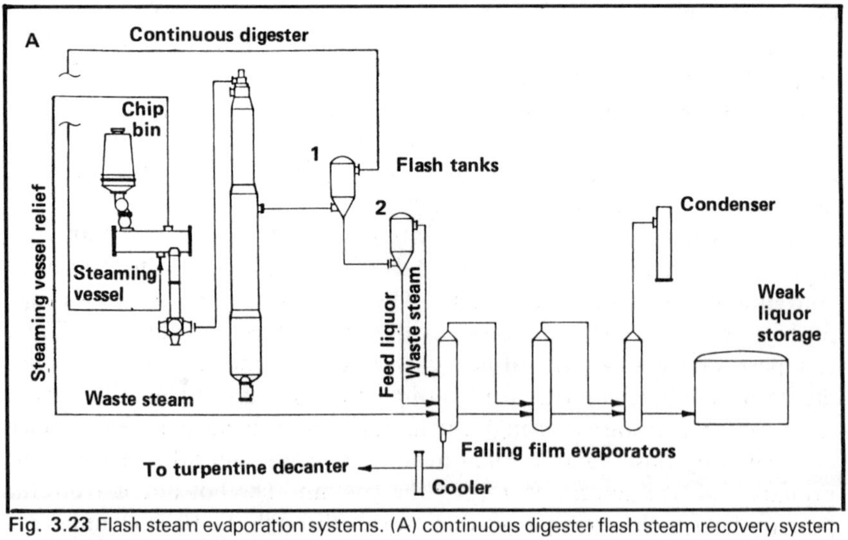

Fig. 3.23 Flash steam evaporation systems. (A) continuous digester flash steam recovery system (courtesy of Kamyr, Inc.)

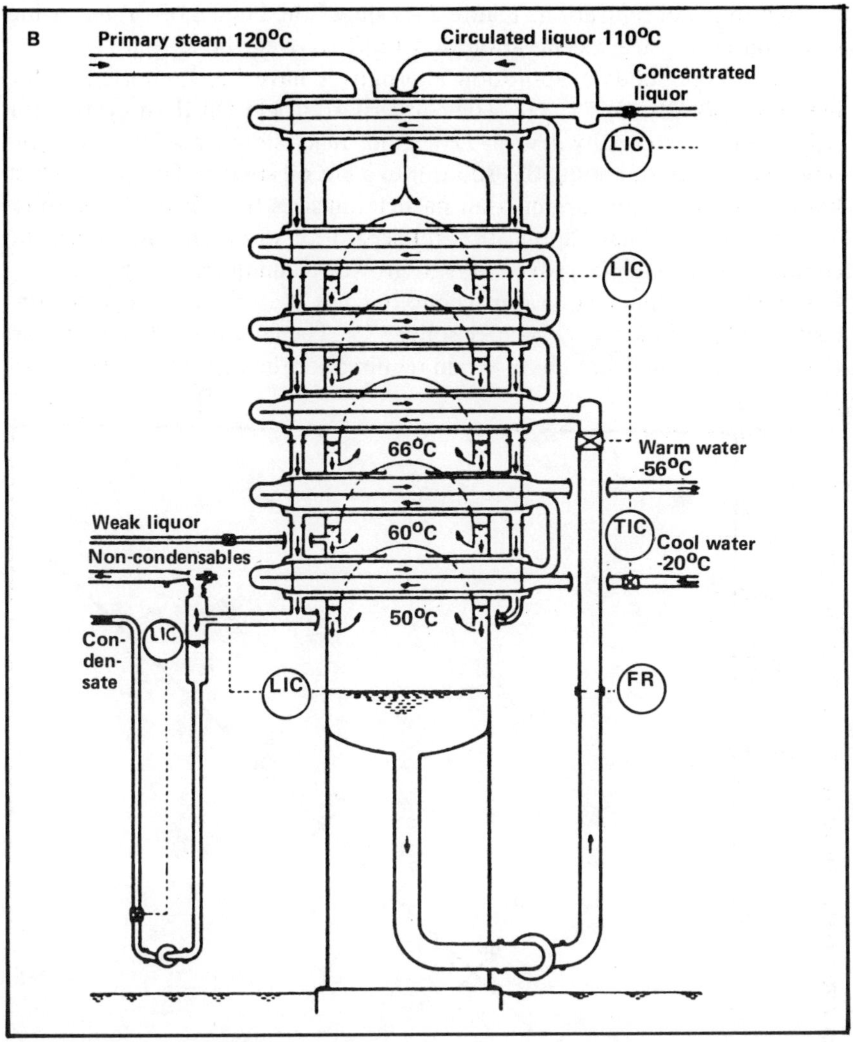

Fig. 3.23 (B) lockman multiflash evaporator.

of the total energy consumed for water removal from black liquor is used in the recovery furnace to evaporate the remaining 10% of the water at 1:1 economy.

Approximately 60% of this energy is used at multiple effect evaporators to remove 90% of the water at about 5:1 economy. Thus, concentration to higher solids prior to the furnace may save a considerable amount of energy. Black liquor throughput to recovery boilers could also potentially be increased by high solid concentration. Figure 3.24 shows the impact of raising firing solids on boiler steam generation in a kraft recovery boiler.

Several high solids evaporation alternatives have been reported in the literature (*60, 61, 65, 72, 73*). The Carver-Greenfield Oil flash evaporator (*73*) produces essentially dry 95–99% solids black liquor particles by injecting black liquor of about 50% solids into a hot oil stream. The dry product has free inorganic and organic salt particles in sizes from 7 to 700 microns. The method could have application for separation of organic and inorganic compounds through improved crystal growth techniques.

Wiped (thin) film evaporator technology has been demonstrated to concentrate black liquor to 80%–90% solids (*72*). The pilot plant data indicated that the incremental amount of steam required for the wiped film evaporator

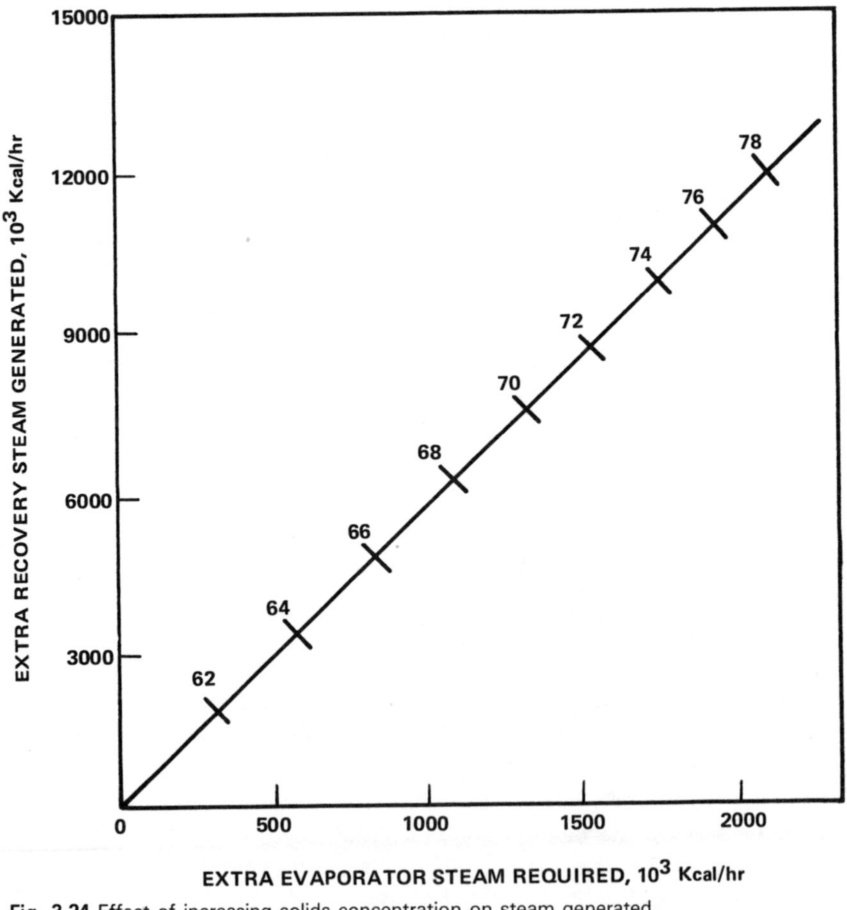

Fig. 3.24 Effect of increasing solids concentration on steam generated.

to concentrate 65% solids black liquor to 80% was almost double of that used in the multiple effect evaporator. The process economics was, however, found to be only marginally attractive.

The plate type falling film evaporator (*60, 65*) and the evaporator/crystallizer (*59, 61*) discussed earlier both appear to be viable alternatives to achieve 75–80% final solids. A major thrust in both approaches has been the ability to operate with high heat transfer rates at low steam-liquor temperature difference. Efforts have also been directed to either minimize scale formation (*58, 59, 61*) or to enable easy descaling without loss in capacity (*60, 65*).

Technology in the high solids evaporator area is being continually developed and merits further consideration, since high solids black liquor combustion not only increases recovery boiler thermal efficiency, but could be an alternative to obtain increased capacity from existing recovery boilers in terms of higher solids throughput.

Evaporator troubleshooting

The ability to quickly and accurately troubleshoot an evaporator is often critical to a mill's production because of the time involved in shutdown, repair, and startup. If a shutdown is required to diagnose a problem, it is essential to know where the problem is. Multiple shutdowns to correct evaporator problems obviously are very costly and should be avoided. Techniques and basic fundamentals for LTV evaporator troubleshooting have been reviewed (*74–76*) and are summarized below.

Problems in multiple effect evaporators can be broadly divided into four categories: low capacity, low steam economy, contaminated condensates, and instability.

Low capacity

Low evaporator capacity often limits pulp production since many mills have insufficient evaporator capacity and limited weak and heavy liquor storage. The symptoms of low evaporator capacity are obvious when sufficient liquor cannot be fed to the evaporator to maintain levels in the weak liquor tanks and, at the same time, produce acceptable solids to the recovery boiler. This problem is directly related to the working temperature drop and heat transfer coefficient of the evaporator system. Reduced heat transfer coefficient is often caused by scaling. The nature of evaporator scaling and cleaning methods have been discussed in detail in the previous section. Figure 3.25 gives a quick reference chart for low capacity problems.

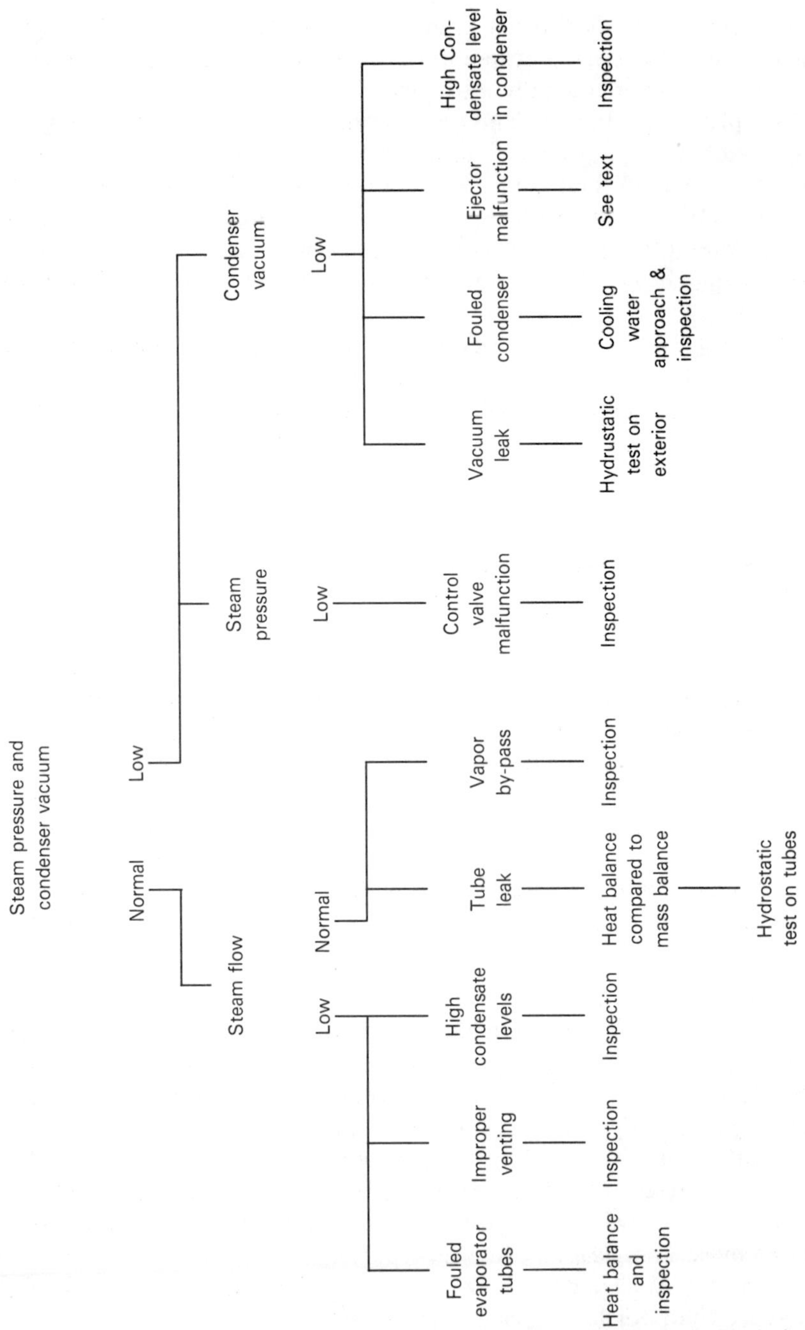

Fig. 3.25 Troubleshooting quick reference chart for low capacity (courtesy of Goslin-Birmingham, Inc.).

Low steam economy

Steam economy is important because of increasing energy costs. Recalibration of steam and liquor flowmeters periodically is a good practice to obtain accurate information pertinent to the evaporator steam economy. The following are possible causes of low steam economy:

1. Excessive venting especially from the first, second, and third effects;
2. Vapor blowing or liquor piping into downstream effects and product flash tanks. This may be due to improperly sized orifices, lack of loop seals in the piping, and leaks in bypass valves;
3. Fouling of after heaters so that liquor from the colder effect is not properly reheated before entering the hotter effect. Afterheaters are important because exclusion of them imposes high preheat duties on the upstream effects. If malfunctioning of afterheaters are suspected, approach temperature can be determined and compared to that of the clean heater.

Evaporator instability

Instability during normal operation is often a difficult problem to analyze because it can result from a number of causes, and heat and material balances are not helpful in these situations. Instability in multiple effect evaporators can often be characterized by fluctuations in steam flow and product solids concentration. During normal operation, steam flow and product solids should be reasonably steady. Evaporator instability can be mainly attributed to unsteady liquor flow. The liquor moves in surges from one effect to the next by the action of the liquor pump from the liquor downleg. As the liquor is pumped, the level of the liquor in the downleg is decreased and the pump NPSHA is reduced below NPSHR, at which time the pump cavitates. The liquor level rises again in the downleg, and the pump reprimes and starts to move the liquor. If the time between surges is too great, the next effect is momentarily dried out. Normal time between surges is about a few seconds. Other causes could be vacuum leaks, tube leaks, and excessive turndown. Obvoiusly, process control systems would greatly improve stability.

Contaminated condensate

Contaminated condensates are often produced by a nonfunctioning entrainment separator or evaporator tube leakage. Figure 3.26 presents a quick checklist for this problem with centrifugal separators. In impingement type separators or mesh pads, problems with entrainment are usually associated with the dislodgement of the pad due to high pressure drop under upset conditions.

Instrumentation and process control

Evaporator instrumentation has enabled mills to not only troubleshoot and improve evaporator performance, but also to schedule boilouts and institute

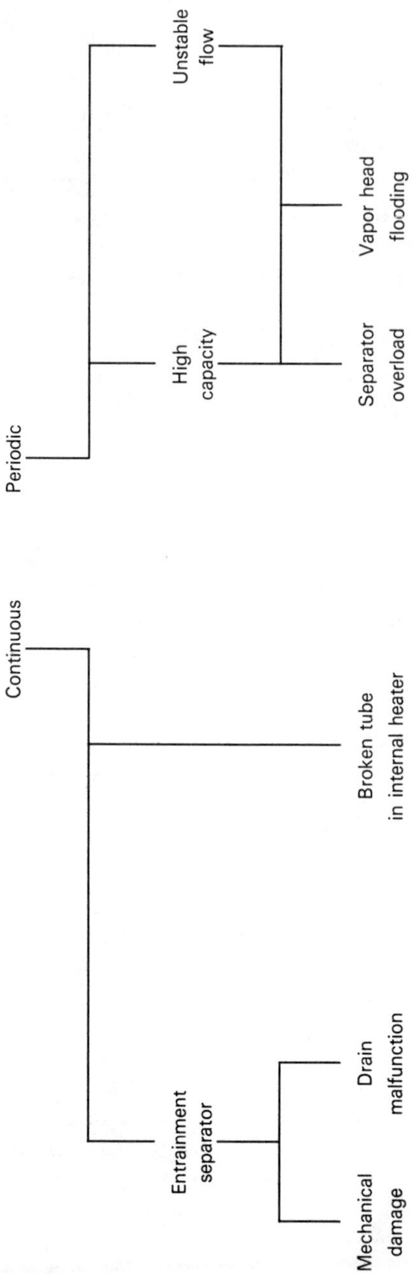

Fig. 3.26 Troubleshooting quick reference chart for contaminated condensates (courtesy of Goslin-Birmingham, Inc.).

preventive maintenance programs. Distributed control systems are replacing the conventional instrument panel in newer mills. Key instrumentation features are briefly highlighted below.

Steam supply

Steam pressure and flow recorders are needed. (Steam pressure is usually controlled to maintain temperature difference). Pressure controller setpoint is adjusted according to the boiling point rise.

Weak black liquor

Feed liquor flow is measured with orifice plates/magnetic flow meters (for feed flow split to two effects, ratio controller is also used). Baumé recorders/Gamma gauges/refractometers are used for solids concentration. Level controller in soap separation tank is used to regulate feed flow to the next effect. Liquor level recorder is used in the weak liquor storage tank.

Strong liquor

Liquor level recorder for strong liquor storage and Baumé recorder for solids concentration are needed.

Condensates

Level controllers are needed to hold a condensate seal in each effect and allow condensate flashing. Conductivity meters with alarm system are used on combined condensate and steam condensate lines to dump contaminated condensate and also warn operators.

Temperature/pressure

Liquor temperature in each effect is recorded with an electronic-multipoint temperature recorder. Local pressure gauges should be mounted in the vapor dome of each effect.

Surface condenser

A cooling water flow meter, outlet water temperature indicator, and vacuum recorder are generally provided.

Several workers (*77–80*) have discussed computer simulation models describing the dynamics and control of multiple effect evaporators and the control strategies proposed have relied on classical feedback/feedforward algorithms. Actual industrial application of these computer models in distributed control systems have, however, not been disclosed.

Several vendor-packaged systems are currently available (Measurex, Nokia, Sentrol, and AccuRay), wherein the vendor supplies the necessary hardware, software, and support programs for maintenance, etc. Typical analog inputs to the process control computer would include signals from most of the instrumentation listed above.

The basic objectives for process control of black liquor evaporators are:

1. Stabilize evaporator operation and maintain uniform product concentration to the recovery boiler.
2. Maximize black liquor solids throughput.
3. Improve overall evaporator efficiency in terms of steam economy.
4. Respond to swings in production rate.

Control of product liquor concentration in a steam-limited situation is achieved by adjusting the weak black liquor flow. If steam supply is not limited, the steam flow and pressure together are the primary control variables. The important properties of an evaporator control system are to quickly damp oscillations in product solids; adjust steam flow; compensate for upstream changes in header pressure; and correct feed liquor solids concentration.

Auxiliary operations in the evaporation plant

Besides evaporation of black liquor in multiple effect evaporators, other operations to minimize the malodorous nature of evaporator gases and condensates and to enhance the evaporator performance with fiber and soap removal are equally important. These operations are briefly discussed in the following sections.

Fiber removal

Weak black liquor from the brown stock washers generally contains some pulp fiber and shives. The trend, in recent years, has been to increase the width of the extraction screen slot in the digester and to use a more open mesh cover screen on the pulp washers. While this has reduced screen and wire blinding and improved throughput in the digester and washing areas, it has resulted in a general increase in the fiber content of the black liquor to the evaporators. Moreover, it has increased the chances for occasional fiber bundles and shives to be present in the liquor stream. Total fiber concentration in the black liquor stream under these circumstances can attain levels of 150–200 ppm.

While such fiber concentrations may not cause difficulties in the back bodies of the evaporator line, fiber plugging can occur in the mist eliminators on some designs, particularly if flushing is not employed or if mesh pad type eliminators are used. Also, fouling in the front bodies and particularly in the concentrators will be more severe with such fiber inclusions in the liquor streams. Most mills have, therefore, installed black liquor filters with sufficiently fine screens to capture all coarse fibers and shives and minimize weak liquor fiber content. Moreover, such filters provide a safeguard against fiber shock loads that can occur when mechanical damage or wear results in breakage or rupture of the extraction screen or the face wire of a pulp washer.

Numerous designs of black liquor filtration equipment have been used in the past. These range from the very simple "side hill" screen to the traveling wire type drum filter, which employs a precoat technique to assist in providing a very clear liquor to the evaporators. While the more complex design provided a better clarity, the degree of operator attention and required maintenance also increased.

The black liquor filter design most commonly used today is of a gravity type, which is automated to a large degree to minimize operator attention. The basic construction consists of a drum cylinder usually fitted with a 100 mesh screen and mounted in a high submergence vat. The inside of the cylinder is open and the driving force is created by the difference in level in the vat and inside the drum. Since only small amounts of fiber are present in the feed, there is little restriction to flow and a relatively small size filter, approximately 6 ft (2 m) diameter, can handle the black liquor flow from an average size pulp mill.

Under normal operation the filter rotates at a fixed speed with the incoming liquor passing through the screen and discharging from the filter to the weak black liquor storage tank. Fibers filtered out on the surface of the screen accumulate in the vat and are removed in a small bleed stream back to the blow tank or first stage pulp washer. Fine-sized fibers may accumulate in the cover screen and reduce the flow volume handled by the unit. When this occurs, the level in the vat will rise and the higher liquor level will activate a controller which is set to actuate a solenoid valve that opens a steam line to an internal steam cleaning spray. The passage of steam from within the cylinder toward the outside effectively cleans the screen. With a clean screen, the filter will handle a higher flow rate and the vat level will drop to a normal operating position. The liquor level sensor will in turn close the solenoid valve and stop the internal steam cleaning.

A normal fiber content in the weak liquor from a black liquor filter should be in the range of 5–25 ppm. This low fiber level can be usually handled by the black liquor evaporators, particularly since any fiber bundles and shives would be eliminated.

Economic justification for the installation of the black liquor filter can often be made on the basis of the fiber recovery values alone. The primary purpose, however, is the increase in on-line time for the black liquor evaporators due to reduced fouling.

Soap recovery

The alkaline pulping of softwoods liberates resin acids, fatty acids, and unsaponifiable compounds, which exist as sodium salts of soaps that are carried with the black liquor from the pulp washing system to the evaporators. The quantities of soaps produced vary considerably with the species of wood pulp, the percentage of total solids of the black liquor from which soap is skimmed, and the retention time. Soap yields are considerably higher in the southern portion of the United States as compared with other areas of the country.

While conversion of black liquor soaps to refined soaps or tall oil has long been practiced in the South, increased interest in soap skimming in the northern and northwestern states has been largely brought about by the rise in the market price of resin. This has stimulated considerable research into the most economical manner in which to obtain the maximum recovery of tall oil.

While the recovery of soap has proven to be a very valuable byproduct of the alkaline pulping operation, a highly important and corollary benefit was gained through improved evaporator operation due to soap removal. This is particularly true for rising-film evaporators which discharge at 48–50% solids. The presence of soap in the first effect body alone, but often together with pulp fibers, formed a deposit on the tube walls that severely limited capacity by increasing the temperature differential in that body, which reduced the temperature differential available for the balance of the evaporator. The decreased overall heat transfer rate necessitates more frequent evaporator tube cleaning. The skimming of soap has contributed to high overall evaporator capacity with the elimination of the need for frequent tube-descaling operations.

The crude soap is recovered from the black liquor after it has been partially concentrated by the multiple effect evaporator (*81–83*). The concentrated black liquor is cooled and allowed to settle where the fatty acid and rosin soaps separate as a brown curdy mass. It is then skimmed off for further processing into tall oil. The recovery of tall oil with skimmings depends on the retention time and black liquor solids content; 3 hours and 25–29% are recommended respectively. About 10% liquor solids and 30–40% water are occluded in the skimmings. The soap skimmings containing the tall oil components as their sodium salts are reacted with H_2S0_4 to convert the sodium soaps into free fatty and resin acids. The reaction mixture is then allowed to settle for the organic acids to separate from Na_2S0_4 solution formed during the acidification and from the lignin and other impurities.

The crude tall oil is further washed with hot water, allowed to settle, and drawn off the storage tanks. The tall oil compositions change according to the wood raw material. In the case of scots pine it may consist of about 50% fatty acids (linoleic, oleic, linolinic, stearic acids); 40% resin acids (levopimaric, dihydroabietic, abietic acids); and some 10% unsaponifiables (ligno ceryl alcohol, B-sitosterol, B-sitostanol). Tall oil can be further refined either by acid refining or by distillation to remove odorous sulfur components and color bodies, which are mostly oxidized resin acids (*82*).

Drew (*81*) has studied a general tall oil material balance, from the time resin acids and fatty acid esters are in the living tree until it is recovered as crude tall oil. This result indicates that less than 45% of the tall oil available in the living tree is recovered. The greatest single loss may be from deterioration in storage, but during processing, about 30% of the remaining is also lost. It is essential that high brown stock washing efficiency is attained to enhance tall oil recovery. Average loss of tall oil following the washing step is about 3–25 lb/ton (1.5–13 kg/metric ton). There are usually some unsaponifiable rosin and fatty acid esters in the pulp that cannot be washed out with water. Negative tall oil soap displacement ratios may be observed in the later stages of washing. This indicates that the soap may be readsorbed on the pulp. Crude tall oil recovery during washing may be enhanced by: 1) pressing the pulp to as high consistency as possible prior to dilution with filtrate liquor; 2) use of wetting agent, dispersant, or solubilizing agent.

Other considerations in the handling of weak liquor and skimming practices to enhance the tall oil recovery are summarized here:

1. Since long storage time of black liquor may deteriorate the tall oil, especially if it is maintained at high temperatures, weak liquor should be pumped directly to the evaporators using the storage tank only for surge.
2. Up to 60% of the tall oil can be separated from weak black liquor storage tanks by sweetening the weak liquor with recycled strong liquor to about 17% solids. This results in improved overall recovery efficiency of the tall oil from black liquor.
3. Complete separation of hardwood and softwood black liquors is desirable since mixing of these liquors cause a decrease in tall oil recovery efficiency. This is probably because of the change in saturation of the tall oil. It has also been pointed out that the higher concentration of calcium in hardwood liquors may inhibit soap separation. At 75% hardwood and 25% pine black liquor there is no separation of soap; at 25% hardwood liquor, the loss in tall oil is about 5%.
4. One of the main reasons for incomplete tall oil separation in the skimmer is attributed to the solubility of tall oil. The solubility of tall oil in pine black liquor is highest in weak black liquor, about 8 lb/1000 lb of solids (8 kg/1000 kg), lowest at intermediate solid concentrations, about 3 lbs/1000 lbs solids (3 kg/1000 kg), and increases with increased concentration. The minimum point of tall oil solubility exists because

inorganic salts tend to salt out the soap, while lignin salts tends to solubilize the soap. The addition of salt cake before skimming may be used to enhance the recovery of tall oil.

5. Parameters affecting recovery of tall oil in the skimmer include concentration of solids, temperature, retention time, composition of tall oil, pH, sulfidity, and reactor configuration. Generally, smaller losses of soap result when the temperature is lower; retention time is greater; fatty to resin acid is higher; sulfidity is lower; concentration of solids has little effect in the range of 22–28%; and the pH value between 11–13 is optimum. The skimmer should have the following characteristics: minimal short circuiting (baffles), minimal distance for soap to rise (small H/D ratio), minimum turbulence, and no areas in the tank that has the downward movement greater than rising velocity of soap, about 2.5 ft/h (0.8 m/h).
6. Separation of tall oil soap components generally has the following efficiency: 70% for fatty acids, 55% for resin acids, 55% for unsaponifiables, and 20% for oxidized compounds.

Black liquor oxidation

Black liquor oxidation (BLO) has been widely accepted as a method to prevent the generation and stripping of hydrogen sulfide and other malodorous organic sulfur compounds from black liquor in direct contact evaporators for odor control in kraft recovery boilers. H_2S is released by reaction between the CO_2 in the flue gas and Na_2S in the black liquor. Newer "low odor" boilers equipped with concentrator evaporators do not require black liquor oxidation for odor control. However, as many older boilers continue to employ black liquor oxidation, a brief discussion appears warranted. The Institute of Paper Chemistry is exploring the potential application of BLO as a means to debottleneck capacity-limited recovery boilers. The overall stochiometry and reaction kinetics of black liquor oxidation have been reported by Bentvelzen (*84*) . At conventional oxidation temperatures, normally above 160°F (70°C), the major reaction is the oxidation of Na_2S to non-volatile sodium thiosulfate as follows:

$$2\ Na_2S + 2O_2 + H_2O \rightarrow Na_2S_2O_3 + 2\ NaOH.$$

There is some heat evolved as the reaction is exothermic. Also, several side reactions occur involving the complexing of mercaptans with the lignin. The oxidation of mercaptan to dimethyl disulfide also takes place, but is inhibited when the sodium sulfide is more than 0.2 g/l concentration. The dimethyl disulfide and lignin-mercaptan formed may undergo reversion reactions to regenerate the mercaptan and hydrogen sulfide. As the sulfide is converted to thiosulfate, contacting the oxidized black liquor with the boiler flue gases minimizes the release of H_2S.

Oxidation of black liquor can be conducted either with weak or strong black liquor. The earlier weak black liquor oxidation systems were used mostly with nonfoaming liquors. Foaming is caused by the tall oil components in the weak black liquor. Chemical defoamers (oil, turpentine, kerosene) and mechanical foam breaker systems have been used to combat severe foaming in liquors with limited success. In the southern United States, mills pulping southern pine consequently adopted strong black liquor oxidation, as the tall oil is removed during soap skimming. Several advantages have, however, been claimed for oxidizing weak black liquor rather than strong black liquor (*85, 86*): a) evaporator corrosion is reduced and heat transfer is, therefore, better; b) malodorous noncondensible gases such as H_2S evolved with evaporator condensates are substantially reduced; c) sulfur losses in multiple effect evaporator can be reduced significantly; d) lower blower horsepower requirement of about 0.08 hp/ton pulp (65 watts/ton), compared to about 0.5 hp/ton pulp (400 watts/ton) for strong liquor oxidation; and e) avoids sparger nozzle pluggage.

Besides foaming, another drawback of weak liquor oxidation is the partial reversion of the oxidized sulfur compounds to hydrogen sulfide during black liquor evaporation. Several types of black liquor oxidation equipment have been used in kraft pulp mills. These include packed tower, bubble tray tower, and air-sparged units with and without stirring. Air or oxygen have been used to accomplish oxidation. Systems using oxygen tend to be tubular plug flow reactors. The air-sparged design predominates for strong black liquor oxidation systems. A typical air-sparged oxidation system is shown in Figure 3.27. In several mills, two-stage strong black liquor oxidation is practiced to ensure the 99% sulfide oxidation efficiency that is required to meet current boiler odor release standards. In some mills, an oxygen system is used as a final "polishing" stage to reduce sodium sulfide concentration in the oxidized black liquor to less than 0.2 gm/L. Black liquor oxidation systems can be operated with liquor from an intermediate evaporator effect since strong liquor at 50% solids tends to cause air nozzle pluggage.

NCASI has recently published a survey of the current design criteria and performance of new black liquor oxidation systems (*87*). Stringent recovery boiler odor control standards require the continuous operation of black liquor oxidation systems at high sodium sulfide oxidation efficiencies. Table 3.12

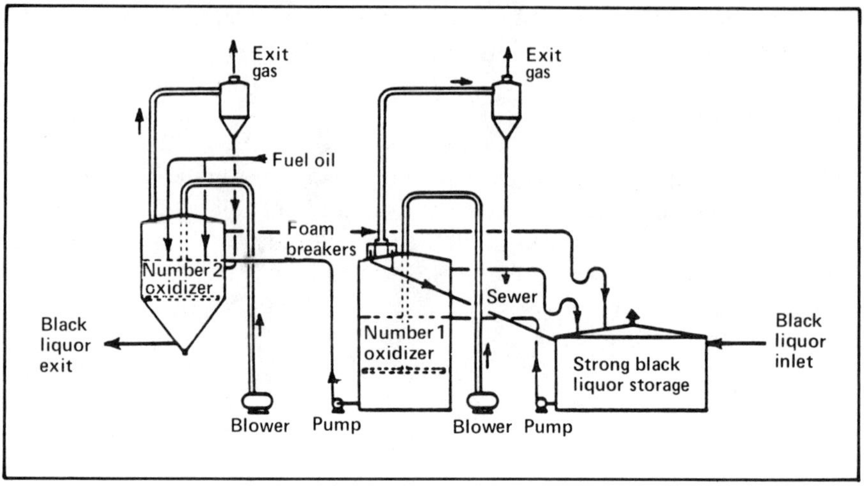

Fig. 3.27 Champion two-stage unagitated strong black liquor oxidation system.

is reproduced from an Environmental Protection Agency (EPA) report (*88*) and gives typical operating and performance data for black liquor oxidation systems.

Treatment of contaminated condensates

Foul condensates in a kraft mill are produced from two major sources: digester area and multiple effect evaporator. The total condensate flows vary from mill to mill. Typical flows for a 500 ton/day kraft mill are 19,000–23,000 L/h (5000–6000 gal/h) digester condensates, and 130,000 L/h (35,000 gal/h) evaporator condensates. Table 3.13 shows representative amounts of contaminants in the condensates for batch and continuous digester plants (*89*). It is evident from Table 3.13 that BOD (attributed mainly to methanol) and TRS (due to malodorous sulfur compounds) are the major polluting components in evaporator condensates. Methods proposed or used for treatment of condensates include carbon adsorption, air stripping (*90*), stripping with flue gas (*91*), and steam stripping (*92, 93*).

Air and steam stripping processes are the two methods which have gained general commercial acceptance. Environmental standards require the removal of TRS, and, in recent years, BOD from these condensates. Condensate stripping is common in Scandinavia. In North America several units have been installed in the last few years.

Air stripping can be achieved by passing the condensates downward countercurrently with air either through a packed or a sieve tray tower. Air stripping is generally sufficient for removing TRS compounds. The stripped condensates can be reused in the mill or discharged and the air containing

Table 3.12. Typical ranges in operating variables reduced sulfur emissions and cost factors for black liquor oxidation systems

	Air oxidation systems			*Molecular oxygen systems*			
Item	*WBLO* only*	*SBLO** only*	*WBLO & SBLO*	*Digester only*	*WBLO only*	*SBLO only*	*WBLO & SBLO*
Operating variables:							
Oxygen requirement, act/theor.	4-6	5-8	7-10	2-3	1.1-1.3	1.4-1.7	1.2-1.5
Power requirement, kW/daily	0.2-0.8	0.4-1.2	0.6-1.5	0.1-0.2	0.02-0.08	0.02-0.04	0.04-0.12
(hp/daily ton)	(0.3-1.0)	(0.5-1.5)	(0.7-1.8)	(0.1-0.3)	(0.03-0.10)	(0.02-0.05)	(0.05-0.15)
Na_2S to direct contact evaporator, g/L	0.3-1.5	0.02-0.1	0.01-0.5	0.05-0.5	0.3-1.5	0.1-1.0	0.01-0.10
Na_2S oxidation efficiency, %	97-99	95-99	99-99	90-99	98-99	96-98	99-100
Reduced sulfur emissions:							
Evaporator gases, kg/L	0.05-0.5	0.5-5.0	0.05-0.5	0.025-0.25	0.05-5.0	0.05-5.0	0.05-0.25
Tall oil vent, kg/L	0.05-0.1	0.5-0.75	0.05-0.1	0.025-0.1	0.025-0.1	0.5-0.75	0.05-0.1
BLO tower vent, kg/L	0.05-0.25	0.05-0.15	0.05-0.3	0	0-0.1	0-0.005	0.025-0.25
Recovery furnace, direct contact evaporator, kg/L	0.1-1.5	0.05-1.0	0.05-0.5	0.025-0.5	0.05-1.5	0.1-1.5	0.05-0.25
Economic factors:							
Capital cost, $/daily	330-880	550-880	660-1100	11-55	27-137	55-220	55-165
($/daily ton)	(300-800)	(500-800)	(600-1000)	(10-50)	(25-125)	(50-200)	(50-150)
Annual operating cost, $/daily	11-110	27-220	55-275	33-550	11-330	22-440	17-385
($/daily ton)	(10-100)	(25-200)	(50-250)	(30-500)	(10-300)	(20-400)	(15-350)

* WBLO = Weak black liquor oxidation
** SBLO = Strong black liquor oxidation

Table 3.13. Pollution load of the volatile compounds (excluding turpentine which is recovered)

	Unit	*Continuous digester process*	*Batch digester process*	*Variation*
Digester area condensates				
BOD_7	kgO_2/ADT	3	5	3.9
TRS	kgS/ADT	0.1	0.2	0.1-0.9
Flow	t/ADT	0.1-0.4	1-2	
Evaporation plant condensates				
BOD_7	kgO_2/ADT	9	7	5-15
TRS	kgS/ADT	0.3	0.3	0.1-2
Flow	t/ADT	6-9	6-9	
Total				
BOD_7		12	12	7-15
TRS	kgS/ADT	0.4	0.5	0.1-2

odorous sulfides may be burnt to SO_2 in the lime kiln, recovery furnace, or power boiler. To prevent explosion when burning these malodorous gases, flame arresters and safety diaphragms are provided.

Steam stripping is effective in removing both TRS and BOD constituents from contaminated condensates. Figure 3.28 from an EPA report (*88*) shows that stripping steam requirements for BOD (methanol) removal are much higher.

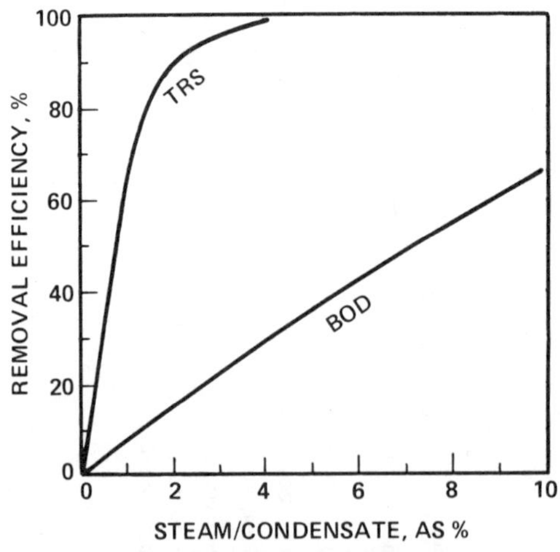

Fig. 3.28 Stripping efficiency for different steam condensate ratios with 10 theoretical plate column.

Several papers have discussed the design of condensate stripping systems (*94–96*) and the following approaches are identified:

1. Stripping in a packed or tray tower with live steam (*94, 95*).
2. Condensate stripping in a distillation column (*62, 94*).
3. Operating the stripping integral with the multiple effect evaporator (*60, 94, 95*).

Stripping steam requirements are about 9–13% of the contaminated condensate flow to achieve 80+% BOD removal and 95+% TRS removal.

As steam stripping can have high operating costs, only selected condensate streams are subject to treatment. For evaporators operating on the 5-6-4-3-2-1 sequence (as in North America), more than 70% of the TRS and BOD are generally concentrated in the surface condenser and 6th and/or 5th effect condensates, which together constitute less than 40% of the total flow. For a typical Scandinavian sequence (3-4-5-2-1), over 80% of the TRS is concentrated in the secondary condenser and vacuum pump condensates and more than 80% of the BOD is present in the 4th and 5th effect condensates. BOD containing condensates can be reused for brown stock washer showers and lime mud washing. Malodorous condensates, however, would require treatment prior to reuse. The current trend is to integrate stripper operation with the multiple effect evaporators. In some integrated systems, by condensate segregation, steam stripping is required on only 3% of the total combined condensates (*60*). Net steam requirements for an integrated system of this type are less than 2% of the total evaporator steam requirement, compared to 5% or more for stand-alone condensate strippers.

Literature Cited

1. Han, S.T., *Tappi*, **40**(11):921(1957).
2. Regestad, S.O., *Svensk Papperstidning*, **54**(2):36(1951).
3. Frederick, W.J., Sachs, D.G., Grady, H.J., and Grace, T.M., *Tappi*, **63**(4):151(1980).
4. Grace, T.M., and Funk, M.S., Proceedings International Conference on Recovery of Pulping Chemicals, Vancouver, B.C. (1981).
5. Hultin, S.O., Proceedings of Symposium on Recovery of Pulping Chemicals, EUCEPA, Helsinki (1968), p.180.
6. Harvin, R.L., and Brown, W.F., *Tappi*, **36**(6):270(1953).
7. Kobe, K.A., and Sorenson, A.J., *Pacific Pulp and Paper Industry*, **13**(2):12(1939).
8. Kobe, K.A., and McCormack, E.J., *Ind. Eng. Chem,*, **41**:2847(1949).
9. Hedlund, I., *Svensk Papperstidning*, **12**:408(1951).

10. Harvin, R.L., "A Study of the Thermal and Physical Properties and Heat Transfer Coefficients of Sulfate Paper Mill Black Liquor," Ph.D. Thesis, University of Florida (1955).
11. Laurola, H., and Wallendahl, V., Proceedings of the 1981 TAPPI Pulping Conference, Denver, p.97.
12. Ghalke, S.N., and Veeramani, H., TAPPI Nonwood Plant Fiber Pulping Progress Report, (8):33(1977).
13. Othmer, D.F., and Conwell, J.W., *Ind. Eng. Chem.*, **37**:1112(1945).
14. Co, A., Kim, H-K., Wright, M.O., and Fricke, A.L., *Tappi*, **65**(8): 111(1982).
15. Liem, A.J., Proceedings of the International Conference on Recovery of Pulping Chemicals, TAPPI/CPPA, Vancouver, B.C. (1981), p.133.
16. Sandquist, K., *Pulp Paper Canada*, **84**(2):30(1983).
17. Stenuf, T.J., and Agrawal, M.L., *AIChE Symposium Series*, **207**(77): 13(1981).
18. Backteman, E.H., and Marr, R.Y., *Pulp Paper Canada*, **72**(12):79(1971).
19. Gudmundson, C., Aloholm, H., and Hedstrom, B., *Svensk Papperstidning*, **75**(19):773(1972).
20. Gudmundson, C., *Svensk Papperstidning*, **75**(22):901(1972).
21. "Pulp and Paper Science and Technology," (Libby, C.E., Ed.) Vol. 1, McGraw Hill, N.Y., p.194.
22. Arhippainen, B., and Jungerstam, B., Proceedings of the Symposium on Recovery of Pulping Chemicals, EUCEPA, Helsinki (1968), p.119.
23. Swartz, A., *Chemical Engineering*, **85**(11):175(1978).
24. Beckwith, W.F. et al., Proceedings of the International Recovery Conference on Recovery of Pulping Chemicals, TAPPI/CPPA, Vancouver, B.C. (1981), p.49.
25. Passinen, K., Proceedings of the Symposium on Recovery of Pulping Chemicals, EUCEPA, Helsinki, (1968), p.183.
26. Burdett, J.W., and Holland, C.D., *AIChE Journal*, **17**(5):1080(1971).
27. Lavery, H., Ph.D. Thesis, Institute of Paper Chemistry, Appleton, WI. (1982).
28. Fosberg, T.M., and Claussen, H.L., *Tappi*, **65**(8):63–66(1982).
29. Logsdon, J.D., *Chemical Engineering Progress*, **79**(9): 36-40 (1983).
30. Coats, G.S., *Tappi*, **56**(4):124(1973).
31. Wetherhorn, D., *Tappi*, **56**(6):88(1973).
32. Clay, D.T., *Pulp and Paper*, **49**(9):106(1975).
33. Itahara, S., and Stiel, L., *Ind. Eng. Chem. Proc. Des. Dev.*, **5**(3):309–315 (July, 1966).
34. Edwards, L., and Baldus, R., *Tappi*, **61**(3):75(1978).

35. Bolmstedt, U., and Gudmundson, C., *Svensk Papperstidning*, **77**(1): 27(1974).
36. Jernqvist, A., Olgard, G., and Hedstrom, B., *Svensk Papperstidning*, **69**(15):477(1966).
37. A. E. Johnson & Associates, "GEMCS" General Engineering and Management Computation System, University of Western Ontario, London, Ont. (1971).
38. Shewchuck, C.F., and Leaver, E.W., *Pulp and Paper*, **56**(6):129–133 (1982).
39. Venkatesh, V., Kirkman, A., and Mera, F., *Tappi* **61**(3):87 (1978).
40. "A Survey of Evaporator Scaling in the Alkaline Pulp Industry," Institute of Paper Chemistry Report Project 3234, September, 1975.
41. Grace, T.M., *Southern Pulp and Paper*, **40**(8):16(1977).
42. Grace, T.M., and Frederick, W.J., *Southern Pulp and Paper*, **42**(8):22(1979).
43. Westervelt, H.H., Frederick, W.J., Malcolm, E.W., and Easty, D.B., *Tappi*, **65**(5):179–180(1982).
44. Harrison, R.E., *Pulp and Paper*, **54**(3):156(1980).
45. Novak, L., *Svensk Papperstidning*, **82**(8):240(1980).
46. Lamy, E.J., *Pulp and Paper*, **53**(12):92(1979).
47. Heit, A.H., *Tappi*, **65**(5):184(1982).
48. McCollister, R.D., *Tappi*, **62**(12):89(1979).
49. Behr, J.G., Halko, D.L., and Gansler, N.R., TAPPI Pulping Conference, Toronto, (1982), pp. 63–67.
50. Butcher, C.H., *Pulp and Paper*, **49**(5):52(1975).
51. Korhonen, O., *Pulp and Paper*, **53**(10):104(1979).
52. Ericsson, E.O., and Moody, D.M., TAPPI Annual Meeting, Atlanta, TAPPI PRESS, 1983.
53. Kallenfors, R., *Pulp and Paper*, **49**(11):144(1975).
54. Vegeby, A., *Paper Trade Journal*, **158**(3):22–27 (January 21, 1974).
55. Thoen, G.N., DeHaas, G.G., Tallent, R.G., and Davis, A.S., *Tappi*, **51**(8):329–333(1968).
56. Arhippainen, B., and Jungerstam, B., *Tappi*, **52**(6):1095–1099(1966).
57. Streit, D.E., *Chemical Engineering Progress*, **79**(9):41-45 (1983).
58. Cheng, P.J., and Walters, W.Z., TAPPI Engineering Conference, San Francisco, 1982).
59. Barde, D.K., and Patel, J.C., *Chemical Engineering Progress*, **79**(9):27-30 (1983.)
60. Rosenblad, A.E., AIChE Annual Meeting, Los Angeles, 1982, paper 68f.
61. Karoly, J.A., and Stefango, D.G., "High Solids Concentration of Scaling Pulp Mill Liquors," Technical Report, HPD, Inc., Naperville, IL., 1973.

62. Beder, H., and Madrid, L., *Tappi*, **60**(9):94(1977).
63. Hough, G.W., *Tappi*, **61**(7):23(1978).
64. Rosenblad, A.E., *Chemical Engineering Progress (CEP)*, **72**(4):53 (1976).
65. Rosenblad, A.E., Paper presented at the International Conference on Recovery of Pulping Chemicals, TAPPI/CPPA, Vancouver, B.C., September 22–25, 1981, pp. 249–260.
66. Guimaraes, J.C., Edwards, L., and Kelly, M.G., *Tappi*, **65**(6):45 (1982).
67. Jacoby, H.E., U.S. Pat. 3,286,763.
68. Rosenblad, A.E., *Pulp and Paper*, **55**(1):109(1981).
69. Johnson, S.E., "The Lockman Evaporator, A New Liquor Evaporation System," Pulping Chemical Recovery, Helsinki Symposium, 1968.
70. Olauson, L., *Pulp and Paper Canada*, **80**(3):45(1979).
71. Ahlstrom Newsletter, February, 1975.
72. Andrews, R.S., and Roscoe, R.W., *Tappi*, **64**(12):49 (1981).
73. Clay, D.T., and Karnofski, M.A., *Tappi*, **64**(12):45 (1981).
74. Almond, C.B., *Southern Pulp and Paper*, **45**(8):30(1982).
75. Lankenau, H.G., and Badyrka, J.T., *Tappi*, **65**(2):75(1982); **65**(3): 153(1982).
76. Swartz, A., *Chemical Engineering*, **85**(11):175(May 8, 1975).
77. Rastogi, L.K., and Williams, T.J., "Modelling and Control of Multiple Effect Black Liquor Evaporator Systems," PLAIC Report 62, W. Lafayette (1974).
78. Niemi, A., and Koistinen, R., *Papri jaa Puu*, **54**(2):297(1972).
79. Isner, A., and Michaelson, R., *Pulp and Paper Canada*, **79**(6):T223 (1978).
80. Ricker, N.L., and Sewell, T., AIChE Symposium Series, **80**(232):87 (1984).
81. Drew, J., *Chemical Engineering Progress*, **72**(6):64(1976).
82. "Tall Oil and Its Uses," Tall Oil Products Division, Pulp Chemicals Association, McGraw Hill, N.Y., 1965.
83. Douek, M., and Allen, L.H., *Tappi*, **62**(7):47(1978).
84. Bentvelzen, J.M., Ph.D. Thesis, North Carolina State University, Raleigh, N.C., (1973).
85. Collins, T.T., *Paper Trade Journal*, **146**(30):39(July 23, 1962).
86. Bergstroem, H., *Pulp and Paper Magazine Canada*, **54**(12):135–140(November 1953).
87. NCASI Technical Bulletin No. 83, A Survey & Analysis of Design Criteria and Performance of New Generation Black Liquor Oxidation Systems, (1976).

88. EPA Report—Pulp & Paper Industry, Part 1.
89. Sebbas, E., "Evaluation of Condensate Stripping in Kraft Mills," Paper presented at TAPPI Environmental Conference, 1982.
90. Hansen, S.P., and Burgess, F.J., *Tappi*, **51**(6):241(1968).
91. Hisey, W.O., *Tappi*, **34**(1):1(1951).
92. Bergstrom, H.O.V., Swed. Pat. 190,258 (June 30, 1964).
93. Murray, F.E., *Pulp and Paper Magazine Canada*, **69**(1):82(1968).
94. Carter, D.N., and Tench, L., *Pulp and Paper Magazine Canada*, **75**(8):71(1974).
95. Hough, G.W., and Salee, R.W., *Tappi*, **69**(2):82(1977).
96. Venkatesh , V., and Nguyen, X.N., TAPPI Engineering Conference, San Francisco, (1978).

4
Chemical Recovery from Concentrated Liquor

T. M. Grace

Introduction

This chapter is concerned with the combustion of black liquor in the recovery boiler. It includes the recovery boiler, direct contact evaporator (if any), precipitator, and dissolving tank. It does not cover items specifically directed at steam and power generation or boiler feedwater treatment. Those items are appropriately covered in texts on steam and power. The liquor combustion step interfaces with the rest of the recovery cycle with concentrated liquor (from heavy liquor storage or heavy black liquor oxidation) on the input end, and with green liquor from the smelt dissolving tank to the green liquor clarifier on the output end.

The chemical recovery furnace has two major functions: recovery of the sodium and sulfur from the spent liquor in a form suitable for regeneration of cooking liquor, and recovery of the energy values in the liquor in the form of high-pressure steam.

The value of the chemicals recovered exceeds that of the energy. However, since these activities are not mutually exclusive, much of the development in recovery boilers has been concerned with energy recovery. The major influence of the chemical recovery function is the requirement to process the black liquor solids at the average rate they are produced in the digester, and the need to maintain a reducing atmosphere in part of the furnace to recover sulfur as Na_2S.

In some ways, recovery furnaces can be thought of as power boilers being fired with a peculiar fuel, black liquor. However, there are major differences between recovery furnaces and power boilers, so that the analogy cannot be carried too far. The major features that make recovery furnaces unique are:

1. The need to fire an aqueous fuel. Heavy black liquor introduced into the furnace still contains a significant amount of water. This must be evaporated before combustion can begin. Special provisions must be made to permit drying of the liquor while maintaining stable firing.

2. The need to maintain an active reducing zone to recover the sulfur as Na_2S. This is achieved through control of the spatial distribution of air. Only a portion of the air needed for complete combustion is added in the reducing zone. The remainder is added at higher levels in the furnace.
3. The need to remove inorganics from the furnace as a molten smelt. Provisions must be made to permit the smelt to drain from the unit to the dissolving tank. The smelt stream must be broken up by shatter jets to prevent dissolving tank explosions. The construction of the hearth must be such that it is smelt tight so molten smelt does not leak through.
4. The need to handle a fuel of very high fouling tendency. Provisions must be made to avoid plugging of gas passages and heat transfer surface with tacky material. This includes sootblowing, special designs for heat transfer surface, and the use of solids recovery equipment.
5. The hazards introduced by potential smelt-water contact. Pressure part failure or firing weak liquor can lead to destructive smelt/water explosions. This particular problem does not exist with conventional boilers. A forced shutdown of a chemical recovery unit involves more than fuel shut-off and repair, as is the case with a conventional boiler. Such an outage usually requires extensive draining and cleaning work and results in loss of production.
6. The potentials for production of relatively large amounts of odorous reduced-sulfur gases (TRS). Elimination of these emissions requires ensuring complete combustion of gases within the furnace itself, and high-efficiency black liquor oxidation on units with a direct contact evaporator.
7. The potential for very high fireside corrosion rates due to the sulfur compounds which are present. Protective measures have included the use of advanced metallurgies or extensive studding of tubes in the critical regions. Fireside corrosion has tended to limit steam pressures and temperatures obtainable from recovery boilers.
8. The relatively low heating value of black liquor solids due to the high inorganic content. This can lead to difficulties in maintaining autogenous combustion.

Material and energy balances

Material and energy balances for the combustion of black liquor use procedures which are very similar to those used for conventional fuels. Proper allowance must be made for the different elemental constituents of black liquor, the different combustion products formed, and the recycling of material within the recovery unit boundary. The material balance relates the chemical flows through the unit to the liquor solids input. The purpose is to provide a proper framework for an accurate energy balance. An accurate energy balance is essential to an understanding of the recovery process and for optimizing performance. Standard procedures for determining recovery boiler thermal performance are described in References (*1*) and (*2*).

Boundaries

The boundaries of the recovery unit for purposes of material and energy balances are drawn so that internal recycle streams of materials in the combustion process do not enter or leave the boundary and, hence, can be ignored. The following equipment would be included within the recovery unit: furnace, direct contact evaporator, flue gas-to-air air preheater, chemical makeup mix tank, and liquor heater. This definition specifically excludes the multiple-effect evaporator and concentrators from the recovery unit. It also excludes the induced draft fan and the precipitator (except for fume recovery), and the smelt dissolving tank (except for the material balances required to determine smelt composition from green liquor analysis). Sootblowing steam, shatter-jet steam, and steam for air heating are not included in the material and energy balances per se and must be accounted for separately. If a direct-fired air preheater is used, the fuel used in the air preheater is treated as if it were auxiliary fuel fired directly in the furnace, and the air enters at ambient temperature.

Material balances

Material balances for combustion of black liquor must be based on an elemental analysis of the liquor solids. Five elements are always considered: sodium, sulfur, carbon, hydrogen, and oxygen. Under some circumstances, chlorine, potassium, and inerts are also included. An accurate elemental analysis of black liquor is a difficult, time-consuming process. Standard procedures for this determination have not yet been officially defined. Elemental analyses should always be reported as weight % on a dry solids basis. A sample elemental analysis for a kraft black liquor from the evaporators is given below.

Sodium	18.7
Sulfur	3.4
Carbon	38.8
Hydrogen	3.9
Oxygen	35.2

Water entering the combustion system with the liquor solids can be considered as an inert in combustion calculations. This water must be included, as such, in material balances because it has a major impact on energy balances.

In addition to black liquor, the inputs which must be included are air and any chemical streams added to the black liquor at the mix tank or fired separately. The latter would include makeup chemicals, such as salt cake, and any auxiliary fuel fired. It would not include chemicals or waste streams added to the weak liquor or those added to the heavy black liquor before the heavy liquor storage tank, since these would already be accounted for in the elemental analysis of the liquor entering the recovery unit boundary.

Basically, there are two product streams from black liquor combustion: flue gas and smelt. Although there is substantial particulate carryover with the flue gas, air emission control requires that this material be removed from the flue gas, and it is recirculated to the furnace without leaving the recovery unit boundary. Essentially, all of the alkali salts leave as smelt. The major constituents of the smelt are Na_2CO_3, Na_2S, Na_2SO_4, NaCl (when chlorine is a significant elemental input), and possibly some unburned carbon. Potassium salts of the same anions could also be present. The major constituents of the flue gas are N_2, O_2, CO_2, H_2O vapor, and possibly some SO_2 and unburned combustibles (e.g., CO). Odor control requirements preclude the presence of significant amounts of reduced sulfur gases, and hence also large amounts of combustibles.

A schematic diagram of the input and output streams for black liquor combustion is shown in Fig. 4.1.

The most convenient method for formulating the material balance is per unit weight of black liquor solids entering. The air flow is determined by specifying the percentage of excess air over theoretical. Theoretical air is defined as that amount of air which would be required to just form fully oxidized furnace products (except for Na_2S at the prevailing reduction efficiency) without any air left over. It can be calculated from the elemental analysis of the black liquor solids and the reduction efficiency. The amounts of each major furnace product are calculated from the elemental analysis. Corrections can be made for the presence of SO_2, unburned combustibles, and carbon in the smelt. The material balance provides a direct relationship between the excess air and the O_2 content of the flue gas, which is not very sensitive to the elemental analysis of the liquor. This relationship is shown in Fig. 4.2, which can be used to estimate the excess air in an operating unit.

A systematic procedure for carrying out material balances for black liquor combustion is given in Table 4.1. The information required is:

elemental analysis of black liquor solids;
liquor solids content;

excess air (from % O_2 in flue gas and Fig. 4.2); and
reduction efficiency.

If auxiliary fuel is burned, or chemical makeup added to the liquor within the recovery boundary, they must be separately accounted for. This can be done by adjusting the elemental input per unit amount of black liquor, or by carrying through a parallel set of calculations using the same procedures and the elemental composition of the additional streams.

A sample material balance for a black liquor with the elemental analysis listed previously is shown in Fig. 4.3.

Corrections to the material balance for the presence of significant amounts of SO_2 and unburned combustibles in the flue gas, or unburned carbon in

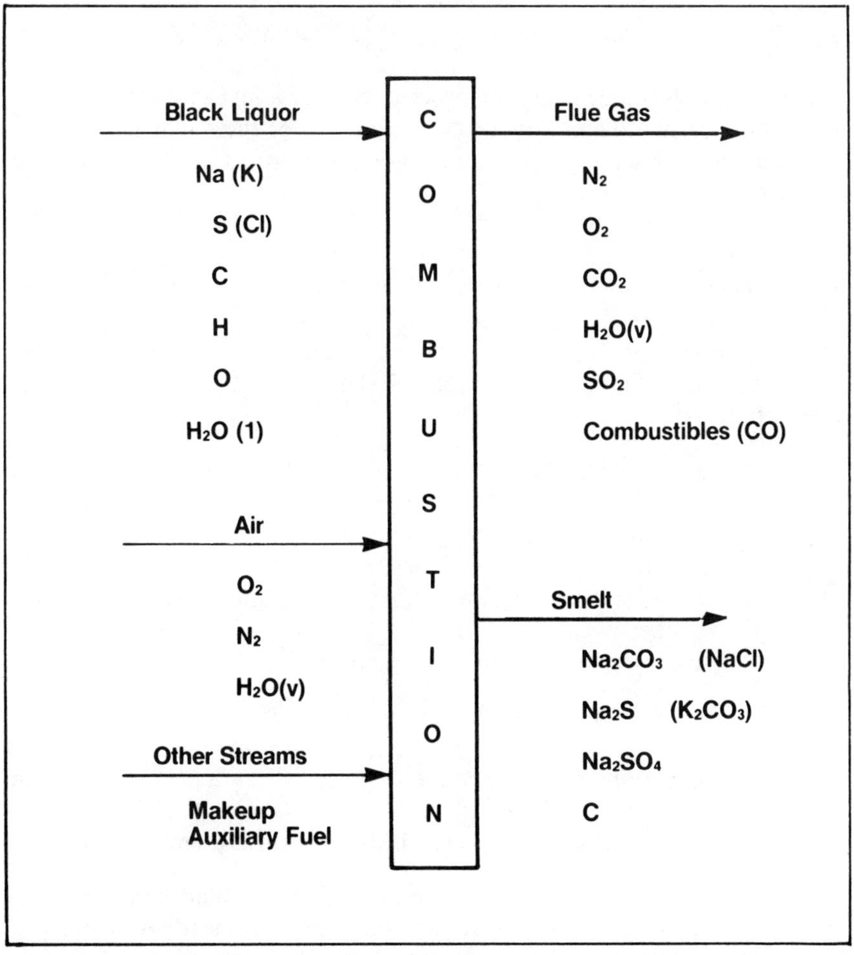

Fig. 4.1. Black liquor combustion - input/output.

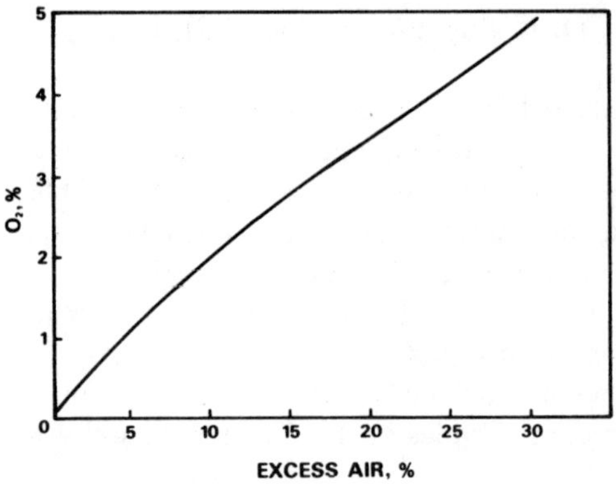

Fig. 4.2. Relationship between excess air and O_2 content of flue gas.

the smelt, can be made using the procedures shown in Table 4.2. These are based on measured values of the concentrations of these materials and on the assumption that their presence does not greatly affect the overall material balance.

Energy balances

Energy balance around a recovery unit can be formulated in a number of different ways. The procedure described herein uses the heat output method. In this approach, the heat to steam is calculated as the difference between the heat input and all other heat outputs. The advantages of this approach are as follows:

1. It avoids the need to measure flow rates of various streams with the attendant inaccuracies. Instead, a closed material balance forms the basis for the heat balance calculations.
2. By quantitatively defining all of the ways in which energy leaves the unit, it points the way toward changes in operation which will improve thermal performance.
3. Relatively few measurements are needed to complete the balance.

The major disadvantage of the approach is that it puts a premium on accurately knowing *all* of the heat outputs. Fortunately, the state of knowledge concerning recovery boilers has reached a state where this is no longer a serious drawback.

The information needed to complete the heat balance is summarized below:

1. Information required for a complete material balance including the corrections for SO_2 in flue gas, unburned combustibles in flue gas, and unburned carbon in smelt.

Table 4.1. Procedure for Carrying out Material Balance for Black Liquor Combustion. Basis: 100 lb dry liquor solids ([illegible] 100 kg dry solids)

	Compound kg	*O_2 Needed, kg*
Na_2S: $\frac{78}{32} \times \frac{\text{\% reduction}}{100} \times (\text{\% S})$	= a	0
Na_2SO_4: $\frac{142}{32} \times \left[1 - \frac{\text{\% reduction}}{100}\right] \times (\text{\% S})$	= b	$\frac{64}{142} \times b$
$NaCl$: $\frac{58.5}{35.5} \times (\text{\% Cl})$	= c	0
Na_2CO_3: $\frac{106}{46} \times [(\text{\% Na}) - \frac{46}{32}(\text{\% S}) - \frac{23}{35.5}(\text{\% Cl})]$	= d	$\frac{48}{106} \times d$
K_2CO_3: $\frac{138.2}{78.2} \times (\text{\% K})$	= e	$\frac{48}{138.2} \times e$
CO_2: $\frac{44}{12} \times [(\text{\% C}) - \frac{12}{106} \times d - \frac{12}{138.2} \times e]$	= f	$\frac{32}{44} \times f$
H_2O: $\frac{18}{2} \times (\text{\% H})$	= g	$\frac{16}{18} \times g$
		$= \Sigma$

$$\text{Theoretical Air} = \frac{[\Sigma - (\text{\% O})]}{0.232}$$

$$\text{Total Air} = \text{Theoretical Air} \times \left[1 + \frac{\text{\% Excess Air}}{100}\right]$$

$$N_2 \text{ in Flue Gas} = 0.768 \times \text{Total Air}$$

$$O_2 \text{ in Flue Gas} = [\Sigma - (\text{\% O})] \times \frac{\text{\% Excess Air}}{100}$$

$$\text{Dry Flue Gas} = N_2 \text{ in Flue Gas} + O_2 \text{ in Flue Gas} + CO_2$$

$$H_2O \text{ in Flue Gas} = g + \text{Water in Liquor} + \text{Humidity in Air}$$

$$= g + \frac{100 - \text{\% solids}}{\text{\% solids}} \times 100 + \text{Total Air} \times \text{Humidity}$$

Note: Elemental analysis of liquor solids given as (wt. % element).

Table 4.2. Minor corrections to material balance for SO_2, CO, and unburned carbon (per unit weight dry solids)

SO_2:	2.08 × (wt. dry flue gas) × (ppm* SO_2 on dry flue gas)/10^6 = Y
	reduce Na_2S by 0.0122 × (% reduction) × Y
	increase Na_2CO_3 by 0.0166 × (% reduction) × Y
	ignore effect on Na_2SO_4.
CO:	9.1×10^{-7} × (wt. dry flue gas) × (ppm* CO on dry flue gas) = Z
	reduce CO_2 by 1.57 Z
	increase O_2 by 0.57 Z
Unburned C:	(wt. % C in smelt) × (wt. smelt)/100 = W
	reduce CO_2 by 3.67 W
	increase O_2 by 2.67 W
	* (ppm by volume)

2. Higher heating values for black liquor and any other stream introduced having fuel value.
3. Temperatures for black liquor and combustion air entering the recovery unit boundary and for flue gas and smelt leaving the unit.

Heat inputs The most significant heat input is the heating value of the black liquor solids. This is essentially the energy released by combustion of the black liquor organics. It is North American practice to define the heating value of black liquor solids as the heat released during complete combustion in an oxygen bomb calorimeter without further correction for products formed. The bomb calorimeter combustion products are CO_2, Na_2SO_4, Na_2CO_3, and *liquid* H_2O. These are not, of course, identical to the products of combustion in a recovery boiler. Specific terms are added to the *output* side of the heat balance to account for these differences. The heating value, per se, is not corrected.

Other heat input terms are the sensible heat in the black liquor and the sensible heat in the incoming air. The sensible heat in the black liquor is the product of the weight of wet liquor (per unit weight of dry solids), the specific heat of the liquor, and the difference between the liquor temperature and a reference temperature of 77°F (25°C). (Note that the proper value of liquor temperature to use is the temperature at which it enters the recovery unit boundary, not the temperature at which it is fired into the furnace.) The sensible heat in the air is given as the product of the weight of air per unit weight of liquor solids, the specific heat of air, and the difference between air temperature and 77°F (25°C). Thus, the proper air temperature is that at which it enters the recovery unit boundary. For a steam coil air heater, this is the combustion air temperature. For a flue gas-air heater or a direct-fired air heater, this sensible heat term is zero.

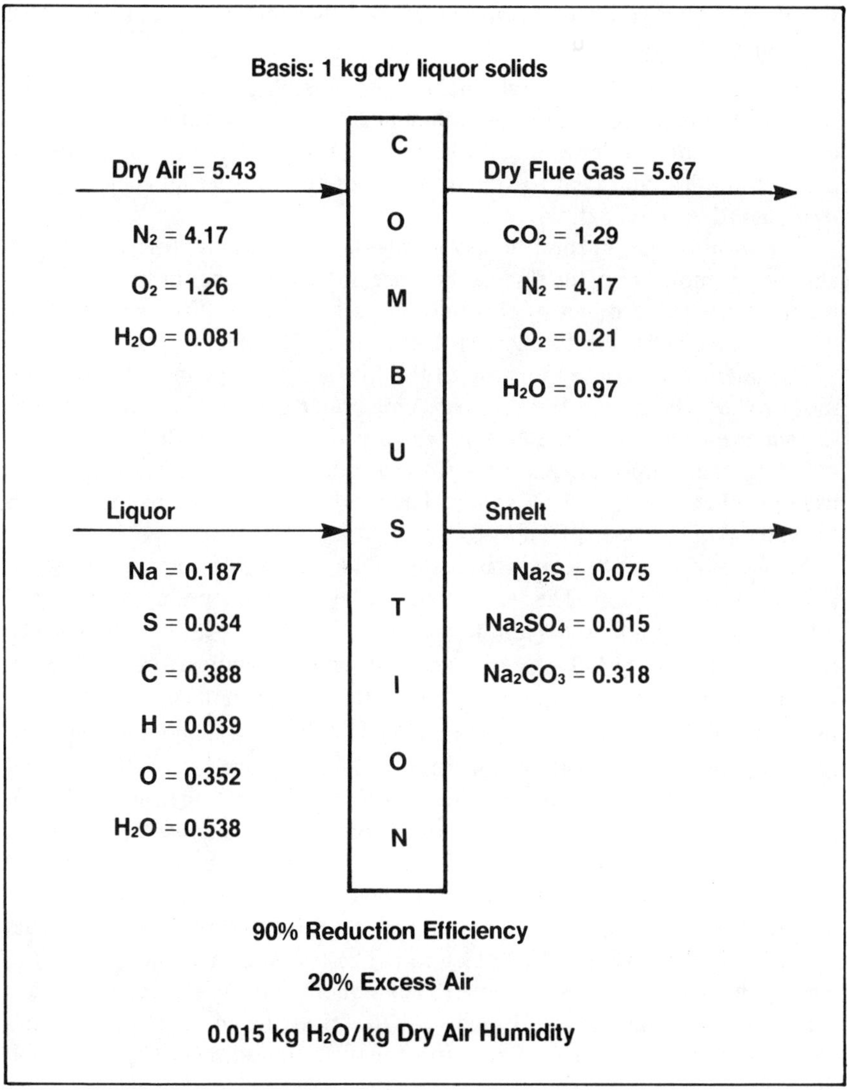

Fig. 4.3. Sample material balance.

If auxiliary fuel is fired, the heating value of the auxiliary fuel must be included as a heat input. This includes any auxiliary fuel fired in a direct-fired air heater. The term would be the product of the amount of auxiliary fuel burned per unit weight of liquor solids and the heating value of the fuel.

A similar heat input for the heating value of any other chemical stream introduced into the recovery unit is also required. Salt cake makeup is Na_2SO_4, a bomb calorimeter product, and has a heating value of zero.

Sensible heat inputs with auxiliary fuel and other chemical streams are small and can be ignored.

Heat Outputs The heat outputs include the sensible and latent heats in the flue gas, the heat content in the molten smelt, heat to form sulfide, heat in unburned combustibles and carbon, radiation and losses to surroundings, unaccountables, and heat to steam. Methods for calculating each of these terms are described below.

The sensible heat in the flue gas is calculated separately for the dry flue gas and the moisture in the flue gas because of the difference in specific heats. Each term is the product of the amount of gas, the specific heat, and the difference between the flue gas temperature and ambient.

The latent heat of the moisture in the flue gas must be included for that moisture which entered the recovery unit boiler as a liquid. This includes the water entering with the liquor, which is evaporated in the furnace and direct contact evaporator, and the water produced by combustion from the hydrogen in the black liquor solids. The term is the product of the amount of water and the heat of vaporization.

The heat content of the molten smelt includes the sensible heat of the smelt and the heat of fusion. The term is the product of the amount of smelt per unit weight of liquor solids and the smelt enthalpy. The latter can be taken to be 580 Btu/lb (1350 $\frac{kJ}{kg}$) at a smelt discharge temperature of 1550°F (1116°K). The temperature dependence of the heat content can be calculated using a specific heat of 0.4 Btu/lb°F (1.67 $\frac{kJ}{kg\ {}^\circ K}$). The dependence of the heat content on smelt composition can be neglected.

The heat to form sulfide accounts for the fact that less heat is released in the furnace when Na_2S is the combustion product than when Na_2SO_4 is produced. This is an energy penalty which is inherent in producing (and using) Na_2S and is properly accounted for on the output side of the heat balance. The term is the product of the amount of Na_2S formed and a heat of reaction of 5550 Btu/lb Na_2S (12,900 kJ/kg). The endotherm applies to all Na_2S leaving the unit regardless of origin (black liquor solids or makeup chemical), since the heat inputs are all based on Na_2SO_4 as the combustion product. The endotherm is the heat of reaction for the reaction, $Na_2SO_4 \rightarrow Na_2S + 2O_2$; i.e., the heat of reduction of sulfate.

Incomplete combustion can result in a very substantial heat loss and must be accounted for. Combustible gases may be present even with an excess of air. These may be considered to be CO for convenience, although other combustibles (such as H_2) might also be present. The heat loss for CO is 4350 Btu/lb CO (10,110 kJ/kg). If other combustibles besides CO are present, a higher heat loss representing the heat of combustion of all combustible gases per unit weight of CO should be used. Otherwise, these will show up as part of the unaccountables. In addition, some unburned carbon is normally present in the smelt. The heat loss for unburned carbon is 14,100 Btu/lb C (32,800 kJ/kg).

Radiation and losses to the surroundings are normally a small fraction of the total heat input. Table 4.3 (*2*) provides a means for estimating this loss.

A final term on the heat output side is unaccountables. The magnitude of this term is dependent on the accuracy of all other parts of the heat balance. It is especially influenced by the degree that unburned combustibles in the flue gas and smelt are determined. A value of 2.0% of input is normally used for design purposes.

Heat to steam is determined by difference between the heat inputs and the sum of all the other heat outputs. This value can be used to calculate the thermal efficiency of the recovery boiler. It represents the gross steam output. Net steam production can be determined by subtracting the steam used in the recovery unit for sootblowing, shatter jets, liquor heating, and air heating. A comparison of the measured gross steam output with that calculated from the heat balance can be used to check procedures and estimate unaccountables.

A systematic procedure for carrying out an energy balance for black liquor combustion is given in Table 4.4.

A sample heat balance for combustion of black liquor is given in Table 4.5. This is for a recovery unit without a direct contact evaporator with the liquor coming from the concentrator at 65% solids. The liquor composition and firing conditions correspond to those used in the material balance shown in Fig. 4.3. The liquor heating value is taken to be 6600 Btu/lb (15,350 kJ/kg). The 65% solids liquor from the concentrator enters at 220°F (105°C) and

Table 4.3. Estimation of radiation loss

Unit size ton/day	*Radiation loss % total input (at rating)*	*Unit size ton/day*	*Radiation loss %total input (at rating)*
100	0.58	700	0.32
150	0.52	750	0.31
200	0.48	800	0.30
250	0.44	850	0.30
300	0.41	900	0.29
350	0.39	950	0.28
400	0.38	1000	0.28
450	0.37	1100	0.27
500	0.36	1200	0.26
550	0.35	1300	0.26
600	0.34	1500	0.25
650	0.33		

Note: Radiation loss is inversely proportional to load when furnace operated off rating.

Table 4.4. Procedure for carrying out energy balance for black liquor combustion (basis: 1 lb black liquor solids)

Heat Input

1. Heating value of black liquor solids, Btu/lb ____________

2. Sensible heat in black liquor ____________
$\frac{(100)}{(\%\ solids)} \times Cp \times (T_l - T_{ref})$

3. Sensible heat in air ____________
(Total air from M.B.*) $\times\ 0.24\ \frac{Btu}{lb\ °F} \times (T_a - T_{ref})$

4. Heating value of auxiliary fuel ____________
(Fuel used/wt. black liquor solids) × HHV fuel

5. Heating value of other chemical streams ____________
(Amount stream/wt. black liquor solids) × HHV chemical

Total Heat Input ____________

Heat Output

1. Sensible heat in dry flue gas ____________
(Dry flue gas from M.B.) × 0.24 Btu/lb °F × $(T_{fg} - T_{ref})$

2. Sensible heat in moisture in flue gas ____________
(Total H_2O from M.B.) × 0.45 Btu/lb °F × $(T_{fg} - T_{ref})$

3. Latent heat of water in black liquor ____________
$\frac{(100 - \%\ solids)}{(\%\ solids)} \times 1030$ Btu/lb

4. Latent heat of water from combustion ____________
(Combustion water from M.B.) × 1030 Btu/lb

5. Heat content of smelt ____________
(Total smelt from M.B.) × [580 + 0.40 $(T_{sm} - 1550)$] Btu/lb

6. Heat to form sulfide ____________
(Total Na_2S from M.B.) × 5550 Btu/lb

7. Heat loss to combustibles in flue gas ____________
(CO from M.B.) × 4350 Btu/lb

8. Heat loss due to carbon in smelt ____________
(C from M.B.) × 14,100 Btu/lb

9. Radiation loss ____________
(Total heat input) × % radiation loss (Table 4.3)/100

10. Unaccountables ____________
(Total heat input) × 2.0%/100**

Sum Items 1→10 ____________

11. Heat to Steam = Total heat input − Sum ____________

Note: To work in SI units $1\ \frac{kJ}{kg \cdot k} = 0.24\ \frac{Btu}{16°F}$ and $1\ \frac{kJ}{kg} = 0.43\ \frac{Btu}{lb}$

*M.B. — material balance.

**Or use % unaccountables found by experience.

Table 4.5. Sample heat balance for black liquor combustion

		Btu/lb Solids	%
Inputs			
Heating value black liquor		6,600	93.8
Sensible heat of black liquor		150	2.1
Sensible heat of air		290	4.1
		7,040	100.0
Outputs			
Sensible heat of dry flue gas		440	6.2
Sensible heat of moisture in flue gas		140	2.0
Latent heat of water in black liquor		555	7.9
Latent heat of water from combustion		360	5.1
Heat content of smelt		230	3.3
Heat to form sulfide		420	5.9
Radiation loss		20	0.3
Unaccountables		140	2.0
	Sum	2,305	32.7
Heat to steam		4,735	67.3
		7,040	100.0

the combustion air at 300°F (150°C). The flue gas discharges at 400°F (205°C) and the smelt temperature is 1500°F (815°C). It is assumed that no auxiliary fuel is being used and that there is no makeup chemical addition. The reference temperature is 77°F (25°C).

This heat balance shows that only about 8% of the heat is lost out the stack as sensible heat. The biggest single loss is the latent heat of water in black liquor, the heat needed to evaporate water in the furnace. This can only be reduced by bringing the liquor into the recovery boundary at a higher solids content. The next most significant output term is the heat used to form sulfide. It can only be reduced by reducing the amount of sulfide produced (lower sulfidity or poorer reduction). Table 4.6 shows the sensitivity of the heat balance to a number of operating parameters. This indicates the leverage factors for improving the thermal performance of recovery boilers. Clearly, it is essential to avoid combustibles in the flue gas. The most important factor for high thermal performance is ensuring complete combustion.

Processes and chemistry

The process of burning black liquor in a recovery furnace is complicated. The chemistry of sulfur retention and sulfide formation is not straightforward and only partly understood. The formation of gaseous combustibles by the pyrolytic decomposition of black liquor solids and char gasification is critical to the burnability of the liquor and not well understood. Material flows follow

Table 4.6. Heat balance sensitivity

Change in parameter	% Change in heat to steam
27°F (15°C) Drop in flue gas temperature	+ 0.8
1% Increase in liquor solids at 50% solids	+ 0.6
1% Increase in liquor solids at 65% solids	+ 0.4
90°F (50°C) decrease in smelt temperature	+ 0.2
10% Increase in reduction efficiency	− 0.6
1000 ppm CO in flue gas	− 0.3
1% C in smelt	− 0.8
200 ppm SO_2 in flue gas	+ 0.1
Saltcake makeup, 1% of solids (90% reduction efficiency)	− 0.4

complicated paths with a number of internal recycle loops. Proper understanding of furnace chemistry and internal processes is essential to the proper interpretation of the effects of operating variables.

The various items considered in this section include the following:

- Drying of black liquor
- Pyrolysis of black liquor solids
- Combustion of gases
- Gasification and combustion of char
- Chemistry of sulfide formation
- Chemistry of fume formation
- Sulfur pathways in the furnace
- Entrainment and carryover
- Corrosion reactions
- Ash fusion characteristics
- Influence of nonprocess elements.

The treatment tends to follow what happens sequentially to a particle of liquor as it moves through the unit, but some exceptions are made in the interests of clarity.

Drying of black liquor

Drying is the first process to occur after the liquor in introduced into the furnace. Liquor temperatures remain near the boiling point until the water is gone. This delays the inception of other higher temperature processes until the liquor is dry. With macroscopic sized particles, pronounced gradients of temperature and dryness can exist within the particle, and higher temperature processes may occur on the outer surface while the interior is still wet.

The evaporation of single drops of liquor has been described by Monaghan and Siddall (*3*). They observed bubbling and violent expansion and contraction of the drop. The viscosity of the liquor in the drop increased gradually. The outer layers of the drop became solid, while the inside was still a highly viscous liquor. The final shape varied greatly from drop to drop but was often partially hollow. These findings are consistent with the concept that drying is controlled by the heat transfer rate, whereas mass transport considerations determine the final structure of the element.

Two different techniques are employed for drying black liquor in a recovery furnace: suspension drying and wall drying. In suspension drying, the liquor is sprayed into the furnace cavity in the form of relatively coarse droplets, and the individual droplets dry as they fall to the hearth countercurrent to the hot combustion gases. In wall drying, the liquor is sprayed onto the walls of the furnace, where it sticks and dries. As the material accumulates on the walls and dries, relatively large chunks break off because of their weight, and fall to the hearth.

The droplet size distribution is very critical in suspension drying. Droplet size controls the surface to volume ratio of the liquor (hence the drying rate) and the terminal velocity of the particle (hence the time spent in suspension). Drying rates are increased by going to finer sprays with large surface to volume ratios, but this also tends to increase entrainment of the liquor droplets. It is necessary to strike a balance in the spray characteristics which will avoid wet liquor reaching the char bed while also avoiding excessive entrainment.

The spray droplet size distribution is controlled by parameters at the liquor gun. Present technology employs a nozzle assembly which gives a conical spray shape and a uniform distribution pattern, while providing large liquor passages within the nozzle to avoid plugging. For a given nozzle size, the spray size distribution is very strongly influenced by liquor temperature, apparently because of the very pronounced effect of temperature on the viscosity of high solids liquor. Normally, spray nozzle pressure ranges between 12-18 psig (180 and 270 kPa) and the liquor temperature is held between 235-250°F (113 and 120°C). A change in liquor temperature of only 1 to 2°F can have a noticeable effect on the droplet size distribution.

The situation is somewhat different with wall drying. The objective is to deposit the heavy black liquor in a partially dried form in a band on the furnace wall where drying is completed. The spray system employs nozzles that produce a jet of liquor which impinges on a flat plate distributor, resulting in a nearly horizontal sheet of relatively large particles of liquor. It is oscillated in a "paintbrush" motion to obtain a uniform distribution on the walls and to control drying. In larger units there is a greater tendency for liquor droplets to dry in flight and never reach the wall. Operational adjustments may be required to prevent this from happening. Typical operating conditions at the gun are 35 psig (340 kPa) and a liquor temperature of 220 to 230°F (104-110°C). The spray characteristics and drying rate are less sensitive to changes in liquor temperature in wall drying.

The amount of heat involved in drying the liquor in the furnace is quite significant, normally 8-10% of the furnace heat release. The effect on temperatures in the bottom part of the furnace is much greater. Drying is a heat sink and tends to cool the region where it is taking place. Thus, the degree of dryness of the liquor solids reaching the bed has a strong influence on bed temperature and on the tendency toward blackouts and partial blackouts.

Pyrolysis

Pyrolysis is the production of volatile combustibles through the thermal degradation of complex organic compounds. Pyrolysis reactions are irreversible degradation reactions caused by thermal effects alone. They are not oxidation reactions, although they may be accompanied by parallel oxidation reactions. Pyrolysis of black liquor can be represented as follows:

$$\text{liquor solids} \xrightarrow{\text{heat}} \text{pyrolysis gases} + \text{char}.$$

Pyrolysis is important in the recovery furnace for three reasons:

1. It governs the production of volatile combustibles and their distribution within the furnace. This is important because homogeneous gas-phase combustion reactions are much faster than heterogeneous char burning, and the greater the extent of pyrolysis, the more easily the liquor solids burn.
2. Pyrolysis is primarily responsible for the release of sulfur gases within the furnace cavity.
3. The residual material from pyrolysis is the starting material for char bed processes.

Dry black liquor solids undergo substantial thermal degradation, beginning at temperatures near 400°F (200°C) (*4*). The liquor solids become converted to a tarry mass, which begins to swell as volatiles are produced. The residue at the end of the pyrolysis is a friable carbonaceous material, somewhat papery in appearance, of high porosity and low bulk density. The volume occupied by the solids increases during pyrolysis.

The major variables affecting pyrolysis are heating rate, final temperature, and liquor composition. The extent of volatile production increases with increasing heating rate and temperature. The pyrolysis behavior of black liquors can be characterized to some extent by laboratory tests, such as thermal gravimetric analysis (TGA). Arbitrary definitions of "volatile matter" and "fixed carbon," based on weight losses over prescribed temperature ranges similar to the proximate analysis of coal (*5*), can be made.

Liquor composition can have a major influence on pyrolysis (*6*). Cellulose pyrolysis begins at about 460°F (240°C). The presence of inorganic compounds, such as alkali carbonates, lowers the temperature at which pyrolysis begins and promotes char formation. Hemicelluloses start to decompose between 300°F and 400°F and evolve more gas than does cellulose. Lignin pyrolysis produces more aromatic products and char than does cellulose. Pyrolysis can be affected by the presence of individual compounds. For example, ellagic acid in liquors from eucalypt pulping has been shown to have an adverse effect on the swelling of the liquor mass during pyrolysis, and hence on the burning characteristics of the liquor (*7*).

Pyrolysis is also responsible for the release of sulfur gases within the furnace. As much as 70% of the total sulfur in kraft black liquor has been volatilized in laboratory tests (*8*). Sulfur release is not limited to organic sulfur. It has been shown that sulfur from inorganic compounds can be converted to gases by pyrolysis with organic substances. Only sulfate and sulfite are resistive to sulfur release during pyrolysis. The sulfur in sulfide and thiosulfate is not stable during pyrolysis, and substantial fractions of it can be converted to reduced sulfur gases. Elemental sulfur also tends to form reduced sulfur gases. The behavior of thiosulfate shows why black liquor oxidation, which converts sulfide to thiosulfate, does not prevent TRS emissions from the furnace itself.

The volatilization of sulfur during pyrolysis has a major effect on potential furnace emissions. The only completely acceptable way for sulfur to leave the furnace is tied up with sodium in the smelt. During pyrolysis some sulfur is separated from sodium. If it does not recombine with sodium somewhere in the furnace, it will leave as either reduced sulfur gases or SO_2. The sulfur pathways in the furnace which lead to its ultimate recovery are discussed in detail later.

Combustion of gases

The final combustion of the volatile combustibles in the oxidizing zone is a gas-phase homogeneous reaction. The main factors required to ensure complete combustion are a sufficient supply of air, an adequate degree of mixing between combustibles and air, temperatures high enough for the reactions to proceed, and a contact time at temperature sufficient to allow the reactions to go to completion. These are the classical three Ts of combustion processes:

time;
temperature; and
turbulence

Mixing is the most difficult problem, particularly in large units. A substantial amount of the air is introduced above the liquor guns and must be mixed with the gaseous fuels coming from the lower part of the furnace. A combination of high injection velocities (to generate turbulence) and a relatively large amount of air is used to achieve this.

Since air is introduced around the periphery of the units, there is variability in the degree of completeness of combustion in any horizontal plane in the furnace. There tends to be a cone of unburned combustibles in the center of the furnace that gradually decreases in diameter, proceeding upward in the furnace. If the cone extends up into the heat traps in the upper part of the furnace, gas temperatures may drop to a level where combustion is kinetically blocked. Under these conditions, unburned combustibles (and reduced sulfur gases) can be present in the flue gas, even though there is a substantial amount of excess air.

In large recovery boilers, special steps are sometimes taken to make sure sufficient air penetrates to the center of the furnace. Some units use special nozzles, referred to as high primary, to direct air into the center of the furnace just above the char bed.

Char combustion

Pyrolysis of black liquor solids produces char as well as combustible gases. The char falls to the bottom of the furnace and forms a char bed which must be burned. Char consists primarily of finely divided carbonaceous material and inorganic salts. Samples taken from active char beds indicate that the char contains 5-10 wt% carbon, with the remainder being Na_2CO_3, Na_2S, and Na_2SO_4. Analyses of material taken from the char band show pyrolysis to be much less complete and the carbon content as high as 25-30% by weight (*10*).

To prevent the char from building up in the furnace, the inorganic salts must be melted so they can run out of the furnace, and the carbon must be burned to CO or CO_2. The main chemical reactions which can be involved in carbon burnoff are as follows:

1) $C + 1/2\ O_2 \rightarrow CO$
2) $CO_2 + C \rightarrow 2CO$
3) $H_2O + C \rightarrow H_2 + CO$
4) $Na_2SO_4 + 2C \rightarrow Na_2S + 2CO_2$
5) $Na_2SO_4 + 4C \rightarrow Na_2S + 4CO$

Carbon is a relatively minor constituent in the hot, actively burning region of the char bed, and the temperatures are above the melting point of the inorganic salts. Under these conditions, the carbon particles tend to be wetted by the molten sodium salts and are not in direct contact with the gas phase. The predominant reaction by which carbon is burned are then the reactions with molten sulfate. The sulfate-sulfide present can act as a catalyst for carbon burnup, with the sulfate reduced to sulfide by carbon and the sulfide subsequently reoxidized to sulfate by contact with air (*11*).

Thus, the conditions needed for stable burnup of carbon in the bed are not only a sufficient amount of air reaching the bed, but also conditions favorable for the reaction of Na_2SO_4 with C. The latter imposes definite temperature thresholds for char combustion.

Blackout A blackout may be defined as a marked reduction in the rate of burnup of the organic material in the black liquor so that autogenous burning is lost. It can occur locally or generally in the furnace. Black liquor combustion is a highly complex process, involving first drying and pyrolyzing the liquor, and then combustion of gases and char. The initial steps tend to be endothermic, with the final combustion reactions the main heat releasing reactions. Many of the processes are temperature dependent. Thus, thermal feedback is a key element in sustaining the combustion process. Critical conditions can exist in which low combustion rates result in a precipitous temperature drop and cessation of combustion, thus causing a blackout.

Blackouts can be caused by a number of factors:

1. Insufficient air reaching areas of the bed (often because of blocked primary air ports).
2. Water or wet liquor reaching the bed.
3. Insufficient combustible gases generated by pyrolysis (often related to poor drying delaying the start of pyrolysis).
4. Insufficient heat release in area above bed.
5. Low bed temperatures.

Conditions leading to blackout are generally those related to marginal combustion such as low solids liquor, liquor with low heating value, or operation well below rating.

A blackout is a potentially dangerous situation. Pyrolysis gases will be released from the blackout area and mix with air without immediately burning. Ignition of these combustible/air mixtures can result in puffing and could lead to damage to the unit. In addition, the rapid accumulation of char and lowering of bed temperature can impede smelt flow and result in plugged spouts and primary air ports.

Incipient blackouts can be monitored by flue gas analysis (TRS or CO) (*12*). As a blackout condition is approached, pyrolysis will occur without follow-up local combustion. A sudden drop in steam flow may also signal a blackout condition.

Sulfide formation

There are essentially two mechanisms by which sodium sulfide is formed in a recovery boiler.

1. Reduction of oxidized sodium sulfur compounds, especially sodium sulfate, by a reducing agent; and
2. Release of sulfur as H_2S during pyrolysis and subsequent absorption in molten Na_2CO_3.

A reducing environment must be maintained for sulfide to be present, because Na_2S is not stable in a high temperature oxidizing environment. However, sulfide formation does not require a strong reducing atmosphere. At 1700°F (930°C), a typical temperature in the char bed, reduction of sulfate can take place in an atmosphere where the CO_2/CO ratio is about 20 to 1 (*13*). This is clearly a mild reducing condition.

Reduction The direct reduction of sulfate is the primary means for sulfide formation within the char bed. Reduction of sulfate within the char bed occurs mainly by reaction with carbon. The main reactions and the associated heats of reaction at 1700°F (930°C) are:

1. $Na_2SO_4 + 2C = 2\ CO_2 + Na_2S$

$$\Delta H = +\ 518\ \frac{\text{Btu}}{\text{lb } Na_2SO_4}\ (1205\ \text{kJ/kg})$$

2. $Na_2SO_4 + 4C = 4\ CO + Na_2S$

$$\Delta H = +\ 1539\ \frac{\text{Btu}}{\text{lb } Na_2SO_4}\ (3580\ \text{kJ/kg})$$

These are both endothermic reactions. The magnitude of the endotherm is three times as great when CO is the reaction product instead of CO_2.

The endothermic reduction reaction acts as a heat sink inside the char bed, and so bed temperatures decrease from the bed surface inward. The bed regions in which significant temperature gradients exist are thus the regions where reduction is taking place at significant rates. Temperature measurements in a number of operating char beds have shown that this region is normally less than 1 to 3 ft (1 m) deep, and that the interior of large beds is essentially inactive (*14*).

Reduction of sulfate by solid carbon is a highly temperature-sensitive reaction. The rate doubles for about a 100°F (55°C) rise in temperature. The combination of an endothermic reaction and strongly temperature-sensitive kinetics causes the reaction to be self-limiting, unless heat is added from an external source. Inside the char bed, the reaction ceases when temperatures approach 1470°F (800°C).

The two parameters which determine the amount of reduction which can take place inside the char bed are the bed surface temperature and the magnitude of the reduction endotherm. The most important of these is the bed surface temperature, which sets the initial rate of reaction, determines the amount of sensible heat available for the endothermic reaction to draw on, and influences the apparent thermal conductivity of the bed. High bed surface temperatures are favorable to reduction on all three counts. The endotherm magnitude is dependent on the relative amounts of CO and CO_2 formed. Very limited testing inside one char bed gave about equal amounts (moles) of each gas. This would give a heat of reduction of about 867 Btu/lb Na_2SO_4 (2020 kJ/kg Na_2SO_4) (*15*).

Sulfide reoxidation by oxygen in air is a very rapid exothermic reaction. Reoxidation is inherently faster than reduction and is totally controlled by rates of oxygen transfer to the melt. Reoxidation proceeds as rapidly as oxygen can be supplied. In effect, smelt burns. A 5% decrease in reduction efficiency caused by reoxidation can raise melt temperature by about 200°F (110°C).

The net reduction efficiency obtained in a recovery furnace depends on the extent of reduction on the bed surface and within the bed, and the extent of reoxidation as the smelt flows out of the unit. Bed temperature is the most important parameter. Hotter beds tend to increase reduction and cold beds tend to give poor reduction. However, it is important that high bed temperatures be obtained witout extensive reoxidation of the smelt. Control of the primary air supply is critical. Relatively large quantities of air are introduced at the primary level and can contact smelt and reoxidize it. A large char bed is not normally required to obtain high reduction efficiencies. An active zone of a foot or so is sufficient if the bed is hot enough. A large bed may have some beneficial effect on reduction efficiency by serving as a shield against reoxidation.

H_2 Absorption Sulfide may also be produced according to the reaction:

$$H_2S + Na_2CO_3 = Na_2S + H_2O + CO_2$$

The equilibrium for this reaction favors Na_2S at higher temperatures and Na_2CO_3 at lower temperatures. Equilibrium calculations (*16*) indicate that optimum conditions for Na_2S formation are temperatures between 1650-1920°F (900 and 1050°C), less than 90% of theoretical air, and little H_2O vapor. Sulfur capture improves noticeably when temperatures are such that a free-flowing smelt is obtained. Moving, turbulent smelt would be expected to give much better absorption of H_2S. Substantial quantities of H_2S are released by pyrolysis of black liquor solids. Much of the pyrolysis will occur in and around the char bed where Na_2CO_3 is available and H_2S absorption can occur. In individual particles of black liquor solids undergoing pyrolysis, the exterior may be at high temperatures with molten Na_2CO_3 present, while H_2S is still pyrolyzing off from the cooler interior. This H_2S could be reabsorbed in the outer layer and never enter the furnace atmosphere.

Fume formation

One of the products of black liquor combustion is fume (very small particles of Na_2SO_4 and Na_2CO_3). Most fume is believed to originate from sodium vapor produced in the high temperature reducing environment of the char bed. This fume can be considered to be the result of the following series of reactions.

1a. $Na_2CO_3 + 2C = 2Na + 3CO$
1b. $Na_2CO_3 + CO = 2Na + 2CO_2$
2. $2Na + 1/2\ O_2 = Na_2O$
3. $Na_2O + CO_2 = Na_2CO_3$
4a. $Na_2O + SO_2 + 1/2\ O_2 = Na_2SO_4$
4b. $Na_2O + SO_3 = Na_2SO_4$

The key reaction is the formation of sodium vapor in the char bed by reduction of Na_2CO_3. The remaining reactions take place in the area above the bed and in the oxidizing zone, wherever free oxygen is present. A very fine particulate is produced because the solid phase condenses from the gaseous state. The tendency toward sodium volatilization, and hence fume formation, increases with temperature and with the intensity of the reducing atmosphere.

The reduction of Na_2CO_3 to yield sodium vapor is endothermic. The heat of reaction according to reaction 1a above (a likely path in the char bed) is about 3700 Btu/lb (8600 kJ/kg) Na_2CO_3 reduced. This is a substantial heat sink. The thermal effects of sodium volatilization parallel those of sulfate reduction; i.e., lowered char bed surface temperature and a temperature gradient within the bed where significant reaction rates are present.

There is very little information available on fume formation kinetics. There is little doubt that Na_2CO_3 reduction in the furnace is kinetically controlled, since calculations assuming equilibrium grossly overpredict the amount of fume formed and also predict a bed surface temperature nearly independent of operating variables.

Reactions taking place in the upper part of the furnace yield the products normally found in the fume (Na_2SO_4 and Na_2CO_3). The sulfate-forming reactions take precedence as long as there is an adequate supply of sulfur present. The relative proportion of sulfate and carbonate in the fume is a reflection of the relative amounts of sulfur and sodium volatilized in the lower furnace.

Another source of fume is NaOH (g). This can be formed by reactions of the type

$$Na_2CO_3 + H_2O = CO_2 + 2NaOH\ (g)$$

Equilibrium calculations by Warnqvist (*17*) indicate that NaOH (g) formation is significant and is the primary factor responsible for sodium emissions in a high-temperature oxidizing atmosphere. Thermodynamic data on NaOH (g) are relatively uncertain, so there is also some uncertainty in equilibrium calculations. However, there is no doubt that some NaOH (g) exists in the furnace atmosphere, because cold side corrosion problems have been traced to NaOH condensing on the back side of furnace tubes.

Sulfur retention

Sulfur enters the furnace with the liquor solids being fired. It is partially tied up with organic compounds and partially in inorganic compounds (associated with sodium). It leaves the furnace either in the smelt as Na_2S and Na_2SO_4, or out the stack with flue gas (as SO_2 or H_2S). Air emissions considerations limit the amounts which may be discharged out the stack. Thus, it is of interest to understand how sulfur flows through the unit and reacts with sodium compounds to ultimately discharge with the smelt.

There is a significant release of sulfur to the gas phase within the furnace by pyrolysis reactions following drying of the liquor. H_2S is the predominant compound. Part of this H_2S will be immediately reabsorbed by Na_2CO_3 in the smelt. Reabsorption is governed by the proximity of the H_2S to molten smelt, local equilibria of the absorption reaction, and the motion and mixing of the smelt. The H_2S which is not immediately absorbed will move on to the upper zones of the furnace. As the gaseous sulfur compounds enter the oxidizing zone of the furnace, they have an opportunity to react with oxygen to form SO_2 (or SO_3), and then with fume to form Na_2SO_4.

Prevention of H_2S emissions from within the furnace itself involves the following steps.

1. Minimization of the release of H_2S to the gas phase at the char bed. Release is inhibited by high bed temperatures.
2. Oxidation of the H_2S which is released within the furnace cavity. Three conditions must be met.
   a) There must be an excess of oxygen over that needed for complete oxidation of all combustible matter.
   b) There must be adequate mixing of the reduced sulfur gases and the combustion air.
   c) The gas temperatures must be high enough for the combustion reactions to go to completion during the time of passage through the furnace oxidizing zone.

The SO_2 produced by oxidation of reduced sulfur gases can be tied up as Na_2SO_4 fume by reaction with sodium compounds in the oxidizing zone of the furnace. The key to this step is the presence of a sufficient amount of volatilized sodium compounds to tie up the SO_2. Sodium compound volatilization is favored by increased bed temperature.

Bed temperature plays a key role in gaseous sulfur emissions control for two reasons. Higher bed temperatures reduce sulfur release and increase the release of sodium which reacts with SO_2. Field testing and analysis of gas samples taken just above the char bed (*18*) have shown a pronounced dependence of H_2S levels on bed temperature.

Bed temperature, °F	H_2S (ppm) above bed
1335 (725°C)	5250
1515 (825°C)	750
1580 (860°C)	250

This sensitivity to bed temperature is one reason why a TRS monitor showing an increase in TRS emissions may give warning of an incipient blackout.

The sulfur that is volatilized and reacts to form Na_2SO_4 in the oxidizing zone of the furnace ultimately returns to the char bed by internal slag falls or to the black liquor feed from the dust hoppers and the precipitator. Since sulfate is reasonably stable during black liquor pyrolysis, the great bulk of the recycled sulfur reaches the bed to be reduced and discharged as smelt.

Entrainment

Entrainment of liquor or char particles occurs when they are caught up in the moving gas stream and carried up in the furnace. Most entrainment results from the small sized fraction of the spray droplet distribution and is directly associated with liquor firing. Since the product of pyrolysis is a light, fluffy carbonaceous char, some entrained material may also be swept from the char bed by high gas velocities.

Very little quantitative information is available on entrainment. A particle will be entrained whenever the upward gas velocity is greater than the characteristic settling velocity of the particle. The settling velocity is a complicated function of the size, shape, and density of the particles and the density and viscosity of the furnace gases. Entrainment increases with any of the following:

1. Decrease in particle size;
2. Decrease in particle density;
3. Increased gas viscosity (increased temperature);
4. Decreased gas density (increased temperature); or
5. Increased air flow to lower zones of furnace.

Liquor spray characteristics (mean particle diameter and the distribution of sizes around the mean) have a major influence on entrainment. Very little quantitative information is available on spray characteristics of black liquor fired in furnaces.

Entrained liquor droplets dry, pyrolyze, and burn in suspension and in the oxidizing zone of the furnace. Gasified sulfur has less opportunity to react with sodium fume, and there often is sufficient time for complete carbon burnup before gas temperatures drop below those needed to sustain combustion. Thus, carbonaceous material can be carried into the upper part of the furnace where it can be incorporated into the slag or contribute combustible material to the precipitator dust. Entrainment significantly increases the particulate load and increases the likelihood of sulfur emissions with minimal beneficial effects. Entrainment should be minimized.

Corrosion reactions

Fireside corrosion can occur in a number of different locations in the unit and may be caused by a variety of mechanisms. Some potential corrosive reactions are:

1. Direct attack of molten smelt on iron.
2. Corrosive attack by elemental sulfur generated by Na_2S-CO_2 or H_2S-O_2 combinations.
3. Attack by pyrosulfate, $Na_2S_2O_7$.
4. Acidic corrosion below the dew point of gases.
5. Cold side attack by reaction of NaOH and iron in nonmembrane wall units.

Molten kraft smelt will react very rapidly with iron and steel. Continued direct contact of molten smelt and metal will result in total destruction of the metal. Smelt is contained by cooling the metal so as to form a layer of frozen smelt at the metal surface to act as a protective barrier. Normally, molten smelt does not come into direct contact with the metal in the furnace.

The most common type of fireside corrosion in recovery boilers occurs in the lower part of the furnace and is apparently caused by elemental sulfur reacting with iron to form FeS.

$$Fe + S = FeS$$

The elemental sulfur is produced *in situ* either by reaction of H_2S and O_2 (*19*)

$$\text{e.g., } H_2S + O_2 = 1/2\ S + 1/2\ SO_2 + H_2O$$

or by reaction of CO_2 and Na_2S (*20*)

$$\text{e.g., } Na_2S + 2\ CO_2 = S + Na_2CO_3 + CO$$

The corrosion rate is low at temperatures below about 590°F (310°C), but it increases rapidly above this temperature. Keeping metal surface temperatures below this limit is critical to control of this type of corrosion. Tube wall surface temperatures are dependent on the local heat flux, internal deposits within the tube, and the boiler water saturation temperature (dependent on operating steam pressure). Regions of high heat intensity or areas where internal deposits have formed can be regions of rapid corrosion. A frozen smelt layer on the tube surface is protective. Tube vibrations or exfoliating scale can crack the frozen layer and lead to increased corrosion.

The H_2S-O_2 reaction seems to predominate in wall drying units, because relatively large amounts of H_2S are generated directly at the walls by pyrolysis of liquor solids in the char band. In suspension drying units, the Na_2S-CO_2 reaction is believed to be the major source of the elemental sulfur which attacks the metal. In either case, a reduced sulfur species and an oxidant are involved. Intermittent exposure to oxidizing and reducing environments tends to aggravate corrosion, because protective layers formed under one regime tend to be attacked in the other regime.

Tube wastage in the lower furnace can be controlled by the use of composite tubes, studding, or various repairable coatings. The trend appears to be toward composite tubing, especially on high pressure units, but the use of high density 1/2-in.-diameter studding can also provide protection (*21*).

Pyrosulfate corrosion occurs when liquid pyrosulfate, $Na_2S_2O_7$, is present and tends to be located in the boiler bank, superheater, and economizer sections. The major reactions involved are:

$$Na_2SO_4 + SO_3 \rightarrow Na_2S_2O_7$$

and $Na_2S_2O_7 + H_2O \rightarrow 2\ NaHSO_4$

Sodium pyrosulfate has a melting (freezing) point of 750°F (400°C). Potassium salts can lower the freezing point to 510°F (280°C). Pyrosulfate corrosion is prevented by avoiding pyrosulfate formation. Conditions favoring pyrosulfate formation are those which lead to high SO_3 concentrations. These are excessive sulfur gasification in the furnace, insufficient sodium volatilization, and excessive O_2 concentrations in the flue gas.

Direct sulfidation by molten smelt has been observed in the superheater region. Such an attack by reduced sulfur species can occur if unburned entrained black liquor solids or char are incorporated into the deposits. In addition, sodium sulfate can become aggressive at metal temperatures above 1020°F (550°C). This can become a serious problem in the absence of a protective frozen slag layer. Chlorides can lead to increased attack on superheaters by lowering the melting point of the slag and decreasing or eliminating the protective frozen layer.

A number of flue gas components can form acids when the flue gas is cooled below its dew point. These include SO_2, SO_3, and HCl. The acids can cause corrosion in the cold end of the system including ducting, I.D. fan, precipitator, and parts of the economizer.

Ash fusion characteristics

The principal inorganic compounds in the lower part of the furnace are Na_2CO_3 and Na_2S. Pure Na_2CO_3 melts at 1564°F (851°C), and pure Na_2S melts at 2138°F (1170°C). The phase diagram for the Na_2CO_3-Na_2S system is shown in Fig. 4.4 (*22*).

The system forms a low melting eutectic at 40 mole% Na_2S with a melting point of 1405°F (763°C). In the normal range of sulfidities of kraft smelts, the melting temperature decreases with increasing sulfidity.

The phase behavior of carbonate and sulfide is considerably affected by the presence of chloride (as NaCl) and potassium (*23*). Sodium chloride forms a eutectic with Na_2CO_3 at about 63% Na_2CO_3 (Na_2O basis) and a temperature of 1170°F (632°C). It forms a eutectic with Na_2S at 39% Na_2S and 1314°F (712°C). The three-component system Na_2CO_3-Na_2S-NaCl has a ternary eutectic at 3% Na_2S, 62% Na_2CO_3, and 35% NaCl which melts at 1108°F (598°C).

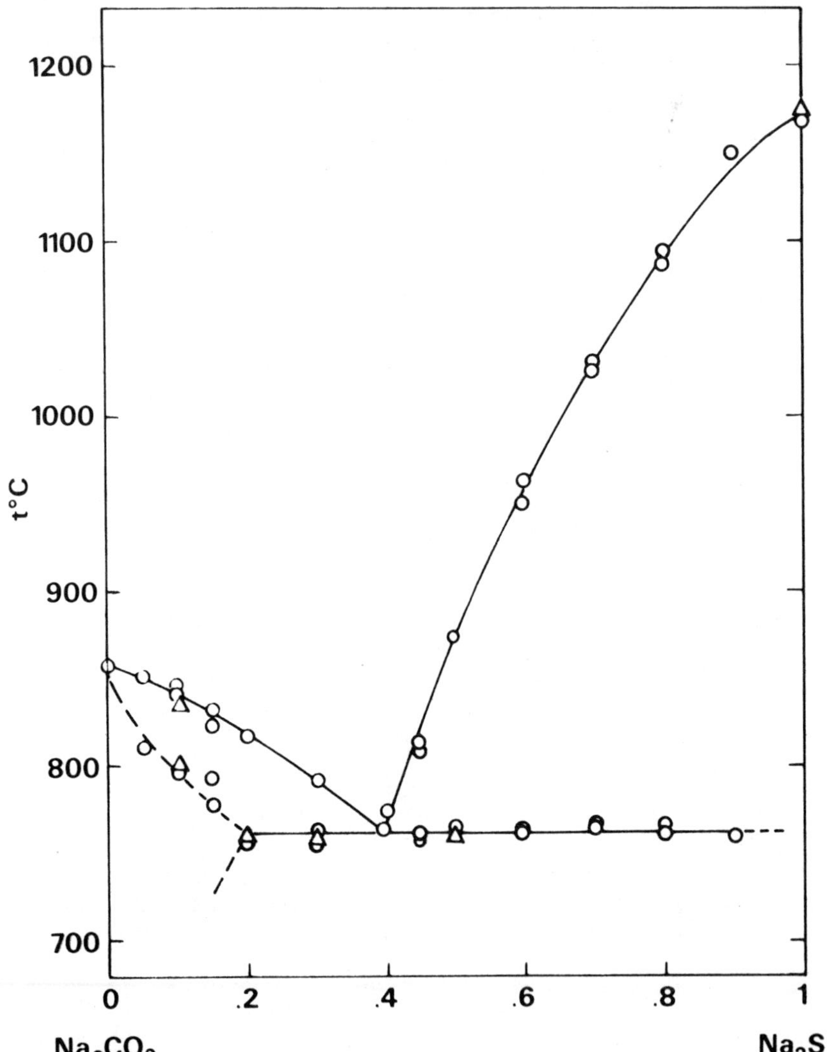

Fig. 4.4. Phase diagram Na_2CO_3-Na_2S.

The effect of potassium on the smelt melting temperature is shown in Fig. 4.5. Up to 25% of the sodium in a smelt consisting of 66.3 mole% Na_2CO_3, 20.8 mole% Na_2S, and 12.9 mole% Na_2Cl_2 was replaced by potassium. The liquidus temperature decreased gradually with increasing potassium content, while the eutectic melting temperature dropped to 977°F (525°C) at 5% potassium replacement. This is a maximum melting point drop of 110°F. The liquidus temperature is the temperature at which crystallization just begins for a given composition. Above this temperature the material is completely molten. The eutectic temperature is that at which final solidification of the smelt takes place. Between these temperatures a mixture of solid and liquid phases will be present.

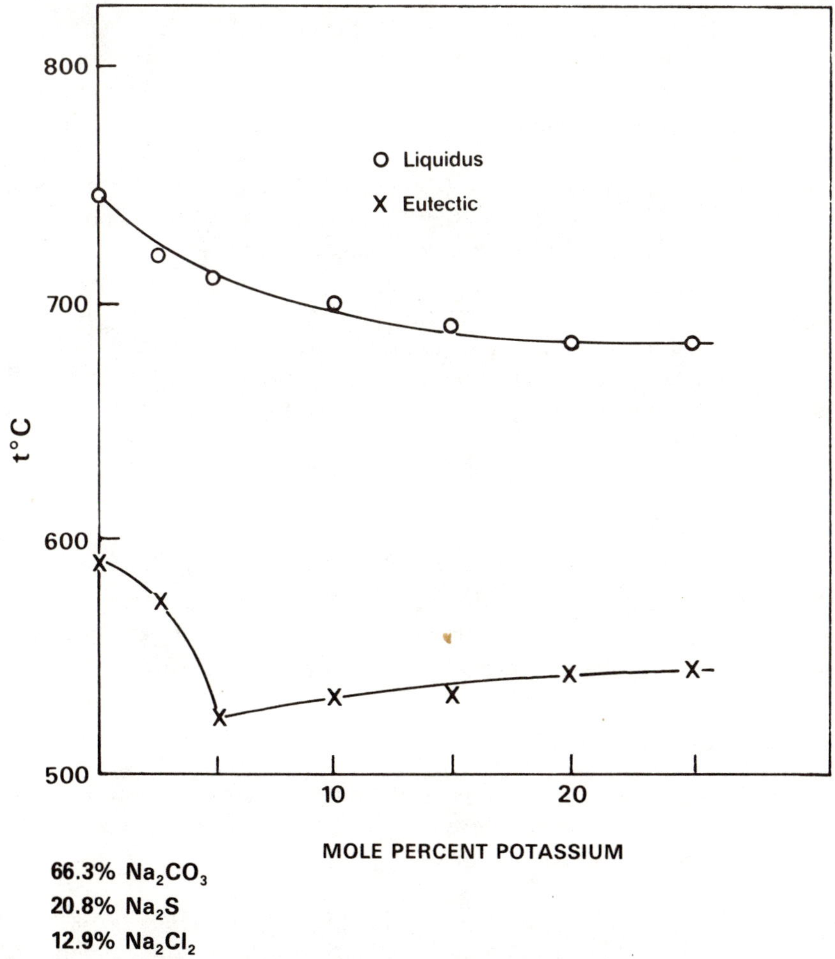

Fig. 4.5. The effect of potassium substitution on smelt thermal events.

Data on the viscosities of various molten salts are shown in Fig. 4.6 (*24*). At any given temperature, the viscosity of the smelt decreases with increasing sulfidity, undoubtedly because increasing sulfidity lowers the melting temperature for sulfidities less than 40%. The viscosities of carbonate-sulfide smelts are a fairly strong function of temperature.

Nonprocess elements

Nonprocess elements are elements other than those which are normally a part of the kraft pulping and recovery cycle (Na, S, C, H, and O). The two nonprocess elements having the greatest effect on recovery boiler operation are potassium and chlorine. This occurs for two reasons:

1. Neither is naturally purged from the cycle, and they can build up to relatively high quantities when the input is significant.
2. They are relatively volatile in the furnace and also have lower melting points of smelt and ash.

During combustion, potassium forms salts which are directly analogous to the sodium salts and chlorine forms NaCl. The influence of potassium salts and chloride on eutectic melting temperatures was discussed in the previous section. Potassium salts are more volatile than sodium salts. The potassium/sodium ratio in the fly ash is 1.5 to 3 times that in the black liquor. Thus, fly ash is enriched in potassium salts with a corresponding lowering of ash fusion temperatures. Sodium chloride is also enriched in the fly ash. Ratios of Cl/Na in fly ash to Cl/Na in smelt range from 3 to 4, and are practically independent of temperature.

Under some circumstances HCl can be formed in the flue gas. The reaction involved is:

$$2\ NaCl + SO_2 + 1/2\ O_2 + H_2O = 2\ HCl + Na_2SO_4.$$

Furnace operating conditions which favor SO_3 formation (high concentration of SO_2 and excess O_2) are likely to result in HCl formation.

Process equipment

The major differences between recovery boilers and fossil-fuel fired power boilers were discussed in the Introduction. The following general features of recovery boilers should be noted:

1. Combustion is carried out in a stationary water-walled furnace, the waterwalls forming part of the boiler (steam generating) circuit. The

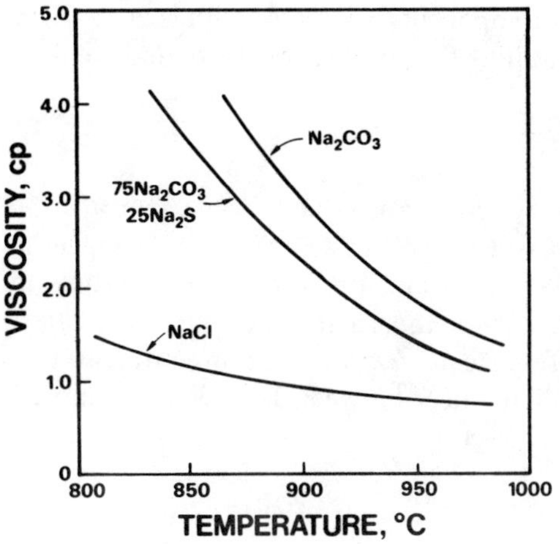

Fig. 4.6. Viscosities of molten salts.

size of the furnace is such as to provide volume compatible with energy release rates, sufficient cross-sectional area to yield reasonable gas velocities, and sufficient heat transfer area to reduce flue gas temperature to desired values. Hot flue gases leaving the furnace cavity flow through a series of heat traps. The most common sequence is screen, superheater, boiler bank, and economizer. Screens have been eliminated in the newest units. Many older units have a direct contact evaporator as a final heat trap. In this case, the economizer is smaller so that there is sufficient heat in the flue gas entering the DCE to achieve the desired concentration of the black liquor.

2. Air is introduced at several levels within the furnace. It is normally preheated to about 300°F (150°C) to help maintain stable combustion. A forced draft fan and adjustable dampers are used to control the total amount and distribution of the air. An induced draft fan is used to remove the flue gas from the furnace and maintain a balanced draft.
3. Liquor is introduced so as to burn on a char bed in the hearth zone of the furnace. The method of introducing the liquor differs in units of different manufacturers. The char bed serves as a reducing medium and a combustion stabilizer.
4. Smelt flows out of the furnace by gravity into one or more dissolving tanks. The hearth may be slanted with drainage to one end, or flat with the smelt removed by decantation.

This discussion of process equipment is divided into three parts for convenience.

1. Basic furnace structure and components associated with heat recovery;
2. Systems associated with handling liquor, air, flue gases and auxiliary fuel; and
3. Equipment for particulate recovery.

In North America, most recovery units in use have been manufactured by either the Babcock & Wilcox Co. (B & W) or Combustion Engineering Inc. (CE). A diagram of a B & W unit is given in Fig. 4.7 and a diagram of a CE unit is shown in Fig. 4.8. For the most part, the equipment can be discussed generically without reference to a particular manufacturer. A specific discussion of the major differences between B & W units and CE units is provided at the end of this section.

Furnace construction

Waterwalls Combustion is contained within a completely water-cooled enclosure, the so-called waterwall furnace. This approach was originally developed to eliminate severe maintenance problems caused by the attack of molten smelt on refractory walls. The heat absorption in the waterwalls currently plays a key role in reducing flue gas temperatures before the subsequent heat traps are encountered.

A number of different wall constructions are used, depending on the age and manufacture of the units. Modern units use a membrane wall or welded wall construction. The space between individual tubes is closed by a metal plate or strip which is welded to each tube. These are normally shop welded, and failures and cracks are rare. Some earlier units used a tangent-tube construction. The boiler wall consisted solely of tubes on close spacing so that the tubes were tangent (touching against each other). The tangent tubes were backed by a heavy skin casing to give a pressure-tight furnace envelope. Another construction, often used on older units, involved the use of flat studs. The space between adjacent tubes was usually 2 in. wide and was closed off with two 1-in. flat studs welded to adjacent tubes. These three constructions are illustrated in Fig. 4.9.

Some recovery boilers have pin studs on part of the tube surface. These provide extended heat transfer area and anchoring for refractory or solidified smelt, which protects the underlying tube metal from corrosion and erosion.

Stack studs are used in very high heat flux areas, such as around air ports and around spout openings where smelt erosion takes place. These are 1/4-in. or 3/8-in.-thick plates welded to the tubes. Consideration is being given to using composite tubes in this area instead of studs.

A large nose baffle or furnace arch is present at the rear of the upper furnace, just below the superheater. The purpose of the nose baffle is to help

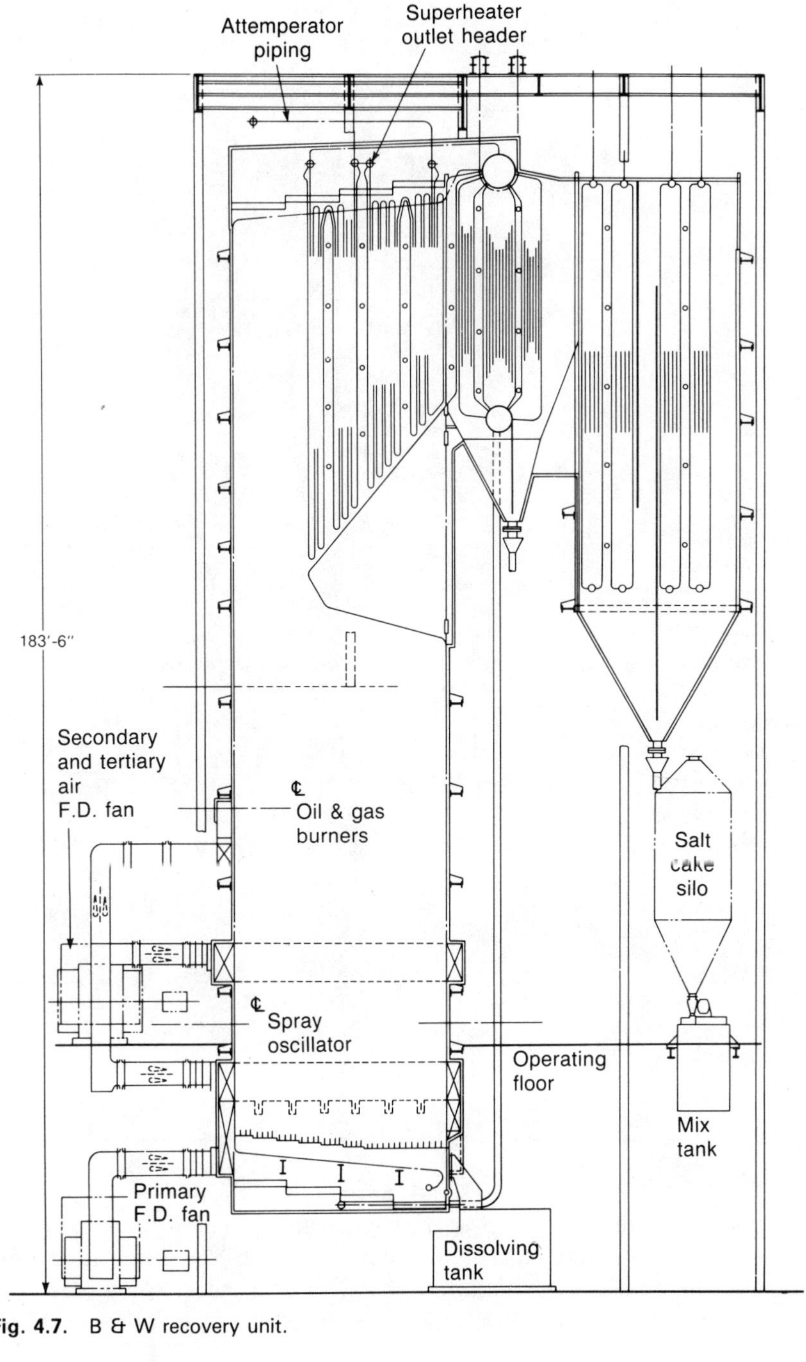

Fig. 4.7. B & W recovery unit.

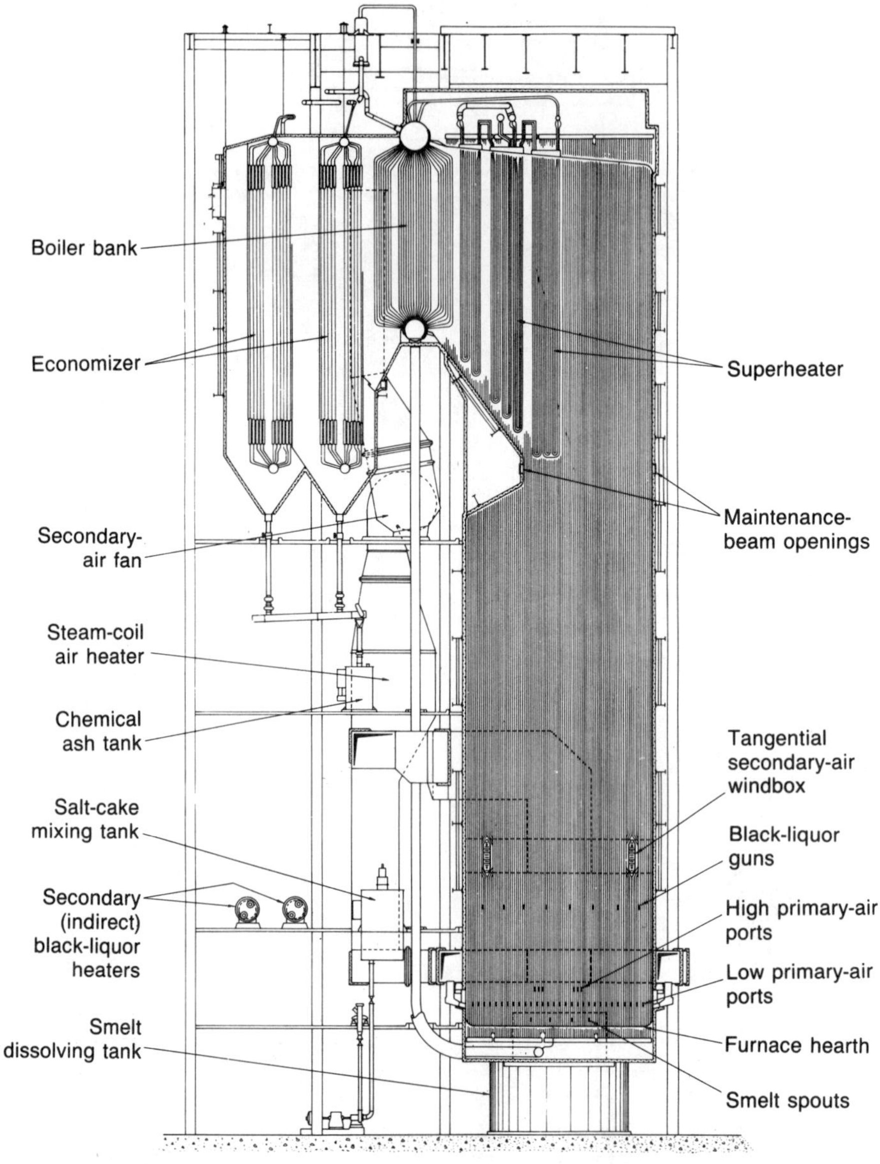

Fig. 4.8. CE recovery unit.

obtain good gas distribution over the superheater and generating bank heat transfer surfaces, and to partially shield the bottom of the superheater loops from the intense thermal radiation from the lower furnace. The furnace arch is contiguous with the rear wall; the tubes merely being bent in toward the

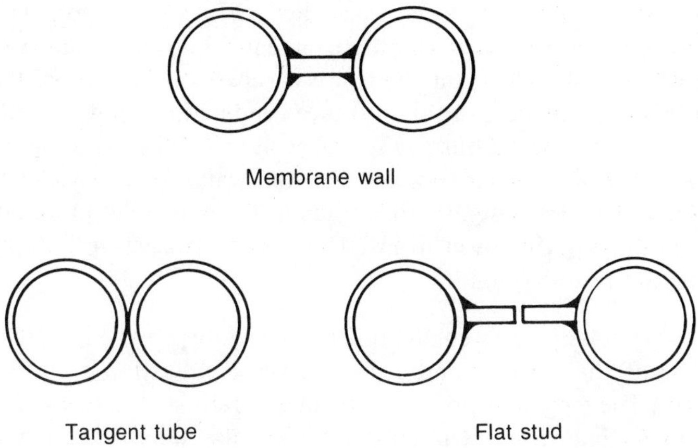

Fig. 4.9. Furnace wall constructions.

combustion space and then bent back. In some furnace designs, the arch protrudes farther into the furnace to give better radiation protection and gas distribution in the superheater.

One of the key factors influencing the design of the furnace waterwalls is the need to protect against corrosion. The normal material used for recovery boiler tubes is ordinary carbon steel. Corrosion of this material in recovery furnace environments occurs at a highly accelerated rate when metal temperature exceeds 590°F (310°C). Several techniques are in use to provide corrosion protection. These include close-spaced pin studs, flame or plasma metallizing, weld overlaying, or use of composite tubing.

Pin studs help to maintain a layer of frozen smelt next to the surface, which acts as a renewable refractory and a gas barrier. They also provide sacrificial heat transfer surface which can be periodically renewed. Conventional refractories, such as chrome-ore, tend to be eroded and are replaced by solidified smelt.

The other techniques operate by introducing a metallurgy which is more resistive to corrosive attack in a furnace environment at higher temperature levels. Metallizing with chromium or chrome-aluminum compounds provides protection but must be periodically renewed. Recently there has been increased use of composite tubes, in which an outer layer of austenitic stainless steel is extruded over a carbon steel inner tube during the tube manufacturing process. This shows promise for reducing furnace maintenance problems and permitting higher pressure (and temperature) boiler operation.

Circulation of boiler water through the waterwalls is provided from headers, which are connected to the lower drum by downcomers. The details of the flow path depend on the nature of the hearth construction. Typically, a unit with flat hearth construction has a center bottom header to supply all four furnace walls. The front and rear walls are tied together by continuous floor tubes, each connected to the furnace header. In sloping hearth units, the floor tubes are contiguous to the rear wall tubes and are supplied from one header. Another header supplies the front wall tubes. Separate headers supply sidewalls. New designs contemplate the use of small modular boiler sections eliminating the lower drum. These are best used in high-pressure, high-temperature operation.

Screens A screen of water-cooled tubes is normally provided at the top and to the rear of the furnace and separates the combustion zone from the superheater. The screen is part of the steam generation circuit. It is becoming common to eliminate the screen tubes entirely in new units. The screen surface is replaced with additional furnace height to compensate for the gas cooling capacity removed along with the screen section.

The screen acts to shield the superheater from direct radiation from the lower furnace and also cools and partially deslags the combustion gases. Formation of deposits in the superheater and in the closely spaced tubes of the boiler bank is dependent on the gas velocity, dust loading, and gas temperature relative to the ash fusion temperature. The combustion products entering the boiler bank must be sufficiently below the ash fusion temperature of the carryover so that cleanliness can be maintained by use of sootblowers. Typically, a gas temperature of 1200°F (650°C) at the front of the generating section (boiler bank) is used for design purposes (*25*). The screen section helps cool the gases down to meet this condition. The amount of heat absorption in the superheaters is limited, in part, by the desired superheat temperature and steam production rate. Cooling in the screen section helps bridge the gap between the gas temperatures leaving the furnace cavity and the allowable gas temperature entering the superheater section.

The screen tubes are spread on wide centers across the width of the boiler to prevent accumulations from bridging across between adjacent tubes. In some units the screen tubes are platen assemblies (tangent tubes in the direction of gas flow) with a wide side spacing. In other units the screens are individual tubes on fairly wide spacing.

Superheaters Superheaters are used to supply the sensible heat to raise the steam temperature from the saturation temperature at the prevailing boiler pressure to the desired superheat temperature, while simultaneously cooling the combustion gases to the temperature at which they enter the boiler bank.

There are two main types of superheaters: platen superheaters and spaced tube superheaters. Platen superheaters are composed of assemblies consisting of tangent tubes with rather wide side spacing. Spaced tube superheaters consist of separate tubes on relatively wide spacing.

The rate of heat transfer per unit circumferential surface area is considerably greater for spaced tube superheaters than it is for platen superheaters. On the other hand, the heat transfer rate of platen superheaters tends to be independent of flue gas temperature, whereas the corresponding rate for spaced tube superheaters decreases with increasing flue gas temperature. This is due to a greater tendency for slag buildup with spaced tube superheaters. The efficiency of superheaters is about 60-70% of the clean tube efficiency in the temperature range between 1470-1650°F (800 and 900°C) (*26*). The superheater must be equipped with sufficient cooling (desuperheating) for metal protection and to provide for the load range specified.

Superheaters are fed from an inlet header directly connected to the steam drum. Steam flow is parallel to the flue gas flow. Superheaters are often divided into sections (primary and secondary) with a spray desuperheater or attemporator located between sections. The intermediate desuperheater is used to control the final steam temperature to the desired value. Steam flow in the secondary superheater can be parallel or counterflow to the flue gas.

Generating section The generating section or boiler bank supplies the latent heat to convert feedwater to steam at the prevailing boiler pressure. It simultaneously cools the combustion products as they flow through the section. The generating section itself is responsible for only a part of the latent heat loads. Additional steam generation takes place in the waterwalls, which are part of the steam generating circuit, and in the screen tubes.

The boiler bank consists of an upper drum (or steam drum) and a lower drum (or mud drum) which are connected by vertical banks of tubes bent along drum radii and expanded into holes in the drum. This is shown schematically in Fig. 4.10. Most North American boilers have no baffles in the boiler bank, and the flue gas makes a single pass through the unit. Some Scandinavian designs and some small U.S. designs use two gas passes through the tube bank.

The steam drum contains internals for separating the steam from the water. Saturated steam is withdrawn from the steam drum. A schematic diagram of a steam drum is shown in Fig. 4.11.

Feedwater from the economizer enters at the steam drum. Saturated water flows downward through the rear boiler bank tubes to the lower drum. A large portion then flows through downcomers to the waterwall and screen

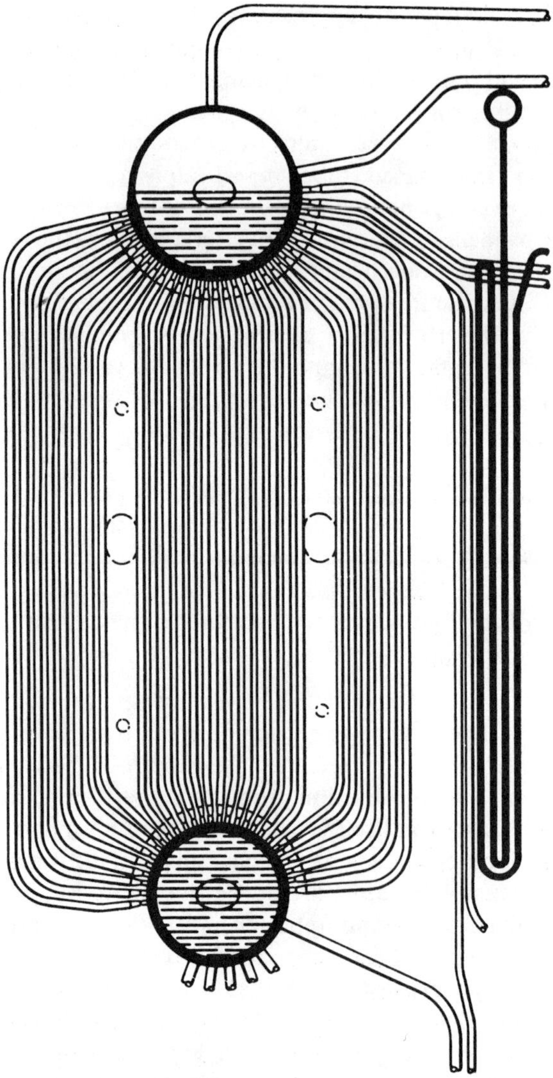

Fig. 4.10. Schematic of boiler bank.

tube headers (if used). The boiling water rises as a steam-water mixture through the furnace waterwalls, screen tubes, and forward boiler tubes to the steam drum where the steam is separated, washed, and removed and the water returned to the lower drum. Flow in this circuit is by natural circulation, driven by the difference in density between the boiling two-phase mixture in the rising section and liquid water in the down flow sections. Feedwater addition ultimately must match the rate of removal of water as steam and

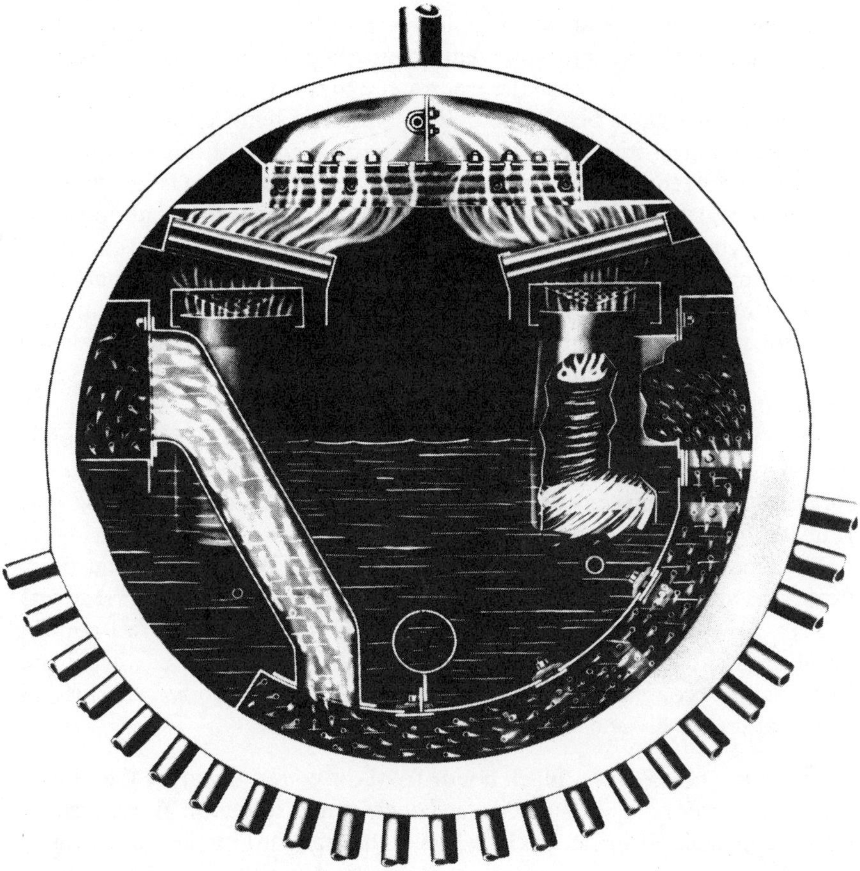

Fig. 4.11. Schematic diagram of steam drum.

as blowdown (the latter is used to control the amount of dissolved solids present in the boiler). Feedwater control is based on maintaining a given water level in the steam drum with due allowance made for changes in voidage (swell or shrink) during transient conditions.

The heat transfer from the gases in the boiler bank is primarily convective. The heat absorption capability of the boiler bank is strongly affected by the buildup of deposits on the external tube surface. This can generally be controlled by sootblowing, but the frequency of sootblowing needed and the extent of degradation in performance between cycles is a function of the amount of particulate carryover, the chemical nature of the carryover, and the prevailing temperature levels.

Economizer The economizer functions as a boiler feedwater heater while simultaneously serving to further cool the flue gas. It brings the feedwater from the temperature at which it leaves the deaerator to a temperature approaching the boiler saturation temperature (a function of boiler pressure). Countercurrent flow between the feedwater and the combustion gases is used to cool the flue gases below the boiler saturation temperature.

The economizer normally consists of a vertical bank of tubes connecting a lower inlet feedwater header to an upper outlet header. Gases enter the economizer at the top and make a number of passes across and down the tubes, directed by baffles, before leaving at the bottom. The feedwater flows upward through the tubes and passes directly from the outlet economizer header to the boiler steam drum. The economizer tubes may be finned or bare tubes. Bypass dampers are normally provided, if a direct-contact evaporator is used, to control the gas temperature entering the direct-contact evaporator.

Some units have long tube-type finned economizers arranged in separate tube banks without any baffles in the tube banks. The casing is arranged to give a certain amount of cross-flow in the tube banks to increase heat transfer.

The amount of heat removed in the economizer depends on what other heat traps are used. In recent years, so-called "low-odor" or NCE recovery boilers have become popular. These are units without direct-contact evaporators with a larger economizer usually used as the alternate heat trap. These larger economizers have generally been designed to reduce flue gas temperatures to around 400°F (200°C).

Direct-contact evaporators Black liquor from the conventional LTV multiple-effect evaporators is not sufficiently concentrated to permit firing directly into the furnace. Prior to the late 1960s, the standard method of bringing black liquor to firing concentration, in North America, was with a direct-contact evaporator. Since that time, indirect evaporators (called concentrators) for achieving firing concentrations have been developed.

A direct-contact evaporator is a device in which hot flue gas is brought into direct contact with black liquor. The flue gas cools, and the sensible heat given up supplies the latent heat to evaporate water from the liquor. The process is essentially adiabatic humidification of the flue gas, and the gas leaves the direct-contact evaporator at a lower temperature and higher humidity than when it entered. Under typical conditions it takes about a 250°F (140°C) drop in flue gas temperature to increase the black liquor solids content from 50 to 65%. Two types of direct-contact evaporators are used: the cascade evaporator and the cyclone evaporator.

The cascade evaporator is a device in which the liquor is picked up on a rotating bundle of steel tubes and moved through the hot flue gases. It is shown in Fig. 4.12. The cascade evaporator may consist of one or more wheels arranged side by side or vertically. Each wheel consists of two circular side

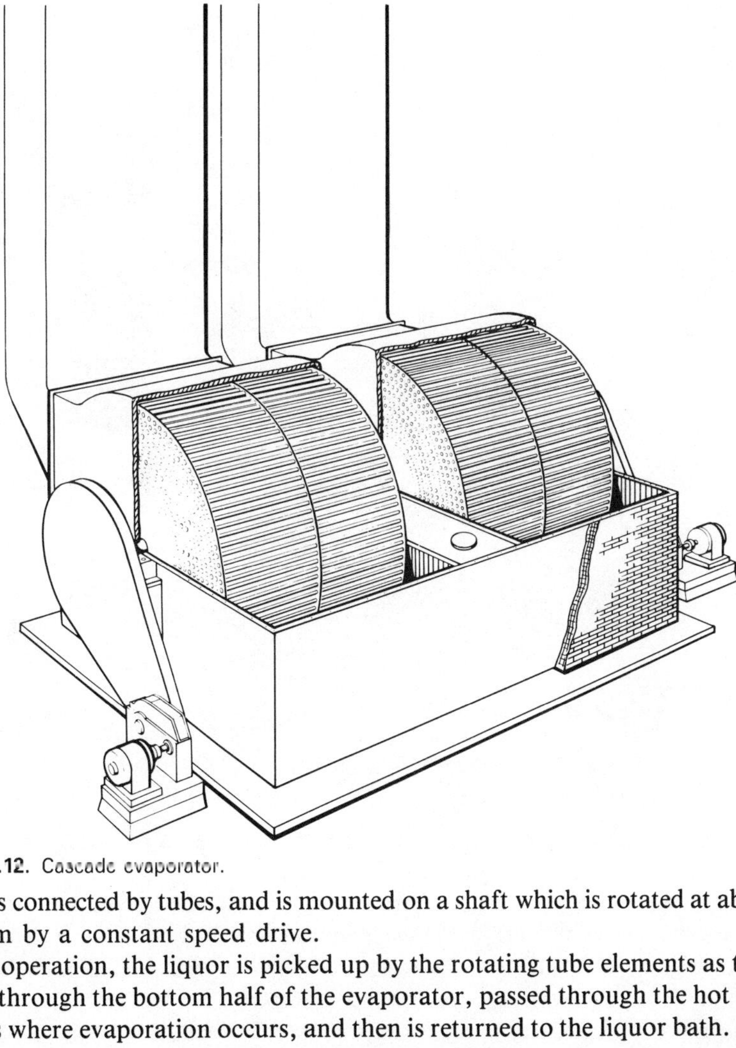

Fig. 4.12. Cascade evaporator.

plates connected by tubes, and is mounted on a shaft which is rotated at about 6 rpm by a constant speed drive.

In operation, the liquor is picked up by the rotating tube elements as they pass through the bottom half of the evaporator, passed through the hot flue gases where evaporation occurs, and then is returned to the liquor bath. The entire body of liquor in the tank ultimately reaches a steady state concentration which is suitable for firing. Flow channels are provided within the tank for return circulation of the liquor as it is moved toward one end of the tank by the rotation of the wheel. The concentrated liquor is withdrawn through a screen in a flow box at the side of the evaporator. The screen is equipped with a moving arm for scraping lumps to a separate chamber where they are redispersed.

The cyclone evaporator is shown in Fig. 4.13. It is a vertical, cylindrical vessel with a conical bottom. Flue gas is admitted through a tangential inlet near the bottom and flows in a whirling, helical path to the top where it leaves

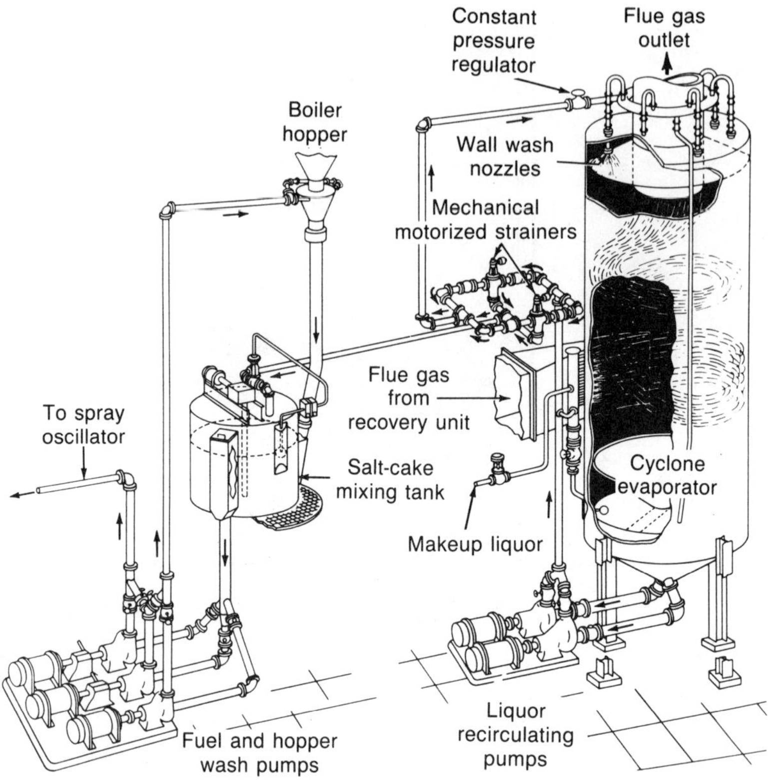

Fig. 4.13. Cyclone evaporator.

through a concentric, re-entrant outlet. Liquor to be evaporated is sprayed across the gas inlet. The liquor droplets mix intimately with the gas and are then thrown by centrifugal force to the cylinder wall. Recirculated liquor to prevent wet-dry conditions flows down the wall of the cyclone and carries the droplets (and any dust removed from the gas) to the conical bottom, out through the drain, and into an integral sump.

Liquor is pumped to spray nozzles located at the tangential gas inlet. Evaporation occurs mainly from the droplets in the flue gas and, to a lesser extent, on the walls of the cyclone. Adjustment of the quantity of spray liquor at the inlet affords control of the degree of evaporation within a limited range. Recirculated liquor is pumped from the sump through strainers to wall wetting nozzles at the top of the evaporator to wet the walls. A distributing skirt arrangement is used to discharge the liquors evenly onto the walls. The walls must be kept continuously wet to prevent localized drying of the liquor. Liquor feed into the units is controlled on the basis of sump level. Concentrated liquor is removed from the unit by bleedoff from the recirculated liquor line. One or more of these units may be used in parallel depending on the size of the recovery boiler.

Flue gas air heaters An alternate heat trap used on a few installations is an air heater. Sensible heat is removed from the flue gas and used to increase the temperature of the combustion air. The specific heats and flows on flue gas and air are not greatly different from each other, so the temperature drop in the flue gas is approximately the same as the rise in the air temperature.

On a few units a regenerative type of air heater is used. It consists of a rotating wheel composed of elements with considerable thermal mass and heat transfer area. The rotating wheel is contained within an assembly which is divided by partitions that confine the flue gas to one side of the apparatus and the air to the other. As the wheel slowly rotates, the elements first store thermal energy by increasing in temperature as they pass through the flue gas side and then release it to the air as they are cooled while passing through the air side.

Flue gas air heaters have met with very limited use on recovery boilers because of the high particulate load and potential corrosion and plugging problems.

Sootblowing Keeping boiler, superheater, economizer, and air heater surfaces in open and clean condition has always been one of the major problems in recovery boiler design and operation. Black liquor combustion leads to rapid buildup of dust or slag on the tubes which must be removed if the boiler is not to become plugged. Recovery units are equipped with automatic sootblowers so that hand cleaning is not required during normal operation at rated capacity.

Current practice tends toward the use of long retractable-type sootblowers throughout the boiler, such as shown in Fig. 4.14. The lance tube remains outside the boiler when not in service and is automatically inserted and traversed across the boiler while being rotated. High-pressure, superheated steam is generally used as the blowing medium. There are usually two large nozzles at the end of the lance tube which give concentrated high-pressure, high-volume blowing. Translation and rotation speeds may be varied within limits to ensure proper coverage. Controls normally permit automatic operation of all blowers in an established sequence.

There is continuing interest in the use of additives to condition the deposits on the heat transfer surfaces to facilitate cleaning. The composition of these materials is usually proprietary. The effectiveness of additives depends on their ability to make deposits more friable and less cohesive. Effectiveness also depends on the method of application, so that the additive gets to the critical areas in sufficient quantities. Caution must be taken to use materials which are compatible with the entire recovery cycle.

Fig. 4.14. Retractable-type sootblower.

Combustion systems

Black liquor firing system This system provides for handling the black liquor within the recovery unit envelope, conditioning the liquor prior to firing, and introducing the liquor into the furnace. Since black liquor is the only fuel fired during normal operation of a properly designed and loaded recovery boiler, the liquor firing system is a key component of the unit.

There are many variables which influence the black liquor flow path in a recovery unit. These include:

1. Direct contact evaporator vs. concentrator;
2. Wet bottom vs. dry bottom precipitator; and
3. Type of liquor spray gun used.

In systems with a direct contact evaporator, the liquor from the strong black liquor storage tank is pumped first to the wet-bottom precipitator (if present) where collected particulate is picked up, and then flows to the direct contact evaporator where it is brought to the solids content desired for firing. From the direct-contact evaporator, the liquor flows to a salt cake mix tank where makeup salt cake (Na_2SO_4) is added and stirred into the black liquor by motor-driven agitators. Makeup salt cake is fed into the mix tank from the salt cake silo, usually by means of a variable speed screw conveyor. The primary liquor heater is normally located in the mix tank. After makeup chemical addition, the liquor goes through a furnace feed pump, liquor heater, flow meter and solids monitor, to the spray guns, and into the furnace.

In systems employing suspension drying, multiple spray guns connected to a common header (ring header) are used. These guns may be located on opposite walls or on all four walls depending on the size of the unit. Large units may use 12 or more liquor guns. The nozzle is designed to give a conical spray of optimum size particles (coarse) for rapid drying and easy burning, while minimizing production of fine particles which are carried along by the gas stream. A diagram of a liquor gun and spray nozzle is shown in Fig. 4.15. The nozzle contains a swirl generator plate and has large helical passages to avoid plugging. It normally requires liquor which is slightly superheated to avoid "roping" of the liquor. The spray guns are normally oscillated in a vertical plane (from horizontal to about 45° downward) to distribute liquor solids over the whole hearth area.

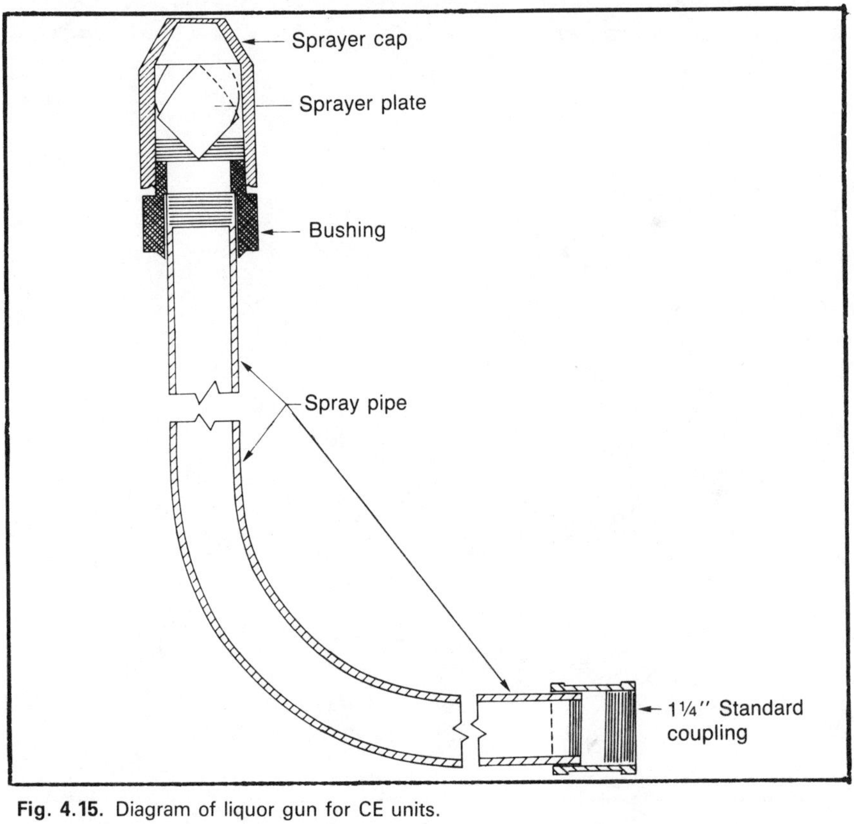

Fig. 4.15. Diagram of liquor gun for CE units.

In systems employing wall drying, between one and four liquor guns are used. A single gun is normally located on the front wall, and the second gun (when used) is located on the rear wall at the same elevation. The third and fourth guns (when used) are located on each side wall at the same elevation. A splash plate type of spray nozzle is used, as shown in Fig. 4.16. The nozzle provides a flat sheet of coarse spray droplets, which is projected across the furnace to strike and adhere to the opposite and side walls. The nozzle is given a coordinated rotational and vertical oscillation. The oscillation tilts the flat spray sheet and causes a 'paint spray'' effect on the walls. This provides a time interval for the liquor to dry between successive spraying on a given spot. The dried liquor accumulates in a char band on the wall before it falls of its own weight to the furnace hearth.

In-line refractometers, as recommended by the Black Liquor Recovery Boiler Advisory Committee (BLRBAC), are the most commonly used means for continuous monitoring of the solids content of liquor to the guns. In some cases, two refractometers are used in series, and the liquor is automatically diverted from the furnace if the liquor solids content falls below some preset value.

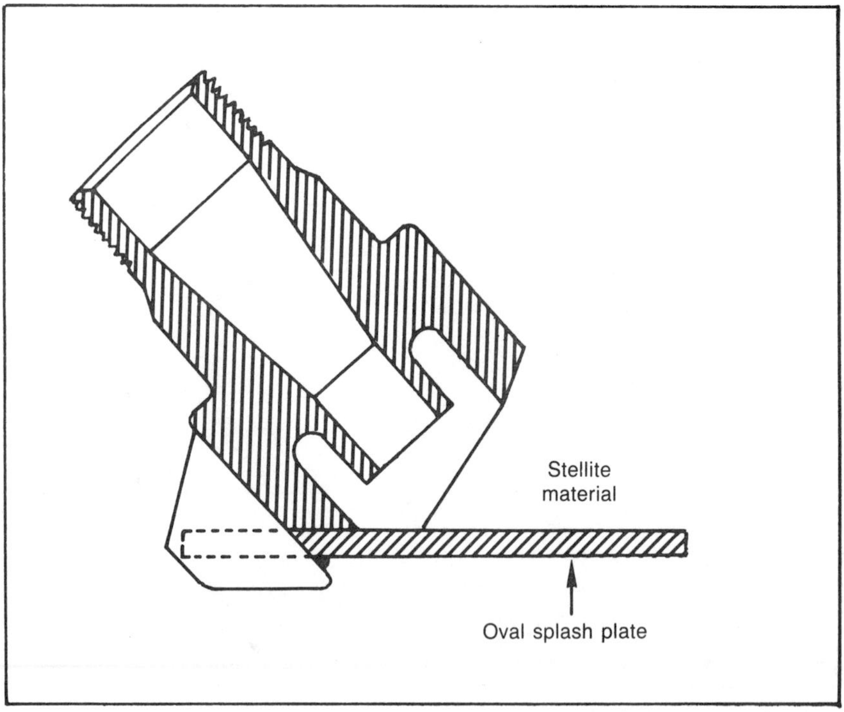

Fig. 4.16. Diagram of liquor gun for B & W units.

Two black liquor heaters are normally installed in the heavy liquor flow sequence. The primary heater is located before the feed pump(s) (normally in the salt cake mix tank). Its function is to heat the liquor to the optimum value for pumping. The secondary heater is located downstream of the feed pump(s). Its function is to heat the liquor to the desired firing temperature. Direct steam heaters are generally used because of their simplicity and resistance to fouling. Indirect liquor heaters (with heating steam and liquor physically separated) have been successfully used as an energy conservation measure (*27*).

Special precautions are taken in the layout of the black liquor piping to prevent settling out of entrained particulate and to provide for the very high viscosity of cold, high-solids black liquor. Steam blowout connections are provided at intervals so that the lines can be blown clear of liquor after a shutdown. Permanent water connections to the black liquor piping are avoided to prevent inadvertent dilution of the black liquor before firing.

Air and flue gas systems Recovery furnaces operate essentially open to the atmosphere. The combustion space is maintained at a very slight negative pressure so that a slight amount of air flows inward through furnace openings (spouts and gun ports). Separate systems for supplying air and removing combustion products are used.

Air for combustion is supplied through one or more forced draft (F.D.) fans. The air is normally preheated to 300-350°F (150-175°C) by an air heater before passing through the air ducts to the wind box and out the air ports into the furnace. Appropriate dampers are used to control the total air flow rate and the distribution to the different levels. Individual dampers are provided for each air port windbox to control the detailed air distribution and the injection velocity.

Combustion products are removed from the furnace by one or more induced draft (I.D.) fan(s). The I.D. fan is located after the recovery unit heat traps. It may be located upstream or downstream of the particulate collection systems. The fan is regulated to maintain the desired furnace draft. The buildup of deposits on the tubes in the recovery unit increases the flow resistance of the combustion gases and results in a larger pressure drop over the flue gas flow path. If this cannot be kept within bounds by periodic sootblowing, the gas flow rate (and hence production capacity of the unit) must be reduced.

There are significant differences in the air supply systems in recovery units of different manufacture. Most Combustion Engineering (CE) units have two levels of air admission. Primary air is admitted to the lower furnace through ports located in the walls directly above the hearth. The remainder of the air is introduced tangentially through the secondary air registers located above the liquor guns. Newer CE units are equipped with two levels of primary air (low primary and high primary). Low primary air is introduced through ports about 1 meter above the floor, whereas the high primary air is introduced through ports about 0.67 meter above the low primary ports.

Babcock and Wilcox units have three levels of air admission. The primary ports admit air through the furnace walls near the bottom of the char bed. Secondary air is admitted through a series of ports at a fixed elevation below the liquor guns and about 3 meters above the spouts. Tertiary air is supplied above the liquor guns and is directed across the furnace from ports in all four walls.

Details of air distributions and injection conditions for these units are discussed later in the operations section.

There are other recovery boiler designs in which the air distribution system is different from either of the above. One design uses four levels of air supply: low primary, high primary, low secondary, and high secondary (or tertiary). The high primary is injected at higher velocities and is intended to facilitate air penetration to the center of the furnace without disrupting the char bed and increasing entrainment. Air penetration to the center of the furnace becomes an increasingly greater problem as the size of units increases, and higher windowbox pressures are needed.

Several different types of air heaters have been used. These include steam air heaters, flue gas air heaters (tubular or regenerative), and direct fired air heaters. The most common system is a steam coil air heater. There are advantages to a steam air heater on recovery units which produce high pressure, high temperature steam for use in turbogenerators. If steam is supplied to the air heater at 125-150 psig (1-1.15 MPa) from one of the extraction points on the turbine, the cycle is thermally closed except for minor radiation losses. In effect, the steam flow to the turbine throttle is increased by an amount equal to the steam consumption of the air heater. Flue gas air heaters were discussed earlier. Direct fired air heaters are simple devices where fossil fuel burners are fired and the hot combustion products mixed with the air to increase the air temperature.

Smelt removal The inorganic salts formed by combustion of black liquor accumulate on the hearth, melt, and flow by gravity out through water-cooled spouts to the dissolving tank. Two floor constructions and smelt removal techniques are used. In units with a slanted hearth, the spout openings are essentially flush with the low point of the hearth. Smelt drains by gravity to the front side of the furnace and out the spouts. Other units have a flat bottom construction, and spouts are located on one or more sides of the unit about 1 ft above the floor. A pool of molten smelt atop a layer of frozen smelt adjacent to the floor tubes accumulates on the hearth and overflows out the spouts.

The spout is basically a water-cooled trough. Adequate flow of suitably treated cooling water is required to prevent overheating the spout metal. The spout cooling system is not pressurized so as to minimize the risk of cooling water entering the furnace in the event of a spout failure.

The smelt flows out the spouts into a dissolving tank. A schematic diagram of a dissolving tank system is provided in Fig. 4.17. It is necessary to break up the stream of smelt before it hits the surface of the green liquor in the dissolving tank in order to prevent dissolving tank explosions. Steam shatter jets directed downward on the smelt stream are used to break up the smelt. Recirculated green liquor is also directed into the smelt stream through a large fishtail nozzle located in front of and well below the spout. The green liquor in the dissolving tank is continuously agitated, usually by a motor driven propeller. It is essential to prevent the accumulation of solid material in the bottom of the dissolving tank and to avoid precipitated material from crusting over the top surface of the tank.

Steam and gases formed in the smelt dissolution operation are removed through a vent pipe and discharged to the atmosphere. A scrubber is usually installed in the vent stack to remove entrained mist and chemicals.

The aqueous input to the dissolving tank is normally weak wash from the causticizing area. Green liquor, at the desired concentration, is removed continuously from the dissolving tank. Automatic controls are normally installed to maintain the desired green liquor density and dissolving tank level by regulating the rates of green liquor outflow and weak wash inflow.

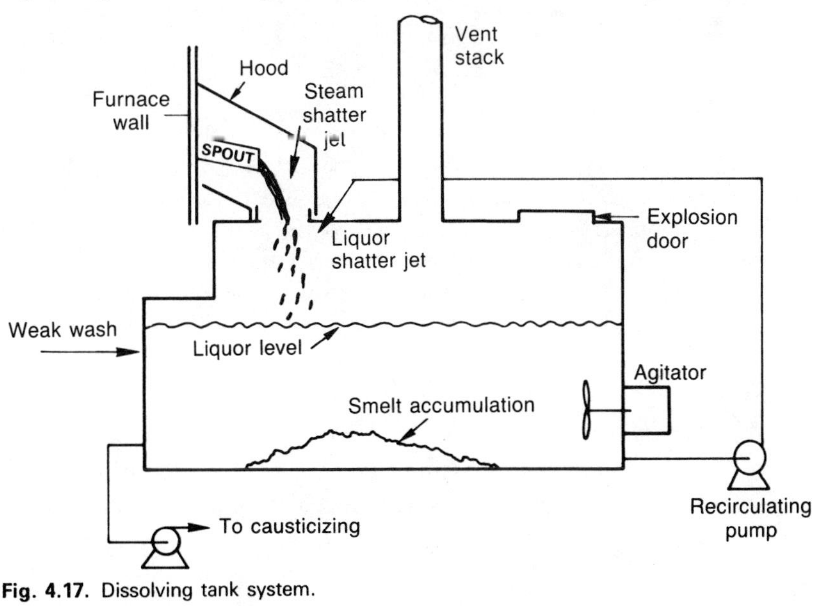

Fig. 4.17. Dissolving tank system.

Auxiliary fuel Auxiliary fuel is always needed for startup and during normal shutdown. It is sometimes used to stabilize combustion during periods of unstable furnace conditions or during low load operation. In some cases, auxiliary fuel is burned in load-carrying burners for purposes of steam generation. Starting burners are provided as hearth burners located in the lower furnace to ignite and sustain ignition of black liquor. Load carrying burners are mounted at the oxidizing zone of the furnace.

Auxiliary fuel burners are provided as integrated combustion systems. Burners are designed for oil and/or natural gas firing and provide for introduction of auxiliary fuel and air into the furnace at the required velocities, turbulence, and proportions to establish and maintain proper ignition and stable combustion of the fuel within the furnace. Each burner is provided with an ignitor to provide ignition energy to immediately light off the main burner. This could be either a continuous gas- or light oil-fueled ignitor or an interrupted direct-electric ignitor. Also included are appropriate flame sensing systems and safety interlocks. These are designed to prevent explosions due to an accumulation of unburned combustibles within the furnace.

Specific types

Most recovery units presently in use in North America are of two types: those manufactured by Babcock & Wilcox Co. (B & W) and those manufactured by Combustion Engineering Inc. (CE). The distinctive differences between these two types of units are discussed below.

B & W units Recovery units manufactured by B & W use the wall drying approach. Black liquor is sprayed on the walls by one to four guns of the splash plate type.

Combustion air is supplied to three zones: primary, secondary, and tertiary. Primary air is supplied through air ports just above the floor. Secondary air is supplied through ports located between the hearth and the liquor guns. Tertiary air is supplied through ports located above the liquor sprays. Tertiary air is injected at high velocity and directed across the furnace.

The furnace floor is slanted so that smelt runs to the low side of the unit (either the front or the rear) and drains out of spouts which are flush with the furnace floor.

Direct contact evaporators on B & W units are normally cyclone evaporators.

CE units Recovery units manufactured by CE use the suspension drying approach. Black liquor is sprayed into the furnace through a number of guns located in all four walls, and the particles dry in flight on their way to the hearth.

In most CE units air enters the furnace at two levels, primary and secondary. Primary ports are located at the hearth level. Some newer boilers employ two levels of primary air. Secondary air is introduced tangentially through ports located above the liquor guns. Tangential injection is used to promote mixing by causing a swirling motion of the gas stream.

CE units have a flat bottom (decanting hearth). The smelt spouts are located about 1 ft above the floor and molten smelt is decanted off. A layer of frozen smelt is present along the floor to seal the floor and protect the floor tubes.

Direct contact evaporators in CE units are normally cascade evaporators.

Particulate recovery

Three types of suspended particles are commonly present in the flue gas:

1. Heavy particles, with char present, large enough to settle by gravity in the boiler hoppers;
2. Fused particles formed in the combustion zone from in-flight burning of entrained liquor droplets; and
3. Fume particles formed by condensation of volatilized inorganic substances. These tend to be submicron in size, although they can agglomerate to some extent to form larger-sized particles.

The dominant factor governing the extent of particulate recovery is air emission limits. Permissible discharge levels have reached quite low values, and high collection efficiencies are now required.

Many of the larger, denser particles are collected in the boiler and economizer hoppers and returned to the liquor system. Additional collection occurs in the direct contact evaporator (when used). Both of these tend to collect the larger sized particles. "End-of-pipe" treatment is necessary to collect the fume particles effectively. The most commonly used system is the electrostatic precipitator.

Electrostatic precipitators The basic action in an electrostatic precipitator is the use of electrostatic forces to remove particles from the gas stream. A high-voltage direct-current corona emanating from a negatively charged discharge electrode, placed some distance from a grounded collecting electrode, ionizes the gas between these elements. The ions attach themselves to the dust particles which then migrate to the surface of opposite polarity under the influence of the electrostatic field.

Collection of particles involves four steps:

1. Charging the particles;
2. Particle transport through the gas;
3. Discharging the particles; and
4. Removing the collected particles from the collecting surface.

A negative potential ranging from 30,000 to 80,000 volts is applied to the discharge electrodes to establish the corona current and the electrostatic field. Excessive voltages lead to sparking, which places an upper limit on operation. The high tension negative current is achieved with a transformer and set of rectifiers normally forming one unit, and monitored by a spark rate control unit designed to give a certain number of flashovers per second. Corona and breakdown are highly dependent on geometry and gas composition.

The forces acting on the charged particle are the electrostatic forces tending to move it toward the collecting electrode and the hydrodynamic forces tending to carry it along with the gas. The resultant of all of these forces determines whether or not a particle reaches the collecting surface.

Three factors are primarily responsible for affecting the removal of a given particle:

1. gas velocity;
2. electrode geometry; and
3. applied voltage.

The gas velocity is initially fixed by the design of the unit relative to load. However, the uniformity of the gas distribution is very important to precipitator performance. Areas of low velocity inevitably mean other areas are operating at higher than average velocity. Another factor influencing overall collection efficiency is "sneakage," gas passing below the fields or at the sides of the fields. The particulate load in this fraction of the gas will not be collected and will contribute significantly to lower overall collection efficiencies for the unit. Since a precipitator is a velocity-sensitive device, any factor tending to increase volume throughput will tend to decrease collection efficiency. The volume throughput is influenced by firing rate, liquor heating value, firing solids content, the amount of excess air, and flue gas temperature.

The electrode geometry is fixed by design, but may change during operation. Structural misalignment or warping, which reduces the distance between wires (discharge electrodes) and plates (collecting electrodes), will reduce the maximum voltage obtainable. The applied voltage is influenced by many variables, including gas conductivity, gas temperature, moisture content, and dust loading.

When the dust particles reach the grounded collecting electrodes, they transfer part of their charge, depending on the resistivity of the dust. If the resistivity is high, the potential drop across the dust layer is increased, which tends to reduce the field strength in the gas space. A high resistivity dust layer can also lower the arcing voltage. A high voltage drop across the layer of collected particles also tends to make the particles more adherent by orienting dipoles formed by uneven charge distribution on the particles. Increased moisture content in the gas (in the correct range) tends to lower dust resistivity.

The dust particles are held on the electrodes by electrostatic forces, surface forces, and contact potentials. Particles are removed from the collecting surfaces by rapping the electrodes. The ideal rapping system should cause the collected dust layer to slip on the plates like sand sliding down a sloping steel plate. This minimizes the reentrainment of collected particles into the gas stream. Minimal reentrainment is essential to achieving very high overall collection efficiencies.

The discharge electrodes collect dust particles with a positive charge and also need rapping. The dust collected on the discharge electrodes can vary between needlelike deposits to thick, evenly distributed layers.

An electrostatic precipitator is shown schematically in Fig. 4.18. The gas inlet arrangement must achieve a good gas distribution and avoid buildup of dust in the lower part of the inlet plenum. Turning vanes and perforated plates are used to provide proper gas distribution. These must be rapped periodically to give satisfactory performance for a prolonged period.

The fields are made up of alternating rows of collecting plates and discharge wires, typically on a plate-to-plate spacing of about 10 in. (250 mm). The flat steel plates form the grounded or positive electrodes. Plate heights have been increasing, and the majority of new units have plate heights of 30 ft (9.1 m) or greater. Two types of discharge electrodes are in use. Weighted-wire units have individual discharge wires which hang from insulators on the roof and are equipped with a stabilizing weight on the bottom. The other type is a rigid-wire-frame electrode.

Precipitators are normally sectionalized, both with respect to gas flow and with respect to the electrostatic fields. Sectionalized fields permit optimization of operating voltages. Sectionalizing into parallel gas flow paths allows a portion of the precipitation to be taken off line for maintenance without disturbing the operation of the other portions. Gas flows horizontally through the unit.

The rapping mechanisms should be sufficient to clean the collecting plates and discharge electrodes without being activated for extended periods. Mechanical, pneumatic, and electric rapping devices have been used. Current

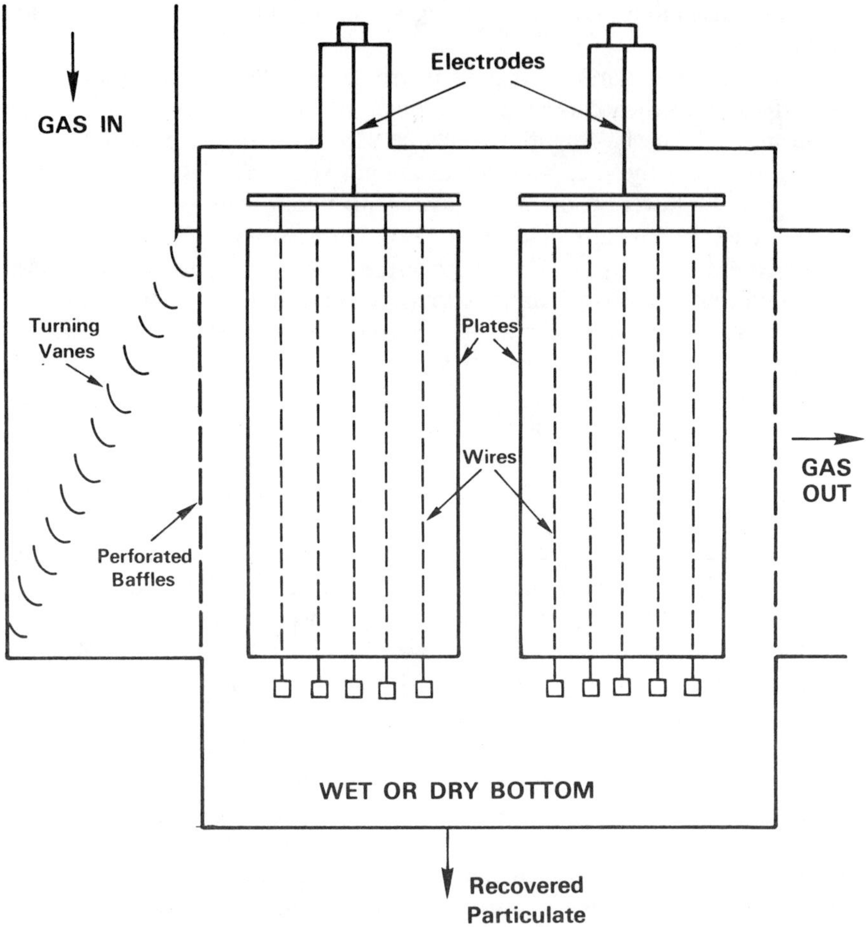

Fig. 4.18. Electrostatic precipitator.

trends are toward the swinging hammer and pneumatic cylinder devices and away from electric impulse devices.

Dust is removed from the bottom of the precipitator. Both "wet-bottom" and "dry-bottom" designs are used. Prior to about 1970, the "wet-bottom" design was standard. In these units, black liquor is circulated through a pan under the precipitator's entire collection zone. Rotating agitators mix the collected dust into the black liquor, and it is transported back to the recovery boiler. When recovery boilers without direct contact evaporators came into use in North America in the late 1960s, the "wet-bottom" design came into disfavor because of suspicion that the gas-liquor contact at the pan surface was a potential odor source. Thus, a trend toward dry-bottom units developed. The dry bottom units have a flat pan and either an all-dry drag-type conveyor or a ribbon-mixer (wet mixing drum). Early dry-bottom units tended to have serious problems with the dust conveyors. Great progresss has been made in reducing dry-bottom downtime (*28*). Because of problems with the early dry-bottom units, the wet-bottom design has again been applied to precipitators on NCE recovery boilers. This application has had some success.

The housing or shell of the precipitator is now normally made of steel. Earlier units used a tile shell construction, but this is no longer in use on new precipitators (*29*). Some of the units have heated shells, but the majority do not. The duct work, roof, and any precipitator steel which is in communication with both the inside and the outside of the unit should be insulated to prevent condensation and corrosion.

The temperature of the incoming flue gas should be high enough to prevent reaching the sulfuric acid dewpoint inside the precipitator. Normally, a temperature of 50°F (30°C) above the dewpoint temperature is considered a safe operating temperature (*30*). The dewpoint temperature is a complex function of the moisture content and SO_3 content of the flue gas. Increasing moisture and increasing SO_3 concentration raises the dewpoint temperature.

The initial application of electrostatic precipitators to NCE recovery boilers was not very successful. Many of the early units had to be retrofitted or replaced. To a large extent the problems have been overcome (*31*). Important progress has been made in precipitator performance and reliability. Problems of wire failure, plate alignment, and collector "snowing" have been reduced. Major problems still exist with collector corrosion and dry dust conveying.

Other methods Although the precipitator is the dominant particulate collection technology used on kraft recovery boilers, other approaches are used in some cases. A crossflow scrubber system in conjunction with a flue gas scrubbing system for odor control is a recent development (*32*). Filtration techniques (bag houses) have not been applied to recovery stack emission to the author's knowledge.

Recovery boiler operation

Operating objectives

The objectives in recovery boiler operation include:

1. Providing continuity of operations;
2. Maintaining desired black liquor firing rate;
3. Maintaining desired steaming rate;

4. Meeting air emissions restrictions;
5. Maximizing thermal efficiency; and
6. Maximizing chemical reduction efficiency.

The relative importance of these objectives can vary in specific mill situations.

Continuity of operations is paramount, since none of the other objectives can be met without it. There are three main elements in providing for operating continuity:

1. combustion stability;
2. minimizing plugging of gas passages; and
3. avoiding hazardous conditions.

Stable combustion without blackouts or partial blackouts is essential. Unstable combustion is a dangerous condition. Combustible pyrolysis gases may form and mix with air, and delayed ignition can lead to puffing and possibly explosive gas mixtures. In addition, the decreased furnace heat release during incipient blackout conditions could result in wetter, less pyrolyzed liquor reaching the bed. This could accelerate the blackout process and would result in rapid bed growth, potential plugging of air ports, increased concentrations of TRS gases, and possibly smelt-wet liquor explosion hazards.

Plugging of convection gas passages with slag and fume is one of the prime factors which determines the actual operating capacity of the unit. The furnace is open to the atmosphere at the liquor gun ports and spouts, and a slight negative pressure (draft) is maintained within the furnace cavity with the I.D. fan to keep flame within the unit. Thus, the resistance to gas flow through the various heat traps is very important. Operation of the furnace must be such that furnace draft can be maintained by regular use of sootblowers. This puts restrictions on the amount of material carried upward in the furnace, temperature levels in the heat recovery zones, and gas flow rates through the unit.

Safety is another important factor. Recovery boilers are faced with some unique hazards because of the possibility of smelt-water explosions. Certain hazardous conditions must therefore be avoided. These include the following:

1. Firing liquor at too low a solids content;
2. Excessive smelt runoff rates or plugged spouts;
3. Localized conditions of excessive temperature (hot spots) such as are caused by the use of air lances; and
4. Operating under conditions which lead to accelerated corrosion of furnace tubes.

The chemical recovery function of the recovery boiler is primarily met by maintaining the desired black liquor firing rate. If the recovery unit is unable to process liquor solids at the rate at which they are being produced in the digester, it will ultimately force a curtailment in the rate of pulp production.

Recovery boilers interface with the mill steam and power system as base load units. The steaming rate is maintained at a particular level, and little attempt is made to adjust the recovery boiler steaming rate to meet short-term changes in mill steam demand. In particular, black liquor firing rates are not continuously adjusted to meet variations in steam demand.

Restrictions on allowable atmospheric emissions have put an additional constraint on recovery boiler operation. The substances for which emission restrictions exist are TRS, particulate, and sometimes SO_2. Control of TRS emissions normally requires essentially complete combustion of the black liquor organics. Particulate emission restrictions are normally met through the use of high efficiency electrostatic precipitators on the recovery stack, and do not usually impact as strongly on recovery boiler operation as do TRS limits. Control of SO_2 emissions is complex, and sometimes presents a problem, especially with higher sulfidity operations.

Maximizing the thermal efficiency is often not a high-priority objective on a recovery boiler. This is in marked contrast to power boilers, where the amount of fuel used is directly related to costs of steam and power. The relatively low importance placed on thermal efficiency simply reflects the fact that recovery boilers are process units as well as steam generators.

Maximizing the reduction efficiency (sulfide to sulfate ratio) is also normally not a high-priority short term operating objective. There are two main reasons for this:

1. The main effect of a low reduction efficiency is an increase in sulfate dead load throughout the cycle, and the economic and operational consequences of poor reduction are diffuse and hard to quantify.
2. Reasonably high reduction efficiencies can normally be obtained without undue attention.

In prioritizing operating objectives it is helpful to distinguish between two situations:

1. The desired black liquor solids firing rate is well below the inherent capacity of the unit (under-loaded furnace); and
2. The black liquor solids firing rate is constrained by furnace capacity (bottleneck).

In the first case the solids firing rate is simply set, and the operation can be optimized around this rate. Auxiliary fuel may be needed to maintain desired furnace temperatures or for higher steaming rates. Operating stability criteria would still be paramount, but objectives such as improved thermal efficiency and reduction efficiency would become more important. In the bottleneck case, the prime objective would be to maximize solids firing rate subject to reasonably stable operation and permissible emission limits. Auxiliary fuel usage would be minimized or eliminated. Thermal efficiency and reduction efficiency would be relatively unimportant.

Operating systems

Operation and control of recovery boilers is most conveniently discussed by considering the overall operation as consisting of a number of distinct systems. These include:

1. black liquor firing system;
2. air supply system;
3. flue gas removal system;
4. auxiliary fuel system; and
5. dissolving tank.

The equipment involved in these systems was discussed previously. This section will be concerned with operating and control aspects.

Although each of these systems has its own control function and equipment, they interact strongly with each other. The basic requirements are to introduce fuel and air into the combustion space in the correct proportions and in such a manner that the fuel burns completely and stably, and to provide for the removal of the combustion products.

Black liquor firing The variables involved in the firing of black liquor can be classified as follows.

1. Input rate (firing rate)
    a. Liquor flow rate
    b. Liquor solids content (as fired)
        i. Dissolved solids
        ii. Entrained solids
2. Spray variables
    a. Number and location of guns
    b. Nozzle characteristics
        i. Type of nozzle design
        ii. Nozzle size
    c. Gun oscillation angles and frequency
    d. Nozzle pressure

e. Liquor properties
   i. Viscosity (temperature dependent)
   ii. Surface tension
f. Liquor superheat (if any)

3. Burning characteristics
   a. As-fired heating value
      i. Heating value of liquor to unit
      ii. Recycled dust load
      iii. Makeup chemical addition
   b. Water load (solids content)
   c. Burnability

The objectives of the black liquor firing control system are to control the rate of input (liquor solids or Btu) to the furnace and to safely control the spray so that the liquor burns stably at the rate at which it is fired without excessive carryover. The processes needed to achieve these objectives are not completely straightforward, and to some extent, black liquor firing has remained an art. Some of the reasons for this are discussed below.

The as-fired solids content of the liquor has a very strong influence on the overall operation. For a given firing rate (flow), it determines the amount of liquor solids entering. It influences the spray characteristics through its effect on liquor properties (such as viscosity), and, since water evaporation is a heat sink, it also affects temperature in the lower furnace. The relevant solids content is the total black liquor solids content (dissolved and undissolved). Refractometers, the most commonly used on-line solids monitor, respond only to the dissolved solids content. Hence, this critically important operating parameter is not directly measured. The common practice of periodically calibrating refractometer readings against moisture balance readings (which will measure total solids) is only valid if the entrained solids fraction remains constant. Since the recycled dust load is normally not known and is uncontrolled, it would be fortuitous if the entrained solids fraction were constant.

The as-fired solids content also influences the safety of the operation. BLRBAC currently recommends the use of a system for automatically diverting black liquor from the furnace when either of two refractometers in series indicates the solids content is below a preselected value.

Once the equipment is installed, (number of guns, location, nozzle type and size, etc.) there are only a few variables available for manipulating spray characteristics. The liquor discharge velocity, an important spray parameter, tends to be proportional to the square root of the nozzle pressure. However, this is not an independently manipulatable variable. The product of density, discharge velocity, and nozzle throat area must equal the liquor flow rate; so velocity and firing rate are coupled. Typically, liquor guns are provided with a relatively large number of tip sizes which are to be used within given ranges of flow rate per gun. These are changed manually when changing firing rate to a value outside the range of the existing nozzle. Within the range, the nozzle size is fixed. In some operations, the liquor is superheated and flashes in the gun before it is discharged. The resulting two-phase flow has a significantly lower effective density, and a higher discharge velocity can be maintained for a given nozzle cross section and liquor flow rate per gun.

Liquor temperature can be used to control the spray characteristics. Liquor temperature affects liquor properties, especially viscosity. These properties influence spray characteristics, especially the droplet size distribution. Higher liquor temperatures tend to give finer sprays and lower liquor temperatures give coarser sprays. Manipulation of spray characteristics by small changes in liquor firing temperature is very important in units using suspension drying. It is less critical in units employing wall drying.

The specific heating value of the as-fired black liquor can vary considerably because of variations in the amount of recycled inorganic solids and chemical makeup, as well as from changes in the liquor supplied to the recovery unit. Thus, the heat input to the furnace is not necessarily fixed by fixing the solids firing rate. The heat input is never directly measured.

The rate of burning of the liquor solids depends on the heating value, the spray size distribution, and those properties that influence the drying rate and combustible gas formation during pyrolysis. Spray manipulation can help counteract a poor burning liquor, because smaller size particles will tend to dry more rapidly, pyrolyze more extensively, and have a greater surface area for carbon burnup. The liquor properties which influence burnability are presently very poorly understood, and no measurements on this factor are available.

The most common operating practice is to control black liquor flow from a manually established setpoint. Liquor solids content is monitored but not controlled directly. (With direct contact evaporators, liquor density can be controlled through manipulation of the flue gas bypass damper and by weak liquor dilution. In "low-odor" systems, the liquor is fired at the solids content at which it comes from the concentrator.) Combustion rates are controlled by adjustment of spray characteristics. Control of liquor temperature to influence droplet size distribution is a critical aspect of suspension firing. Liquor temperature is controlled automatically (by adjustment of direct steam addition) to a manual set point.

Air input The variables associated with the combustion air supply can be classified as follows.

1. Total air input
   a. Stoichiometric air requirement
   b. Desired excess air
   c. Infiltration air
2. Air distribution
   a. Primary-secondary-tertiary splits
   b. Horizontal distribution
      i. Windbox pressures
      ii. Port plugging
3. Air temperature
4. Air for auxiliary burners

The main concern with total air is the proper balance between fuel and air supply. The total air input is tied to the liquor solids firing rate through the stoichiometric air requirement (theoretical air). In addition, a certain amount of excess air must be used to overcome the effects of incomplete mixing between fuel and air. Insufficient excess air leads to incomplete combustion and TRS emissions. Too much excess air can result in lower thermal efficiencies, lower capacities, and cold end corrosion problems.

The demand for combustion air rests almost entirely with the organic constituents in the liquor. Liquor inorganics do not require significant amounts of air. Recycled dust, consisting mainly of Na_2CO_3 and Na_2SO_4, has no oxygen demand. Thus, the proper balance between air and fuel requires tying the total air input to the "organic" input. When the total air input is tied directly to liquor flow rate, disturbances in liquor solids content or in the recycle solids addition will change the balance between air and combustibles.

The composition of the flue gas provides information on the air/fuel balance. When combustion is complete, the O_2 content of the flue gas is a direct measure of the amount of excess air used. It is also possible to have significant levels of combustibles (such as CO) and O_2 present simultaneously in the flue gas because of poor mixing of air and fuel. Infiltration air (air which leaks into the furnace through openings because of the negative draft within the furnace) does not normally mix well with the fuel and does not take part in combustion. It simply dilutes the combustion products. Insufficient velocities in the forced air supply can also lead to this problem.

The char bed contains a substantial inventory of combustibles. Changes in this inventory could divorce the consumption of oxygen from the input solids rate on a shortterm basis. This could confuse the interpretation of flue gas composition information when the char bed is not stable.

The distribution of air in the vertical direction is important to furnace operation. A reducing environment is established in the lower part of the furnace by limiting the amount of air to this zone. The amount of air entering the lower furnace influences the rate of burnup of char bed material, helps control char bed inventory, and affects all temperature-dependent bed processes.

Historically, two different air distribution patterns have been used in North America, one type in CE units and the other in B & W units. Since 1979, CE has been using three levels of air instead of two on all new units. The air distribution patterns are summarized in Table 4.7.

The nomenclature on air distribution can be confusing. Secondary air in a B & W unit is part of the air supply to the reducing zone. In effect, the reducing zone air is split into two levels, high and low. Secondary air in a B & W unit is not analogous to secondary air in a CE unit (which is used to complete combustion). There is close correspondence between tertiary air in a B & W unit and secondary air in a CE unit.

Windbox pressures are directly related to the air injection velocity, and have a pronounced effect on turbulence generation, air/fuel mixing, and the penetration of air to the center of the unit. Air is introduced around the periphery of the furnace through a multiplicity of relatively high velocity rectangular jets. The air must have sufficient momentum for at least some of it to reach the central core of the furnace cavity and mix with combustibles there. Injection velocities for primary air are kept relatively low to avoid air

Table 4.7. Air distribution patterns

	B & W units	*Old CE units*	*New CE units*
Name	Primary	Primary	Low primary
Location	Just above floor	Just above floor	Just above floor
% of total air	45 – 55	65 – 70	40 – 50
Pressure, in. H_2O	3 – 4 (0.75 – 1 kPa)	3 (0.75 kPa)	3 (0.75 kPa)
Name	Secondary		High primary
Location	8 ft above floor.		5.5 ft above floor.
% of total air	20 – 35		15 – 25
Pressure, in. H_2O	5 – 8 (1.25 – 2 kPa)		5 (1.25 kPa)
Name	Tertiary	Secondary	Secondary
Location	Above spray (horizontal)	Above spray (tangential)	Above spray (tangential)
% of total air	15 – 30	30 – 35	30 – 35
Pressure, in. H_2O	5 – 12 (1.25 – 3 kPa)	6 (1.5 kPa)	6 (1.5 kPa)

jets striking the bed and generating local hot spots, undercutting the bed, or entraining bed material. High velocities and turbulence are especially important in the oxidizing zone of the furnace to assure complete burnout of combustibles and TRS gases.

The relationship between windbox pressure and injection velocity is given by $V = 80\sqrt{h}$, where V is in ft/sec when h is in inches of water (V, m/sec $= 15.3\sqrt{h}$, cm H_2O). Thus, injection velocities run around 140 ft/sec (40 m/sec) in the primary zone and from 180 and 270 ft/sec (55-80 m/sec) in the secondary or tertiary zones. Windbox pressure cannot be controlled independently of air flow rate unless air port openings are variable, since the continuity relationship, volume flow = velocity × area open to flow, must be satisfied.

Air port plugging with slag or black liquor can change the peripheral air distribution. If a number of neighboring ports become blocked, a localized area could become starved for air and black out. The reduction in port open area will also increase air injection velocities for the same total air input rate.

Air temperature can be an important variable. Air constitutes 80-90% of the mass flow into the furnace, and increases in the combustion air temperature tend to result in nearly proportional increases in temperature levels within the furnace. High air temperatures can be important during periods of marginal combustion when the lower furnace is tending to run cold. On the other hand, reduced air temperature could be called for when high furnace temperature limits throughput. Although air temperature could be used as a combustion control variable, this is not commonly done. Normal practice is to control air temperature to a preselected value, usually between 300-400°F (150 and 200°C).

Air is also required for the combustion of auxiliary fuel. Since this air supply is normally an integral part of the auxiliary fuel burner, it is discussed in the section on auxiliary fuel.

Flue gas removal At any given moment, the furnace gas pressure, p, is governed by the gas law, $pV = mRT$, where furnace volume, V, and gas constant, R, are fixed. The furnace temperature, T, and gas mass inventory, m, must be balanced to maintain the desired furnace pressure (draft). The rate of change of gas inventory depends on the difference between the rate of air supply plus the rate of formation of combustion products, and the rate of removal of combustion products. Since the air supply is controlled by combustion requirements and the internal combustion rate is not directly controlled, gas inventory and furnace pressure (draft) are regulated by controlling the rate of removal of combustion products.

Combustion products are removed from the furnace by one or more I.D. fans. These are normally operated with variable-speed drives, and fan speed is adjusted to regulate furnace draft. In some cases, variable dampers may be used to regulate draft. Draft is measured at a tap in the side of the furnace, near the roof, before the gases pass through the generating bank. Furnace draft is set at a low value [e.g., 0.1-0.25 in.H_2O (0.025-0.06 kPa)] consistent with the limitation that normal pressure swings do not result in frequent excursions to positive pressure (blowback).

The flow resistance to gas passing through the unit is an important parameter. The product of gas flow rate and the flow resistance must equal the head developed by the I.D. fan. At any given fan speed, the headflow characteristics of the I.D. fan are such that the head increases with decreasing flow. Thus, as the flow resistance to the flue gases increases, the gas flow rate will automatically tend to decrease. This must be compensated by an increase in fan speed to maintain the flue gas flow rate. When the maximum fan speed is reached, a further increase in flow resistance will result in a decrease in flue gas flow and therefore a loss in burning capacity.

The main factor influencing the flow resistance is the tendency for gas passages to plug with slag, carryover, and fume. Sootblowers are used to maintain open gas passages and clean tube surfaces in the superheater section, generating bank, and economizer. Normally, the sootblowers on a recovery boiler are controlled with an automatic sequencer following a preset order. This practice of timed, sequenced blowing tends to use excessive amounts of steam. Recently, systems for monitoring the degree of buildup by measuring the pressure differential over different sections of the gas passage, and using this information to determine when each sootblower should be operated, have been developed. These have often resulted in significant savings in sootblower steam.

A few large recovery boilers have experienced excessively low gas side pressure (high draft) excursions which resulted in damage to the boiler setting. These have been termed "furnace implosions," since the walls of the furnace were sucked in. All implosions have occurred on units with I.D. fans with high-head capability operating, at least momentarily, under low-flow conditions. Two conditions can lead to implosions:

1. Maloperation of equipment resulting in the furnace being exposed to the full I.D. fan head capability, usually at the low flow-high head range and often combined with an interruption of the forced air flow; and
2. The rapid decay in furnace gas temperature (and hence pressure) following a rapid reduction in fuel input or a main fuel trip.

The possibility of such occurrences should be taken into consideration in the design of flue gas removal control systems. NFPA 65G regulation now sets the design pressure of the furnace and pressure parts to correct this situation.

Auxiliary fuel burners Auxiliary fuel burners are used for startup, shutdown, supplying localized heat to the char bed, and sometimes for additional steam generation. To prevent combustible gas explosions, units are equipped with supervised auxiliary fuel burners and combustion safeguards. These include positive ignition of pilot and main fuel, flame supervision, and appropriate interlocks to prevent introduction of fuel into the unit under conditions where its complete combustion cannot be assured.

Recommended practices for the design, installation, and operation of auxiliary fuel systems in recovery boilers are described in detail in "Recommended Good Practice for Safe Firing of Auxiliary Fuel in Black Liquor Recovery Boilers" published by BLRBAC (*33*).

The decision to use auxiliary fuel can be an important aspect of the operation of recovery boilers. Auxiliary fuel is essential during startup and for burning down the char bed during a normal shutdown. It can also be used to provide additional steam generation (load carrying) during periods when insufficient liquor is available or to assist the combustion process during periods of marginal combustion stability.

Auxiliary fuel combustion for purposes of steam generation should be carried out in load-carrying burners located in the upper furnace levels above the liquor guns. These burners may have a modulating control valve to vary the rate of fuel supply to satisfy changing load demands. Load carrying by auxiliary fuel firing in the hearth zone is not recommended. Extensive combustion of auxiliary fuel in upper-level burners (with their attendent combustion air supply) can distort heat release and air flow patterns from those obtained when firing black liquor alone. This can influence the black liquor burning behavior and significantly lower reduction efficiency.

Auxiliary fuel can be used to assist the combustion of black liquor under conditions when autogenous combustion is difficult or impossible. It may be used to clear up wet spots on the char bed, reestablish burning after blackout or partial blackout, help maintain bed temperatures at a reasonable level, and prevent bed buildup. It is sometimes used to aid firing of low solids liquor, although this is a hazardous practice and is not recommended. Auxiliary fuel to assist black liquor combustion is fired using burners located at the hearth zone (hearth burners) near the bottom of the furnaces.

Smelt dissolving system The molten inorganic chemicals flow out of the furnace through water-cooled smelt spouts to an agitated dissolving tank. There are two main operating objectives associated with smelt dissolution:

1. Avoiding smelt-water explosions in the dissolving tank; and
2. Producing a green liquor of suitable concentration for causticizing.

It is necessary to disperse or break up the smelt stream before it reaches the surface of the green liquor in the dissolving tank. Steam shatter jets are the primary means used to accomplish this. Sufficient shatter steam must be used to safely disperse the smelt. Excess steam, beyond that needed to quiet the dissolving tank, should be avoided, as it is wasteful and results in a larger vent plume. The shatter steam valve should be accessible in a safe location so that steam flow can be regulated during periods of heavy runoff. Shattering is often aided by discharging a stream of recirculated green liquor into the smelt stream through a large fishtail nozzle. Because of the need for keeping the recirculation and shatter sprays constantly going, spare recirculating pumps are strongly recommended.

An important precaution in preventing dissolving tank explosions is to be certain that the tank is adequately vented and a safe volume maintained in the air space between the top of the tank and the liquor to act as an explosion cushion. It is also necessary to prevent the buildup of undissolved salts at the bottom of the tank. The level of green liquor in the dissolving tank must also be controlled.

Steady flow of smelt out the spouts is essential to safe operation of the smelt dissolution system. Excessive runoff rates can exceed the capability of the shatter jets to break up the smelt stream adequately and can lead to "noisy" dissolving tank operation and possibly damaging explosions. Plugged or partially plugged spouts can be a serious problem, because the smelt will accumulate in the furnace. This can lead to heavy runoff when the spouts are opened, and even to smelt entering the primary air ports. Frequent observation of the hearth and spouts is recommended to help prevent these problems.

The concentration of the green liquor from smelt dissolution must be adequately controlled. Too low a concentration gives a weak white liquor and results in increased evaporator loads. Too high a concentration can lead to lower causticizing efficiencies, and possibly precipitation problems in the dissolving tank and green liquor processing system. (Under severe conditions the green liquor in the dissolving tank could crust over and molten smelt accumulate on top. This is a very hazardous situation.) Finally, variations in green liquor strength are a serious disturbance to the causticizing system, and will inevitably result in lower causticizing efficiencies and higher Na_2CO_3 dead loads in the liquor system.

Most smelt dissolution installations operate on a continuous basis with green liquor being pumped out and fill liquor (normally weak wash) continuously added. Both green liquor density (concentration) and dissolving tank level must be controlled. In most cases automatic systems are employed to control both of these quantitites by regulating the rates of green liquor withdrawal and weak wash fill back.

Operating and control strategies

Recovery boilers are basically operated as base load units with respect to steam generation. The firing rate in the recovery boiler is kept constant and the variable portion of the total steam demand is handled by varying the firing rate on the power boilers. This approach is adequate as long as a significant demand for steam in excess of recovery boiler capacity exists. Should energy conservation measures ultimately result in a great reduction or elimination of steam demands beyond the capability of the recovery boiler, this strategy would be inadequate. Operation of a recovery boiler as a swing unit with the firing rate adjusted to meet rapid changes in steam demand would be more difficult.

Combustion control There are a number of objectives for the combustion control systems:

1. complete combustion with a minimum amount of excess air;
2. char bed stability;
3. proper char bed temperatures;
4. minimum entrainment and carryover of liquor droplets;
5. blackout avoidance and correction; and
6. emission control

These objectives are achieved through manipulation of liquor firing and air supply variables. The approach that is normally used in black liquor firing is to fire at a constant liquor flow rate. The forced air is governed by combustion requirements, and the I.D. fan is manipulated to achieve a balanced draft. Fine tuning of liquor spray characteristics and the air distribution is used to achieve desired bed stability and temperature levels.

Liquor The black liquor firing rate is established by mill production rates, possibly constrained by the capacity of the recovery boiler and/or emission limits. The black liquor flow rate is normally measured with a magnetic flowmeter and the signal fed back to a flow controller. The setpoint is manually adjusted. The aim is to fire liquor at a constant liquor solids input. However, flow rate is not manipulated in response to changes in solids content. The main reasons for this are the sensitivity of spray patterns to changes in flow rate and the relative unreliability of online solids measurements.

Black liquor solids content is normally measured on an as-fired basis with the solids monitor located downstream from the final liquor heater. The most popular solids monitor is the in-line refractometer, and many mills use two refractometers in series in accordance with BLRBAC recommendations. Refractometers have proven to be reasonably reliable in black liquor service, although they do not sense entrained solids (undissolved solids) and are sensitive to changes in the liquor composition (especially the organic-inorganic ratio) (*34*). The solids content of the black liquor is *not* directly controlled at the point of firing. In units with direct contact evaporators, the black liquor solids content can be controlled to some extent by adjusting bypass dampers and by weak liquor dilution. In units without direct contact evaporators, the liquor is fired as it is received from the concentrator. A number of installations now have systems for automatically diverting black liquor from the furnace when the as-fired liquor solids goes below a preset minimum value, usually between 58 and 60% solids. This is done to prevent explosions from low solids liquor coming in contact with smelt.

Control of the spray characteristics has a strong influence on combustion control, especially in suspension drying units. The selection of the number of guns, location, and nozzle sizes is based on symmetry considerations and manufacturers' recommendations for the firing rate selected. These then become fixed, except for step changes in operation. The drop size distribution is very critical in suspension-fired units. Drops that are too large will reach the char bed without drying and pyrolyzing sufficiently, and could lead to cooling and rapid bed growth or blackouts. Droplets which are too small will be swept up by the gas stream and carried over. The droplet size distribution is not controlled directly. It is adjusted indirectly by controlling the final liquor temperature. Liquor temperature has a strong influence on the spray size through its effect on liquor viscosity. Liquor temperature is controlled by a temperature controller on the rate of steam addition to the final liquor heater. Changes in liquor temperature of a few degrees can have a marked effect on the spray. Wall drying units are less sensitive to spray parameters, although liquor temperature and sweep angles can be adjusted to give desired drying conditions.

Air The first requirement of the air control system is to obtain the proper air-fuel ratio. The normal procedure is to send the setpoint signal from the black liquor flow controller to the total air controller which adjusts the vanes on the F.D. fan. This ties the total air rate directly to the firing rate. The proportionality factor in the controller can be manually adjusted by the operator.

Feedback information from flue gas analysis can be very useful in establishing proper air/fuel ratios. The ideal situation is continuous measurement of combustibles (e.g., CO) and O_2. The minimum amount of air needed is that which just causes combustibles to become insignificant. Thus combustible monitoring can be used to ensure that an adequate amount of air is available. The air controller proportionality factor could be decreased until combustibles just start to rise. This would give the minimum amount of excess air needed for complete combustion. If the O_2 content at which combustibles disappear is too high (>2.5%), it is an indication of inadequate mixing. Steps could then be taken to increase injection velocities and turbulence. Use of a target O_2 value as the only basis for controlling air/fuel values is not completely adequate. It is sensitive to mixing variables and cannot by itself guarantee complete combustion. Effective use of flue gas composition data for air control has been held back by lack of reliable sensors. However, recent developments show promise in overcoming this problem (*35*).

The distribution of air in the furnace is also important. Primary air (primary and secondary in B & W units) plays a strong role in char bed burning. Secondary air (tertiary in B & W units) does not. Thus, primary air rates help govern bed temperatures and bed stability. Control of primary air is primarily based on bed characteristics. Air distribution is controlled by dampers in the supply lines. A number of different control schemes can be used. A common procedure is to manually set a primary air to liquor ratio. Changing the ratio would simply move air from the primary zone to the other zones. In units with more than one F.D. fan, fan speed or inlet vanes can be adjusted to give the desired air distribution.

Recommended windbox pressure levels should be maintained to ensure adequate injection velocities. When flue gas composition data indicate a mixing problem, injection velocities should be increased, particularly at the upper levels. Symmetry of air introduction should be maintained to avoid local areas of air starvation. Ports must be kept clean to ensure an even distribution.

Air is normally preheated to about 300°F (150°C) to aid combustion stability. Air temperature is not normally used as a control variable. It does have some potential in this regard.

Bed A significant amount of the combustion and some of the key process reactions take place in the char bed. There are basically two control problems associated with the char bed: char bed inventory and bed temperature levels.

The inventory of material in the char bed must be kept reasonably constant if furnace operation is to be stable. In general, there is no particular size where the char bed is inherently stable. Bed size is a reflection of a dynamic balance between the rate at which black liquor char is reaching the bed and the rate at which the carbonaceous material is combusted and the smelt flows out. The bed integrates the difference between the rate of supply of liquor solids and the burning rate. Char bed inventory is very difficult to measure, and present control methods are based on visual observation (sometimes with the aid of a TV camera). Attempts have been made to measure bed weight by strain gages on furnace floor beams (*36*). These have not met with great success.

In CE units, two variables influence the size of the char bed — primary air and spray size distribution. Increased primary air (which increases bed burning) and decreased spray size (which reduces the carbon load to the bed) will both tend to lower the bed, and vice versa. A balance between primary air rate and droplet size can be selected to give a bed of desired size. Operators sometimes prefer fine spray–low bed operation because it requires less attention than coarse spray–full bed operation. However, this practice can lead to poorer reduction and increased carryover.

Bed inventories tend to be larger and somewhat more stable in B & W units. They often grow to a level near the secondary air ports and remain fairly stable at that level. This occurs because the additional air input at the secondary level is sufficient to ensure complete gasification of all carbonaceous material present. The density of bed material is also generally higher in B & W units (*37*).

Bed temperature is very important to sulfate reduction, fume formation, and combustion stability. The temperature is highest on the surface of the bed and decreases toward the interior, because of the endothermic reduction and sodium volatilization reactions occurring within the bed. The most critical thermal parameter is the average bed surface temperature, since this determines not only surface reaction rates, but also the amount of heat being carried into the bed and hence the extent of reduction and fume formation within the bed (the volume of bed that is chemically reactive). In general, bed temperatures should be kept as high as possible, consistent with a stable bed inventory. High bed temperatures improve combustion stability and help eliminate blackout conditions. High bed temperatures also give better reduction efficiencies and increase sodium volatilization, which acts as an SO_2 trap. Excessively high bed temperatures could lead to excess fume production and higher dust loads, but temperature levels that would cause great operating difficulties from this source are probably not sustainable without burning down the bed.

It has proven to be very difficult to measure bed temperatures or lower furnace gas temperatures with any degree of reliability. The basic problems are the extreme dirtiness of the furnace atmosphere and interferences from the sodium and water vapor which are present. Recently, a few units have incorporated bed temperature measurement into recovery boiler control loops with apparent success (*38*). An indirect indication of char bed temperature can be obtained by measuring the composition (alkalinity) of the precipitator dust. Volatilized sodium will react with SO_2 in the gas space to form Na_2SO_4.

When enough sodium in volatilized (because of higher bed temperatures), there will be excess sodium present which will form Na_2CO_3. The presence of a reasonable amount of Na_2CO_3 in the fume (10%) is one indication of bed temperatures in the proper range.

Both liquor spray characteristics and air supply (distribution) must be adjusted to achieve simultaneous control of bed inventory and temperature. This balance can vary with the burning characteristics, heating value, and solids content of the liquor being fired, as well as the furnace geometry and load. At present, there are no simple control loops to perform these functions, and there is a general lack of good feedback information on which to base control loops. The present practice can best be described as an art with direct operator intervention (particularly on spray characteristics), based on visual perceptions of the state of the char bed.

Disturbances Recovery furnace control can be greatly improved by minimizing disturbances in inputs to the unit. A listing of sources of disturbances and possible means for minimizing them is given below (*39*).

1. Changes in liquor composition or
   species being pulped,
   pulping yields,
   variable residual alkali levels,
   inorganic recycle variations, and
   changes in makeup chemical addition.

   Liquor feed to the recovery unit should be as homogeneous as possible. Blending liquors ahead of the evaporators can help minimize changes resulting from pulping conditions.

   Digester control loops can be added to minimize variations in residual alkali. (Alternatively, controlled NaOH addition to black liquor can be used.) Sootblowing and precipitator rapping should be done at regular intervals to minimize disturbances in recycle loads. Techniques for interrupting and directly controlling the recycle addition rate could be considered. Makeup chemical addition should be steady.
2. Liquor flow measurement drift. This is apparently an inherent problem with heavy black liquor. Heat release rates and measured air consumption rates can be used to correct for drift in average flow rate.
3. Liquor solids content. Proper attention to the solids control loops in the direct contact evaporator or concentrator will help stabilize this input. If the online solids measurement is considered sufficiently reliable, dilution with strong black liquor could be used to control firing solids directly.

4. Liquor nozzle plugging. Clean liquor nozzles are essential to proper spray characteristics. Operator attention to catch buildup in early stages is required. Homogeneous liquor without lumps or pockets of entrained solids will help minimize plugging. Worn nozzles should be replaced.
5. Nozzle sweep synchronization. Controlled sweep synchronization is preferable to random synchronization, which will occur with unsynchronized sweeps. Alternatively, operation with stationary nozzles can be used and may be preferable.
6. Primary air momentum. Plugging and cleaning air ports results in significant changes in the area available for air flow and will result in significant changes in injection velocities (momentum) for a given air flow rate. Infrequent and sudden cleaning of ports can decrease primary air momentum by over 50%, and can result in blackout or loss of bed control at the center. Port cleaning should be regular and frequent to avoid large changes in open area. Port pressure measurement can be used for control purposes. If the port open area under clean conditions is too great to permit good air penetration to the center, adjustable port openings should be considered. Similar considerations hold for upper air levels, although the sensitivity is not as great since the char bed is not directly involved. Automatic means for cleaning ports are under development.
7. Buildup and falling of large slag deposits from the upper furnace. These create severe disturbances when they strike the bed and can be hazardous. Minimizing carryover and maintaining proper temperature levels in the upper part of the furnace can help to reduce these occurrences.

Air emissions control Air emissions from the recovery boiler are essentially of three types: reduced sulfur gases, SO_2, and particulates. Particulate emissions are normally controlled by a high-efficiency electrostatic precipitator operating on the stack gas and are not strongly sensitive to furnace operation. Sulfur gas emissions are very sensitive to operating parameters, and their control is a significant constraint on operation.

Reduced sulfur gases constitute an odor problem and are subject to very low emission limits (5 ppm on new sources). These are primarily H_2S, although mercaptans and other reduced compounds are also involved. Standards are normally expressed in terms of total reduced sulfur (TRS).

In recovery units with direct contact evaporators, H_2S can be stripped from the heavy black liquor by reaction of acidic flue gas components with sulfide present in the black liquor. Control of TRS emissions in these units requires prior oxidation of the black liquor to give very low residual sulfide levels. High efficiency black liquor oxidation units are now a standard part of the recovery cycle when recovery boilers with direct contact evaporators are employed.

It is also necessary to control TRS emissions arising from within the furnace cavity. This requirement exists for all recovery boilers, regardless of whether or not they have a direct contact evaporator. These emissions originate from the reduced sulfur gases which are formed during the pyrolysis of black liquor. This TRS formation is to some extent inherent in the nature of black liquor and is not prevented by prior black liquor oxidation of sulfide to thiosulfate. Control of TRS emissions from within the furnace cavity rests on two requirements:

1. Maintaining conditions in the lower furnace which minimize the release of H_2S to the furnace atmosphere; and
2. Oxidizing to SO_2 any TRS which is released.

The first item requires high temperatures in the lower furnace. High hearth zone temperatures favor the absorption of H_2S by molten Na_2CO_3, and prevent its release. Minimizing entrained liquor droplets (such as by using coarse spraying) also helps minimize H_2S release.

Any TRS released must be oxidized, and this establishes conditions on the air supply. TRS gases tend to be oxidized at comparable rates to other furnace combustibles. Thus, the complete oxidation of TRS gases requires essentially complete combustion of the black liquor organics. It has been shown that a close correlation exists between TRS concentrations and combustibles (CO) in flue gases, and that combustibles control can be an effective way of controlling TRS (*40*). Complete oxidation requires not only an adequate supply of air, but also good mixing of combustibles and air and adequate retention time.

Oxidation of TRS gases to SO_2 eliminates the TRS problem. If the recovery boiler is not subject to SO_2 emission limits, this is sufficient. Simultaneous control of both TRS and SO_2 is more complicated. Units with a direct contact evaporator have an advantage, because part of the SO_2 will be reabsorbed by the black liquor in the direct contact evaporator. However, both TRS and SO_2 can be controlled simultaneously in units without direct contact evaporators.

SO_2 will react with alkaline fume and be converted to Na_2SO_4 in the oxidizing zone of the furnace. Increasing the volatilization of sodium from the char bed increases the amount of alkaline fume available and reduces the SO_2 concentration in the flue gas. Sodium volatilization is enhanced by increasing bed temperatures while still maintaining a reducing environment. Thus high bed temperatures achieved with a stable bed are favorable for reducing SO_2 emissions.

It is evident that conditions favorable for tying up SO_2 with fume are also favorable for preventing release of H_2S in the first place. Thus, a reasonably high fuming rate is desirable for both TRS and SO_2 control. This requires adequate bed temperature levels. Feedback information from the composition of the flue gas can verify that adequate conditions are being maintained.

Reasonable increases in fuming rate will not necessarily increase plugging problems in the boiler heat traps. Plugging problems are primarily caused by carryover particles from liquor sprays and by excessive temperatures in the upper furnace. Free SO_2 in the flue gas can lower the alkalinity of deposits and increase the sticking tendency. Increased fuming rates will lessen this problem.

If the precipitator efficiency is adequate, increased dust loads could be handled without exceeding permissible particulate limits. An undersized precipitator could make it very difficult or impossible to simultaneously control TRS, SO_2, and particulate.

Overall strategies Until recently, relatively little sophistication was used in overall supervisory control and in performance optimization of recovery boilers. The basic procedure was to establish the firing rate and to keep the unit operating reasonably stably. Extensive use of auxiliary fuel to aid in overcoming marginal combustion was not uncommon. The operation was under manual control with a few automatic loops available on subsystems. Little feedback information on process parameters was available. Operation was essentially an art.

Many recovery boilers are still operated in this mode. A typical control system is as follows:

1. Black liquor flow is automatically controlled from a manual setpoint.
2. Black liquor temperature is automatically controlled from a manual setpoint. This setpoint is adjusted by the operator based on visual observation of furnace brightness and char bed size.
3. The black liquor flow set point is relayed to the set points of the total air flow controller and the primary air flow controller. A change in the black liquor flow setpoint will cause a proportionate change in the setpoints of the air controllers. This provides the only link between the fuel firing rate and the air supply.
4. The total air controller will manipulate vanes at the F.D. fan or adjust fan speed to give the desired air flow. The operator can manually adjust the controller proportionality factor and thus change the proportion of air and liquor. This could be done in response to data on flue gas O_2 content or TRS.
5. The primary air controller adjusts a damper in the primary air duct to give the desired primary air flow. The operator can manually adjust

the controller proportionality factor and thus change the air distribution.

6. Furnace pressure (draft) is automatically controlled by sensing pressure and varying the speed of the I.D. fan. This loop operates independently of the combustion system.
7. Air injection velocities can be changed by manual adjustment of dampers in the windbox.

Complete, automatic supervisory control systems or computer control systems are just beginning to be used on recovery boilers. The degree of sophistication and logic varies considerably from one system to another. Feed-forward techniques with adjustment of controller parameters based on feed-back information are in use. Sensor limitations may be the biggest single factor inhibiting more widespread use of computer control.

It is possible to define a set of conditions needed for optimum furnace performance. These can be used as a basis for establishing control systems.

1. Complete oxidation of the liquor organics with a minimum of excess air. This requires effective use of information on the combustible and O_2 content of the flue gas. Meeting this requirement should automatically eliminate TRS emissions from the furnace as well.
2. High combustion temperatures and combustion as low as possible in the furnace (*41*). Liquor sprays and air distributions should be adjusted to permit high char bed surface temperatures and intense combustion in the region just above the bed and below the spray. This could require two levels of primary air in the primary zone of suspension fired units, so that low velocity air would impinge directly on the char bed with high velocity air available to mix with and burn combustibles in the zone directly above the bed. It would also require measurement of bed zone temperatures for feedback control.
3. Burning rate should be stabilized by operating the unit at a constant heat input rate rather than a given solids rate. The recovery boiler could be used as a continuous calorimeter and the firing rate adjusted (over a limited range) to maintain a fixed heat release. This would require the use of liquor guns in which the spray characteristics were not highly sensitive to liquor flow rate.
4. Liquor sprays which produce particles of the proper size range. The optimum size would be particles which give intense combustion in the lower furnace without the use of auxiliary fuel. The size range would be relatively narrow, since fine particles tend to carry over and large ones bring wet material to the char bed. This may require development of modified gun designs.

5. Moderately sized, active, porous char beds. It has been established that large char beds are not essential for good reduction (*10*). Dense, colder char bed material simply occupies space in the unit and increases burndown and cooldown times. It would be desirable to operate with a stable, porous char bed about 3 ft in height. Measurement techniques need to be developed before such a bed could be controlled on a closed loop basis.
6. Sufficient fume should be produced from the char bed so that SO_2 emissions are minimal. This should also minimize sticky ash and acidic corrosion problems. Na_2CO_3 content of the dust can be used as a check on whether combustion temperature levels are high enough.
7. Minimal disturbances to the furnace. As much as possible, sources of disturbances should be tracked and eliminated. Liquor solids content could be put under a control loop.

There is increasing interest in developing strategies for increasing the capacity (or throughput) of recovery boilers. Tight TRS emission limits require essentially complete combustion of the black liquor solids fired, so that these strategies must incorporate complete combustion. Under these conditions, most of the actual limitations to capacity tend to be heat release related (flue gas flow rates, gas temperature levels, allowable heat fluxes, steaming rates). Capacity optimization strategies require the definition of critical limiting conditions and logic trees for determining when throughput can be increased.

Operating procedures

Startup Before the unit is placed in operation, equipment lubrication should be checked. Cooling water should be supplied to smelt spouts, bearings, and black liquor pump glands. The boiler should be filled with suitably treated water to the proper level. The dissolving tank should be filled with either weak wash or water. Air dampers should be properly set. Access doors should be closed and smelt holes should be free of any obstructions. All personnel not involved in the startup should be cleared from the area.

After the unit has been inspected and preliminary startup steps taken, the lighting-off procedure should follow the manufacturer's recommendations. The following is illustrative:

1. Start I.D. fan and adjust furnace to desired draft.
2. Start F.D. fan and adjust airflow for proper auxiliary burner operation. Purge the unit for the required period of time.
3. Place steam air heater in operation.
4. Start sufficient auxiliary burners to boil out pendant superheater and put unit on the line, following pressure raising curve supplied by the

manufacturer. Maintain the proper boiler water level by blowdown during pressure raising. Maintain the proper boiler water level by feedwater addition when the level drops as the unit goes on the line. As the unit reaches operating pressure, open the main steam stop valve.

*5. Approximately 1 hour before the unit goes on the line, fill the direct-contact evaporator with black liquor from the strong liquor storage tanks. Start the wheels of the cascade or the pumps of the cyclones as the liquor is admitted.
6. Start the green liquor circulating pumps and dissolving tank agitators.
*7. When the liquor in the direct-contact evaporator has reached the desired concentration for firing, open the liquor valve to the mix tank, filling the tank to the operating level.
8. Start the hopper sluice system pump and adjust the liquor to the hoppers.
9. Open the steam valve to the black liquor heater in the mix tank.
10. Install the size liquor nozzle desired for black liquor firing.
11. Start the black liquor fuel pump and recirculate liquor through the header to make sure all lines are flushed free of water or low solids liquor. Start spraying liquor into the furnace after the minimum safe firing concentration is reached. Liquor guns should be installed one at a time with a sufficient period in between to ensure that the firing is stable. The unit should be on the line generating steam prior to firing black liquor.
12. Adjust and put on automatic control I.D. fan and feedwater control.
13. Adjust F.D. fan to compensate for liquor firing.
14. As the liquor firing becomes stable, remove the auxiliary burners one at a time until the liquor firing is self-sufficient.
15. Start salt cake makeup feed.
16. Make final adjustments of oscillator sweep and air distribution.

*Steps 5 and 7 are eliminated in units without direct contact evaporators.

Routine operation

1. Make sure feedwater flow to the boiler is regulated to maintain recommended drum level.
2. Maintain boiler water concentration within prescribed limits by proper blowdown.
3. Maintain black liquor firing rate constant consistent with operating rate of the entire mill.
4. Adjust total air to the unit to give the desired O_2 concentration in the flue gas.
5. Adjust air distribution as needed to maintain desired operating performance and bed shape.
6. Adjust liquor temperature to achieve desired spray characteristics.

7. Check the rate of saltcake addition to the mix tank.
8. Check the chemical ash hoppers to make sure no bridging or accumulation is taking place in the hoppers.
9. If a direct contact evaporator is used, regulate the boiler bypass damper and/or dilution liquor to maintain desired solids content.
10. Check the green liquor density and level control system to make sure it is operating correctly.
11. Keep the air ports, smelt spouts, and black liquor spray nozzles free of smelt accumulations.
12. Make sure the hearth burners are kept clean and ready for service; the burner openings should be kept lanced out and free of smelt.
13. Make periodic checks on all rotating equipment for abnormal operating conditions; check cooling water, gland water, lubrication, bearing temperatures, etc. Check the smelt spout cooling water.
14. Blow all soot blowers as required to keep all gas passes clean. The proper soot blowing operating sequence must be established through experience.
15. Make a habit of studying the instruments so that unusual readings will immediately be recognized. Try to account for all deviations in recording instruments. Take board readings according to the routine set up by the operating departments. Keep the pens of recording instruments supplied with ink.
16. At least once a shift, make an inspection of the upper furnace to see if excessive carryover is taking place. Try to estimate relative temperatures in the screen and superheater zone so excessive slagging and tube fouling will not occur. Know where the various draft readings are taken so that the areas of plugging can be located immediately.

Shutdown The manufacturer's recommendation should be followed for routinely shutting down a recovery unit. The following procedure is illustrative:

1. Shut off salt cake feed to mixing tank.
2. Shut off black liquor feed to mixing tank. As soon as the black liquor flow to the nozzles has stopped, indicating that the mixing tank is empty, turn on steam to spray gun to clean it out. When spray gun assembly has been cleaned out, remove it from the furnace.
3. Flush with water all black liquor lines not in use.
4. When steam flow starts to drop, insert auxiliary fuel burners to burn down bed as much as desired and to maintain steam flow, if necessary. The bed must not be burned down so fast that excessive flow of smelt

is produced, which will cause trouble in the dissolving tank. A burner should be installed near each smelt spout to keep smelt hot as long as there is any appreciable flow. Be sure there is sufficient steam pressure on shatter jets.

5. When black liquor flow to the furnace is reduced or cut off for any reason, air supply to the unit should be reduced accordingly. If air supply is not reduced, char bed will burn out too fast, and increased smelt flow will create problems in the dissolving tank.
6. Pump the black liquor from the direct-contact evaporator back to storage. When the direct-contact evaporator is empty, refill with water, keeping the cascade wheels or cyclone circulating pumps in operation to clean up the system. Caustic should be added during prolonged oil firing. Flush out all black liquor lines not in use.
7. When the char bed has been reduced in size so that the smelt spout openings will be free for the next startup, start removing auxiliary burners from service, one at a time, reducing steam flow from the unit gradually until all burners are out of service.
8. As the control goes out of range, the I.D. fan, F.D. fan, and feedwater controls should be placed on hand- or remote-manual operation.
9. When the last burner is removed, shut off the forced draft fan.
10. Shut the main steam stop valve and open superheater vent valves.
11. Open access doors on the unit to aid in cooling the unit down, running the I.D. fans at a reduced rating, just sufficient to keep the unit under suction.
12. Adjust the drum water level up to within an inch of the top gage glass and secure the feedwater.
13. When the unit cools sufficiently, shut off induced draft fan.

Emergency shutdown An immediate emergency shutdown must be performed whenever (1) water in any amount is known to be entering the furnace and cannot be stopped immediately, or (2) any leak develops in a pressure part except for leaks external to the furnace such as handhole gaskets, gage glasses, etc. The following Emergency Shutdown Procedure (ESP) should be followed (*42*):

1. Sound an alarm to clear the recovery area of all unnecessary personnel.
2. Immediately stop firing all fuel. Secure the unit's auxiliary fuel system at a remote location.
3. Immediately shut off feedwater and all other water sources to the boiler except for smelt spouts.

4. Shut down the air supply to primary air ports immediately. Continue operation of the forced-draft fan to supply as much air flow as possible to the secondary and tertiary air ports (if present). Regulate induced-draft speed or damper to maintain a balanced draft in the furnace.
5. Drain the boiler as rapidly as possible and in accordance with manufacturers' recommendations to a level eight feet above the low point of the furnace floor.
6. Reduce steam pressure as rapidly as possible after the boiler has been drained to the eight-foot level.

Operating problems Many operating problems can develop with recovery boilers. Because of the unique hazards associated with the presence of molten smelt within the unit, these problems can have more serious consequences than they would in a conventional boiler The following are some of the major operating problems, consequences, and the action required (*43*).

1. Electrical power failure. This is one of the most serious problems that can occur in a recovery operation, and it is essential that the boiler be secured in the event of a power failure. Specific procedures will vary depending on the availability of backup power and the nature of drives, controls, and instrumentation. The most critical steps are to maintain drum level as long as possible and to stop firing all fuel.
2. Feedwater supply loss. This is also a very serious operating problem. Without sufficient cooling water, boiler tubes will soon overheat and fail. It is essential that all firing be stopped. Normal draft should be maintained and air supplied to the upper air ports. Extreme care must be exercised when feedwater is again available to avoid quenching hot pressure parts.
3. Low drum level. If water level is low and cannot be explained or corrected, all firing should be stopped immediately. Fast action is important because severe boiler damage can occur if firing continues and water level is excessively low. If the level control valve is wide open and feedwater flow is higher than steam flow, a tube leak should be suspected.
4. High drum level. This can lead to water carryover into the superheater section which will not only cause superheater tube damage due to the thermal shock and chemical deposits, but can also severely damage subsequent steam turbines. If the level cannot be controlled within prescribed limits, the unit should be immediately shut down. If the

level is high but controllable, the level can be brought down by using the manual mud drum blow down valves.

5. Contaminated boiler water supply. While generally not an immediate threat to boiler security, continued operation with contaminated feedwater can result in boiler or turbine damage. Sources of contamination should be traced and eliminated as soon as possible. Blowdown should be increased to purge contamination. Operation should not be continued when contamination exceeds limits set by local management.
6. Loss of main steam header pressure. This will increase drum level and steam flow. The boiler steam production capacity may be exceeded and result in tube damage because of upset boiler water circulation. It is essential to maintain drum level and not to allow the boiler to exceed preset limits for steam production. Be alert for loss of utilities such as power, air, and water.
7. Instrument air failure. This can knock out air-operated controllers. If the boiler cannot be controlled within established limits, it must be immediately shut down. Drum level must be maintained (manually if necessary) while shutdown proceeds. Normal startup procedures are followed when pneumatic power is restored.
8. Smelt spout cooling water jacket failure. This can lead to a smeltwater explosion. Cooling water to a leaking spout should be shut off and the water lines disconnected. The spout opening can then be plugged. Be alert for uncontrolled smelt flow if the plugged spout burns through. *Under no circumstance should cooling water flow be restored to a spout once it has been shut off.*
9. Total loss of smelt spout cooling water. The water supply should be shut off and the lines disconnected. Steam can be used to cool the spouts if the unit is so equipped. The bed should be burned out as rapidly as possible and the boiler shut down for replacement of spouts. Once the cooling water has been lost, it must not be restored to the spouts.
10. Smelt leak through furnace floor. If the smelt flow is uncontrolled or poses a hazard, the unit must be shut down immediately. The area under the boiler should be secured. If the leak can be stopped, the boiler may be left in operation. Air lances or CO_2 lances may be used *external* to the boiler to try to chill and stop the leak. Every effort must be made to keep the molten smelt out of the sewer. Lime, salt cake, or dirt can be used to contain the molten smelt.
11. Loss of all auxiliary fuel. This is usually the result of a fuel interlock trip. Drum level and furnace draft must be maintained while seeking

the cause and correcting the problem. Proper purging procedures must be followed when relighting the burners.

12. Auxiliary burner trip. An attempt should be made to relight the burner. If unsuccessful, a spare burner should be lit off. Then the cause of the burner failure can be diagnosed and corrected.
13. Inadequate combustion air. This will result in a reduction of steam flow, high combustibles readings, and no excess O_2. In this condition, fuel rates should be gradually reduced, starting immediately, until an acceptable O_2 concentration is reached. Air should never be added in this situation. If a high concentration of combustibles is present and air is added, there may be a serious explosion.
14. Induced draft fan failure. The furnace will pressurize and blow smoke and flame out through furnace openings. It is essential to immediately trip the F.D. fan(s) and all auxiliary fuel and then remove the liquor guns. When the I.D. fan is working again, the boiler can be started up according to normal procedures.
15. Flue gas system fires. Fires can occur whenever dry liquor solids accumulate. The hazard is worse by excessive air leakage into ducts. A fire can spread rapidly through the ducts, fans, direct contact evaporators, and precipitators. Procedures for handling fire vary from mill to mill. BLRBAC has recommendations on this subject (*44*).
16. Sootblower problems. Loss of one or more sootblowers may cause dirty gas passages and formation of large slag deposits which can subsequently fall. Sootblowers which blow steam without moving can cut tubes. Deformed or misaligned sootblowers can contact tubes and damage them. Occasionally sootblower lance tubes have been launched causing physical damage where they struck the opposite wall. Stalled sootblowers should be manually retracted and then the steam supply secured. Misaligned sootblowers should not be operated. Periodic visual inspections of sootblowers should be made to prevent problems from developing.
17. Total loss of black liquor supply to furnace. This can be caused by a black liquor fuel interlock trip, plugged gun nozzles, an empty mix tank, a plugged screen on the mix tank outlet, failure of the black liquor feed pumps, heavy black liquor which does not flow easily, an open diversion valve, or the failure of black liquor heaters or controls. If the problem is not easily corrected, drum level should be maintained and auxiliary fuel fired until the problem is corrected.
18. Excessive furnace pressure. Smoke and flame will blow back through furnace openings, and high pressure alarms will be activated. If the unit is not automatically tripped, immediately stop firing all fuels and shut down the F.D. fan(s). If the trip was caused by a tube rupture and water may be entering the furnace, an ESP should be initiated.

19. Low furnace pressure. This is a problem only in those units with sufficient I.D. fan capacity to cause furnace implosions. If the boiler protective interlock system does not function automatically, the protective action in effect at the mill should be initiated manually.
20. Air heater leak. Moisture in combustion air caused by a steam coil leak or rupture is potentially hazardous because of a possible smelt/water explosion. Excessive steam in the bed air supply can also promote a blackout. If water is entering the furnace, all steam to the air heater should be immediately secured and the heater isolated from the condensate system. All duct drains should be open. If water is not entering the furnace, the leaking coil should be located and isolated from both steam and condensate.
21. Low liquor solids. This is a possible smelt/water explosion hazard, and may also cause unstable combustion. If the furnace is equipped with an automatic diversion system, the black liquor will be diverted from the furnace when the refractometer(s) indicates the liquor solids are below a preset standard, typically 58% solids. If the solids fall below the preset minimum and the black liquor is not automatically diverted, it should be manually diverted from the furnace. If the solids content is low, but not below the preset minimum, black liquor firing can be continued (possibly using auxiliary fuel) until the problem is corrected.
22. Partial blackout. If the blackout is caused by a tube leak, an ESP should be initiated. If there is no emergency condition and the boiler can be controlled, the cause of the blackout can be sought. This could be low solids content liquor, improperly operating sprays, inadequate primary air, or plugged primary air ports. The primary air flow should be increased in the blacked-out area. In severe cases, auxiliary fuel burners can be used in the blacked-out area.
23. Complete blackout (loss of liquor ignition). The bed darkens and becomes cooler and ignition is lost. When this occurs, black liquor must be diverted from the furnace and the guns pulled. Auxiliary fuel burners can then be lit to reestablish ignition and to bring the boiler back on line. When everything is in readiness, black liquor firing can be resumed. It is important to *avoid continued spraying of black liquor into a blacked-out furnace*. This practice can lead to explosions.
24. Plugged liquor spray nozzle. Nozzles can plug for several reasons, such as insoluble lumps in the black liquor, failed screens in the flow box or mix tank, low pH liquor, excessively high liquor solids content, low nozzle pressures or liquor flow velocities, and uneven salt cake flow to the mix tank. Nozzles may be unplugged by steaming them out in place, if care is taken to clear the steam-out line of condensate.

If the gun cannot be unplugged, the pressure should be relieved and the gun replaced.

25. High dissolving tank density or low tank level. This condition can lead to crystallization or live (molten) smelt in the tank. If a crystallized or live smelt condition exists, the area should be cleared of unnecessary personnel and all firing stopped immediately. All sources of water to the tank should be shut off and the area around the dissolving tank roped off and cleared of all personnel until ample cooling time has passed. No attempt should ever be made to dissolve live smelt.
26. Loss of smelt shattering. If there is damage to the spout itself, appropriate action must be taken. If no hazard exists, a steam or air lance can be used for temporary shattering.
27. Plugged smelt spout. Smelt will accumulate within the unit and runoff may be heavy when the spouts are opened. Excessive accumulation can result in smelt entering primary air ports. If all spouts are plugged, liquor firing must be stopped before proceeding to trying to unplug the spouts. If only some spouts are plugged, the firing rate should be adjusted to avoid overloading the open spouts. The spouts should be opened as soon as possible with appropriate precautions to evacuate the area if excessive runoff occurs. Air lances, water lances, and sharp pointed rods should *not* be used to unplug spouts. Prior to lancing, the shatter steam should be at maximum to handle the increased smelt flow.
28. Heavy smelt runoff. Heavy runoff is often caused by opening plugged spouts, burning out a large bed, or switching to auxiliary fuel. In severe cases, dissolving tank explosions can occur. Reducing primary air and reducing auxiliary fuel firing can reduce runoff rates.
29. Cascade stoppage. The cascade should be restarted immediately. If successful, find and correct the cause of the stoppage. If the restart fails, all fuel firing should be stopped and the liquor system secured. Precaution against fire caused by dried liquor solids should be taken.

Inspection and maintenance Safety and reliability of recovery boilers can be greatly improved by a regular, systematic inspection and maintenance program. This will help assure continuous operation between scheduled outages and reduce forced outage time due to equipment failure. Forced outages are costly and there is a greater explosion risk during a forced shutdown and subsequent startup. Many forced outages are caused by auxiliary equipment such as fans, sootblowers, precipitators, and pumps. Preventative maintenance programs and careful inspection followed by the proper corrective action can eliminate many forced outages.

Although many system components can be inspected and maintained during normal operations, other items and areas can only be inspected and maintained when the boiler is shut down. Annual inspections of gas side and water side boiler components should be made, and a suitable nondestructive testing program established and maintained. The content and frequency of such programs must be established for each installation. The "Recovery Boiler Reference Manual," produced by the API Recovery Boiler Committee, is a very useful reference on this subject (*45*).

The majority of problems on the gas side of the unit are caused by the following:

corrosion;
overheating;
mechanical damage;
weld failures; and
erosion.

A brief discussion of each follows.

The areas most affected by corrosion are waterwall tubes at the smelt line, smelt spout openings, the area around each air port, the lower superheater bends in the upper furnace, the area around the char band, and (in nonmembrane wall units) the unheated back side. Corrosion protection is provided by close-spaced pin studs, flame or plasma metallizing, weld overlaying, or composite tubing. Corrosion is normally small where protection is intact. Inspections should pinpoint areas where protection has deteriorated and needs to be renewed.

If the heat transfer within the tubes is impaired (from internal deposits or impaired circulation), they can overheat. If metal temperatures are excessive, the tubes can weaken and ultimately fail. Overheated tubes will discolor and can burst if overheating is severe. Tubes showing evidence of overheating should be replaced and the cause determined and corrected.

Mechanical damage is especially likely around air ports and smelt spout openings where rods are used. Floor tubes and screen tubes bent from falling slag have been observed. Misaligned sootblowers and other sootblowing problems have caused mechanical damage.

Numerous welded attachments on furnace tubes are subject to failure. These include welded wall membranes, flat studs or fins, pin studs, stack studs, threaded studs on pads, temporary lifting lugs, and lugs and ties. Pin studs can burn back in high heat input areas and may need renewal. Replacement studs should be closely inspected for possible undercutting.

Areas of localized metal thinning due to erosion and/or corrosion must be identified so that corrective action can be taken. Wastage of metals is often seen around smelt spout openings, sootblower cavities, burner areas, and sometimes at specific areas on the furnace walls or at the smelt line.

Water side inspection should look for deposits or scale, evidence of plugged tubes, and signs of corrosion. Samples of scale should be taken for analysis and the locations identified. Evidence of cracks in tube metal or ligament cracking should be sought. Tube sections can be removed in high heat input areas to show the extent of scale buildup. The degree of scale buildup will determine the need for acid-cleaning the boiler. If evidence of water side corrosion is present, the boiler water treatment program may need to be reevaluated.

The dissolving tank should be inspected. Special attention should be paid to the agitator and the steam shatter jets and recirculation system.

The combustion air system should be inspected for steam coil air heater problems, leaking ducts, condition of F.D. fan and drive, signs of vibration problems, plugged draft taps, airflow measurement problems, and the condition of the windboxes.

The flue gas system should be examined for large deposit accumulations, corrosion, condition of expansion joints, insulation, and lagging. Draft taps should be clear. The I.D. fan needs careful inspection for erosion, corrosion, cracks in fan wheel, deposit buildup, and leakage through shaft seals.

Safety

Recovery boilers are faced not only with all of the safety problems associated with fossil-fuel fired boilers (containment of high-pressure steam, combustible gas explosions, possible fires, etc.) but also are subject to explosion hazards arising from the presence of molten smelt within the unit. The need for safe, reliable operation in the face of special hazards has had a profound effect on operating procedures as was discussed in the previous section. This section will focus specifically on safety aspects of recovery boilers.

Smelt-water explosions

There have been over 90 recovery boiler explosions in North America at this date due to smelt-water contact. These have ranged in violence from nearly complete destruction of the unit to those causing negligible damage. There have also been several hundred emergency shutdowns where the fear of an explosion has led to a forced outage. The frequency of explosions has remained relatively constant over the years and the problem cannot be considered to be solved.

Nature of smelt/water explosions Smelt-water explosions are physical explosions caused by extremely rapid generation of vapor (steam) when liquid water and molten smelt come in contact. The volume expansion on vaporization causes a very rapid pressure rise which generates a shock wave that carries the blast energy. It has been estimated that 1 lb of water vaporizing in 0.001 second (one millisecond) releases the equivalent mechanical energy of 1/2 lb of TNT. Explosions in recovery furnaces are thought to typically involve from 5 to 25 lb (2 to 10 kg) of water and be equivalent to the explosion of several lb of TNT. It has been established that the single most important factor determining explosive violence and furnace damage is the amount of water involved (*46*).

Smelt-water explosions generate shock waves rather than a general overpressurization of the furnace. Severe localized damage can result where the shock wave impinges on the furnace structure. Multiple reflections and reinforcements can also result in general damage to the furnace structure.

Explosions due to sudden vaporization of a liquid are generally termed "vapor explosions," and have become recognized as a cause of accidents in a variety of industries. The nuclear power industry has sponsored a great deal of work on the nature of vapor explosions, and a better understanding of critical factors is beginning to evolve. Vapor explosion theory is continuing to develop and is now approaching the stage where it can be useful in defining conditions under which explosions will and will not occur.

Extensive laboratory investigations of smelt-water explosions have been carried out. These have shown that the explosiveness depends on many variables, including smelt composition, liquid composition, temperatures, and contact geometry. To the present, no method of making recovery furnace operation immune to smelt-water explosions by modification of smelt composition or boiler water chemistry has been developed. The only known way to prevent smelt-water explosions is to prevent smelt-water contact.

Specific hazards Sources of water for smelt-water explosions can be classified as follows:

pressure-part failure;
low solids black liquor;
inadvertent introduction of water; and
failure of other water-cooled parts.

The most common cause of water entering the furnace has been tube leaks (pressure-part failures). Floor tube, wall tube, and screen tube failures constitute the most serious problem. Generating bank failures can result in water entering the furnace depending on the location of the leak and whether or not a baffle is present. Economizer tube failures are normally not an explosion hazard because of the lack of a direct path between the failure point and the furnace cavity. Superheater leaks are also normally not an explosion hazard since steam rather than liquid water is present.

Tubes fail from a number of different causes. Overheating, weld problems, corrosion, mechanical damage, and vibration fatigue are some of the major problems. Much attention has been focused on design and operating practices and on proper maintenance procedures to minimize the occurrence of tube failures.

Low solids black liquor has been the second greatest source of water for smelt/water explosions. The Black Liquor Recovery Boiler Advisory Committee (BLRBAC) recommends that black liquor should not be fired below a minimum concentration of 58% solids. Some mills use a higher minimum value. Increasingly, systems are being installed for automatically diverting black liquor from the furnace when the liquor falls below a minimum solids value. These systems should help eliminate this source of explosions.

Black liquor solids concentrations can become low for a number of reasons. These include evaporator and concentrator problems, excessive dilution or waste stream addition, leaking seals, pressure part leaks which drain into the liquor system through ash hoppers, the direct steam heater, dilution from wash water connections, and dilution when washing the black liquor system or auxiliaries.

Inadvertent addition of water is another common cause of explosions. This has included water in auxiliary fuel, condensate in steam purges, washup hoses, water lances, cross-connections through sootblowers, air heater leaks, and washing out the unit after a shutdown. Special care must be taken when washing equipment around a recovery unit to make sure the water cannot enter the furnace in some way. When starting up after a shutdown, all washup water connections must be blanked off.

Nonpressurized, water-cooled parts may also be a source of water for explosions. For this reason, water-cooled access doors and furnace accessories are no longer used on recovery boilers. The remaining hazard of this type is smelt spouts.

Emergency procedures In the event that a smelt-water emergency does develop (normally from a pressure-part failure), an emergency shutdown of the boiler is required. The BLRBAC-recommended ESP procedure aims at an immediate shutdown of the combustion process while avoiding conditions which could lead to a combustible gas explosion. (The actual procedure is given in the Operating Procedures section.) All firing of fuel is stopped immediately. Feedwater (and any other water source to the boiler except smelt spouts) is shut off. The air supply to the char bed is stopped, but forced air is continued to the other air ports and the I.D. fan is regulated to maintain a balanced draft. The boiler is drained as rapidly as possible to a level eight feet above the low point of the furnace floor.

The rapid drain feature of the ESP procedure is unique for a steam generator. It is designed to minimize the amount of water which can enter the unit and contact molten smelt. The eight feet of water is left to protect the boiler from the heat of the smelt bed. When the rapid drain is completed the steam pressure is relieved as fast as possible. This reduces the rate of leakage if the leak is below the eight foot level, and dissipates boiler energy. Rapid draining of recovery boilers has undoubtedly prevented a number of smelt-water explosions. Several hundred emergency shutdowns have been carried out. The large majority did not result in explosions.

Dissolving tank Smelt-water explosions can also occur in the dissolving tank, since direct contact between smelt and water is an inherent part of this operation. Damaging explosions in the dissolving tank are prevented by effectively breaking up the smelt stream into small particles and by preventing pools of molten smelt from accumulating in the bottom of the tank. In effect, a continuous series of small "explosions" is used to release the energy in the smelt rather than a sudden violent damaging release.

Safe operation of the dissolving tank involves the following steps:

1. Proper operation of the steam shatter jets and recirculating green liquor shatter spray;
2. Steady smelt flow at rates which do not exceed shattering capacity;
3. Adequate agitation of the green liquor in the tank;
4. Maintaining an adequate level of green liquor in the tank; and
5. Keeping the green liquor in the proper concentration range to avoid crystallization and crusting over the tank.

The most dangerous situation is one in which a crystallized or live smelt condition exists in the dissolving tank. If it does, the area should be cleared, the furnace shut down, and all sources of water to the tank stopped. It is then necessary to wait until the tank has cooled before proceeding. No attempt should ever be made to dissolve live smelt.

Combustible gas explosions

The basic cause of combustible gas explosions is the ignition of a dangerous accumulation of gaseous combustibles and air within the confines of the recovery furnace. Mixtures are potentially explosive if the proportions of fuel and air are such that uncontrolled combustion will result from their ignition. The explosive limits for a particular fuel in air are primarily determined by the chemical nature of the fuel and the amount of inerts present. Ignition of a gas mixture within the explosive limits will result in an explosion. The magnitude and intensity of the explosion will depend on the quantity of combustibles which have accumulated and the air/combustibles ratio at the moment of ignition.

Auxiliary fuel Numerous situations can arise in connection with the operation of a boiler-furnace which will produce explosive conditions. The most common circumstances are:

1. An interruption of the fuel or air supply to the burners, sufficient to result in loss of flame, followed by restoration and delayed reignition.
2. The accumulation of an explosive mixture of fuel and air as a result of loss of flame at one or more burners in the presence of other burners operating normally or during lighting of additional burners.
3. The accumulation of an explosive mixture of fuel and air as a result of fuel leaking into the furnace prior to startup, or after a complete furnace flameout and the subsequent ignition of the accumulation by a spark or other ignition source, such as attempting to light or relight burners.
4. Unstable or rich flame, and subsequent dilution of unburned fuel accumulation resulting in a flammable fuel-air mixture with delayed ignition.

The general adoption of monitored burners with appropriate safety interlocks has greatly reduced the frequency of auxiliary fuel explosions in recovery boilers. The design of auxiliary burner systems is such as to ensure proper conditions for combustion and to minimize the possibility of dangerous combustible mixtures from forming. An interlock system is provided to ensure that operators follow a safe, preestablished sequence of operations by excluding improper actions when starting up and operating with auxiliary fuel, and by initiating trip devices when approaching an unsafe operating condition. Recommended interlock system-sequence logic for different burner configurations is described in the recently revised BLRBAC manual on "Recommended Good Practice for Safe Firing of Auxiliary Fuel in Black Liquor Recovery Boiler" (*33*).

Black liquor Gas explosions from black liquor are a possibility. Pyrolysis or partial combustion of black liquor can produce combustible gases which can form explosive mixtures when mixed with air. Continued introduction of black liquor into a blacked-out furnace is a dangerous practice. The char bed retains sufficient heat to partially dry and pyrolyze the black liquor. The resulting combustible gases can mix with air and be ignited when auxiliary fuel burners are lit off to reestablish ignition in the furnace.

Sudden loss of combustion air can also produce a potentially dangerous accumulation of combustibles which could mix with air and explode on ignition. Air should *not* be added to a fuel-rich atmosphere in the furnace. If such a situation exists, fuel rates must be reduced until the mixture is safe on the air-rich side before adding air and reestablishing normal operation.

Other hazards

Fires Fires around recovery boilers can be very hazardous. Water can only be used with great caution in fighting fires because of the possibility of a smelt-water explosion if water enters the furnace. The areas of greatest vulnerability to fires are those in the flow path downstream of the furnace: flue gas ducts, direct contact evaporator, I.D. fan, and precipitator. Dried black liquor solids are combustible with a fairly low ignition temperature. Accumulation of dried liquor solids is the main reason for fires in these areas.

Procedures for combating fires in the flue gas ductwork, direct contact evaporator, I.D. fan, and precipitator vary from mill to mill. Massive low-pressure steam flooding is the preferred fire control method. Points of steam admission would be upstream of the direct contact evaporator, upstream of the precipitator and/or the I.D. fan, depending on the relative position of the latter.

Isolated instances can exist where steam is not available at all times and a secondary means involving water may be necessary. Where this is the only course, the water system should be completely monitored and interlocked with boiler safety systems. Water used to control fires cannot be allowed to enter the furnace directly or through the black liquor system. Since the material burning, in many cases, is black liquor solids; smelting may occur in the fire area. This could explode if contacted by water used to put out the fire.

Procedures can be instituted to minimize conditions leading to fires in the flue gas circuit. These include means for monitoring and removing buildup of dried liquor solids or other combustible material. Temperature monitoring equipment can be used to indicate excessive flue gas temperatures and to initiate the steam flooding action.

Fires have also occurred when auxiliary fuel lines were ruptured and the spilled fuel ignited. Smelt/water explosions have triggered damaging fires by rupturing fuel lines. Readily accessible shutoff valves located outside the recovery boiler area should be provided for the fuel piping system so that in the event of a mechanical break or damage by fire or explosion, the fuel system can be secured.

Implosions Occurrences of excessive low gas side pressure causing boiler setting damage have been termed "furnace implosions." They have occurred in balanced draft, fossil fuel fired (both oil and coal) and recovery boilers where sufficient force was developed to exceed the structural strength of the boiler furnace. All incidents occurred on balanced draft boilers with I.D. fans with high-head capability operating, at least momentarily, under low-flow conditions. All implosions occurred without main fuel firing, either before light-off or following a main fuel trip.

There are two conditions that have caused furnace implosions:

1. Maloperation of the equipment regulating the boiler gas flow path including air supply and gas removal, resulting in the furnace being exposed to the full head capability of the I.D. fan; usually at the low flow, high-head range of a centrifugal fan and often combined with an interruption of the forced draft air flow path.
2. Rapid decay of furnace gas temperature and pressure following a rapid reduction in fuel input or a main fuel trip. At any given moment the furnace gas pressure is governed by the gas law, $pV = nRT$. When a main fuel trip occurs, the temperature drops rapidly resulting in a sharp decrease in pressure, until a compensating increase in furnace mass inventory is achieved.

Combinations of these two events cause the most severe implosions. A detailed discussion of implosions, furnace gas flow dynamics, and solutions to these problems is given by Durrant and Lansing (*47*). As mentioned earlier, regulation NFPA 65G now sets the design pressure of the furnace and pressure parts to correct this situation.

Environmental factors

Air emission control

Sources and requirements The black liquor combustion operation is the source of a number of potential air emission problems. The substances involved are:

1. reduced sulfur gases (TRS);
2. SO_2; and
3. particulates

All three types of emissions are possible from the recovery boiler stack. Particulates and TRS can be emitted from the dissolving tank vent.

National source emission standards for TRS and particulates are summarized in Table 4.8. National source emission standards do not exist for SO_2. Sulfur dioxide emissions are subject to ambient air quality standards and requirements can be highly localized. State or local codes may also be more restrictive on TRS and particulates.

TRS control Control of TRS emissions from the recovery stack is generally based on their elimination from the flue gas within the recovery unit rather than by "end-of-pipe" scrubbing. There are two specific sources of TRS emissions in a recovery boiler:

1. Stripping of TRS gases from black liquor in the direct contact evaporator; and
2. TRS formation during black liquor pyrolysis and its incomplete oxidation within the furnace.

Both of these must be dealt with if emission standards are to be met. TRS emissions from these two processes are essentially additive. There is relatively little scrubbing capability for TRS in the direct contact evaporator.

Emission from the direct contact evaporator can be controlled by either of two methods:

1. Retaining the direct contact evaporator while using high efficiency black liquor oxidation to eliminate sulfide from the black liquor before it is introduced into the direct contact evaporator.
2. Eliminating the direct contact evaporator and replacing it with an alternate heat trap. The liquor is brought to firing strength in a steam heated

Table 4.8. Source emission standards (12 hours average values)

New sources	*TRS, ppm*	*Particulate*
straight kraft recovery	5	0.1 g/DSCM and 35% opacity
cross — recovery	25	0.1 g/DSCM and 35% opacity
dissolving tank	A	0.1 kg/kg dry black liquor solids
Existing Sources		
old-design straight kraft	20	
new-design straight kraft	5	
cross — recovery	25	
dissolving tank	A	

A = 0.0084 g/kg dry BLS

evaporator outside the recovery unit. Recovery boilers without direct contact evaporators are sometimes referred to as "low-odor" units.

Both of these approaches have been successful in controlling TRS emissions. In general, black liquor oxidation has been used on existing units with new recovery boilers employing the "low-odor" configuration. However, a few new units have gone in with direct contact evaporators, and some existing units have been retrofitted to eliminate the direct contact evaporator.

The residual sulfide level in the black liquor entering the direct contact evaporator must be quite low (0.1-0.5 gpl) to meet TRS standards. The exact sulfide concentration required will depend on the direct contact evaporator design and operating variables, residual alkali level in the black liquor, gas and liquor temperatures, and flue gas composition. High concentrations of SO_2 in the flue gas will tend to increase H_2S release because SO_2 is more acidic than CO_2, and the H_2S-releasing reaction is essentially acid hydrolysis.

The most common alternative heat trap is an extended economizer. Regenerative air heaters have also been used to cool the flue gases. Flue gas discharge temperatures in "low-odor" units are typically about 100°F (55°C) higher than they are in units with direct contact evaporators. The moisture content of the flue gas is considerably lower in "low-odor" units.

It is also essential to eliminate TRS that arises from the furnace itself. This is accomplished by maintaining operating conditions which ensure the complete oxidation of any reduced sulfur that is released from the pyrolyzing liquor solids. These conditions are adequate total air (sufficient excess air) and good mixing of the air introduced at secondary and tertiary levels. High bed temperatures help minimize initial H_2S release and aid in controlling TRS levels. In general, operating conditions which give negligible combustibles in the flue gas will be adequate for TRS control. Proper operation of the boiler is essential, even with black liquor oxidation. Black liquor oxidation does not prevent sulfur gasification within the recovery furnace proper.

Flue gas scrubbing systems for TRS control have also been developed (*48*). These are alkaline scrubbers operating at a balanced pH (between 9 and 10) at which the scrubbing solution is in equilibrium with the CO_2 in the flue gas. Sulfide levels in the scrubbing liquor are kept to very low levels by immediately oxidizing the absorbed sulfide to thiosulfate or sulfate. The most promising systems have employed an alkaline suspension of activated carbon as the scrubbing medium. An oxidation reactor on the recycling liquor is used to keep the scrubbing solution sulfide free. These systems are presently in use on only a few kraft recovery boilers.

SO_2 control Control of SO_2 emissions from recovery boiler stacks (where required) is mainly dependent on boiler operating parameters. Direct contact evaporators (when present) have some scrubbing capability for SO_2. However,

they were not designed to be, and do not operate as, high-efficiency SO_2 scrubbers. Even "low-odor" boilers can give low SO_2 emissions if properly operated.

Operational control of SO_2 emissions is based on two factors.

1. Minimizing the amount of sulfur gasified in the furnace.
2. Providing conditions favoring the reaction of SO_2 in the combustion gases with Na_2O and Na_2CO_3 fume so that it ends up as Na_2SO_4 particles.

The major variable influencing both of these factors is the bed temperature. High bed temperatures favor immediate H_2S reabsorption by molten Na_2CO_3 and hence minimal sulfur release. High bed temperatures also tend to increase the fume formation from the bed and hence increase the capture of SO_2 within the furnace. In one operation, the following changes in operating variables were found to give approximately a 20 ppm drop in SO_2 emissions (*49*):

1% increase in liquor solids content;
100°F (55°C) increase in combustion air temperature; and
$0.01 \frac{\text{lb } H_2O}{\text{lb dry air}}$ decrease in moisture of combustion air.

Different operations would undoubtedly have different sensitivities.

The S/Na ratio in the black liquor has a strong influence on the ability to control SO_2 emissions. As the S/Na ratio increases, SO_2 emission levels can increase strongly. It becomes increasingly difficult to control SO_2 emissions, as white liquor sulfidities exceed about 30-35%.

Recovery stack particulates Recovery stack particulates are controlled by "end-of-pipe" methods. Electrostatic precipitators are almost exclusively used for this purpose, although a few installations have wet scrubbers.

The particulates from the recovery furnace consist mainly of Na_2SO_4 with some Na_2CO_3 under certain operating conditions. NaCl may also be a major constituent if there is significant chloride in the liquor. The emission levels without control can exceed 100 kg/ADT. Dust concentrations ahead of the control device may range around 5 gr/SCFDG. The particulate in the recovery stack stems from two main sources; the ash from entrained liquor droplets combusted in the upper furnace, and true fume produced by condensation from the gas phase. The latter, especially, is responsible for the small particle size of the recovery boiler particulate.

Bosch, et al., (*50*) found that 90% (by weight) of the particles entering the precipitator are smaller than 3 microns and 50% are smaller than 1 micron. The presence of a large amount of submicron particles is characteristic of recovery stack particulates. The small size of the particles is a major reason why electrostatic precipitators are used for control of this source.

Gradual tightening of control requirements has led to continually increasing efficiencies for control equipment. For example, precipitator efficiency requirements have gone from 91% in 1950 to 98% in 1965 to 99.6% in 1972. These have resulted in much larger and more expensive precipitator installations.

Electrostatic precipitators on recovery boilers without direct-contact evaporators have presented some problems not previously encountered with precipitators on recovery boilers. Larger units are required to achieve the same collection efficiencies. Part of the reason that precipitators are larger on "low-odor" units is that gas viscosity and volume flows are higher because of the higher gas discharge temperature. Another part is due to the adhesion of the particles and other characteristics associated with their removal from the system after being collected. The problems of obtaining reliable precipitator operation on "low-odor" units have largely been overcome.

Dissolving tank vent Emissions from the dissolving tank vent are mainly small, entrained green liquor droplets formed by the violence of the smelt dissolving operation. Relatively simple devices may be used for control of this source. Wet scrubbers are the most common means of removing particulate from the dissolving tank vent. A wet fan mixes the gases with weak wash (the scrubbing medium) and the liquid/gas mixture discharges tangentially into a cyclonic scrubber. The scrubbed gases are vented and the scrubbing medium returns to the dissolving tank. Materials are normally Type 304 stainless steel to combat corrosion.

External impacts

Attempts to deal with other environmental problems around the mill can impact on the recovery boiler operation. Examples of this are the use of the recovery boiler as an incinerator for malodorous gases, and the introduction of extraneous materials such as bleach plant effluent or process wastes into the recovery loop. Process wastes can be divided into two general categories, gaseous wastes and liquid waste streams. The former would be added directly to the recovery furnace, whereas the latter would normally be added to the black liquor system somewhere before being burned.

Incineration of gases The most common gaseous wastes are TRS-containing streams collected from various sources around the mill. Collection and incineration of TRS streams is an essential step in odor control in kraft mills. These gas streams can be of two types: low volume streams with a high concentration of combustible material, or high volume streams with the TRS and other combustibles diluted in air to a concentration well below the lower explosive limit. Safe incineration of the high concentration streams involves considerations similar to the firing of gaseous fuels. The low concentration streams, on the other hand, simply form part of the combustion air supply to the boiler.

There are a number of problems associated with the handling and incineration of TRS gas streams. They are odorous, toxic in high concentrations, and can be highly corrosive. Production of gases is often intermittent (especially with batch digesters), and they are not easily stored. The high sulfur content of the material being incinerated will tend to increase SO_2 emissions from the recovery boiler and could also increase fireside corrosion problems and dust stickiness. In view of the hazards inherent in the recovery boiler, it would be prudent to incinerate them elsewhere (e.g., the lime kiln), and the recovery boiler should be used only as a last resort.

If high-concentration TRS gases are to be incinerated in a recovery boiler, extreme precautions must be taken to guard against combustible gas explosions in the recovery furnace and in the TRS collection and handling system. Appropriate flame safeguards and interlocks must be used and kept well maintained.

More commonly, TRS gases being incinerated in the recovery boiler are supplied as a low-concentration component of the combustion air. Although this is a less hazardous method of incinerating these odorous materials, it is not without problems. Introduction of a significant amount of sulfur in the air stream will tend to increase SO_2 emissions and possibly fireside corrosion problems. The combustion air supply will tend to be wetter, and this could lead to lower bed temperatures and the operating problems associated with lower bed temperatures. Corrosion problems in the combustion air system could be increased. This could include the F.D. fan, ducting, air heater, and the windbox area of the furnace.

Process wastes in black liquor A more common situation is the addition of process waste streams into the black liquor and their subsequent incineration in the boiler. Brine from tall oil acidulation and chlorine dioxide generator spent acid are normally disposed of in this manner. Oxygen stage bleach effluent and, in the effluent-free mill concept, the entire bleach plant effluent have been incorporated into the black liquor recovery cycle. All of these can influence recovery boiler performance.

The main impact of process waste streams on liquor combustion is through the following factors:

1. Addition of nonprocess elements;
2. Effect on heating value of liquor;
3. Total load of solids to be processed; and
4. Handling characteristics of the liquor.

The most serious impact of waste streams on recovery boiler performance occurs when the waste stream adds an extraneous element that is retained in the recovery cycle and builds up to relatively high concentrations. The most significant of these are chlorine and potassium, with silicon and aluminum also potential problems. Other elements tend to be naturally purged from the system and do not build up; they will not have a major effect on boiler operation unless the input is fairly massive. Chlorine and potassium will tend to be enriched in the ash and slag in the upper furnace and will lower the ash fusion temperature significantly. This will change the slagging and runoff patterns in the upper furnace and could influence plugging and corrosion. It may be necessary to decrease load to reduce gas temperatures entering the boiler bank. The NaCl produced from a chlorine input acts as a dead load chemical, and will result in higher inorganic/organic ratios and lower heating values. Mills with high chloride levels have tended to run with relatively low bed temperatures and high SO_2 emissions. Silicon and aluminum will form compounds which will create havoc around the entire recovery loop. In the furnace, they would tend to make the smelt and slag glassy and viscous.

When the inorganic content of the liquor being fired is increased, the ratio of water to combustible material entering the furnace is also increased (for the same solids content). This will tend to lower temperatures in the hearth zone and increase the difficulty in burning the liquor. High inorganic loads or lowered liquor pH could lead to problems in concentrating and handling the liquor and force operation at lower solids content, thus compounding the problem.

Outlook

Energy impacts

The continually changing world outlook on energy supply and the resultant effect on fossil fuel costs will undoubtedly have a significant impact on black liquor combustion technology. In general, one would expect to see increases in the thermal efficiency of recovery boilers, increased emphasis on by-product power generation, and greater reliance on the recovery boiler as the main steam-power source.

Reasons for relatively low thermal efficiency of recovery boilers are clearly evident in the heat loss method for carrying out recovery boiler energy balances discussed previously. The need to increase the efficiency of this potential source of steam and power is likely to lead to one or more of the following steps being taken:

1. Improved control techniques to insure complete combustion with a minimum amount of excess air;
2. Higher solids black liquor being introduced into the recovery unit boundary;
3. Additional heat traps to lower the temperature of flue gases being discharged;
4. Addition of systems to recover the heat in the smelt stream being discharged;
5. Use of end-of-pipe systems to recover latent heat in flue gas as low-level heat for process purposes; and
6. Minimization or elimination of direct steam usage in the recovery system for liquor heating, sootblowing, and smelt shattering.

The main effect of an increased emphasis on by-product power generation would be a desire to increase steam temperatures and pressures and to eliminate use of high pressure, superheated steam around the boiler when saturated low-pressure steam would suffice. The desire for higher operating pressures and temperatures is expected to increase the use of composite tubes and other metallurgies which permit high metal temperatures without increased corrosion rates.

Greater reliance on the recovery boiler as the main source of steam and power at the mill will increase the need for effective control procedures that will permit a better response of operating loads to changes in steam demand. It will increase the undesirability of forced outages and complicate the handling of emergencies.

Capital intensity

The recovery boiler is capital intense because it is both a steam (and power) generator as well as a chemical processing unit. No foreseeable changes in liquor burning technology are likely to result in greatly reduced capital requirements.

The challenge, from a capital standpoint, is to provide means for achieving incremental capacity increases on existing units for incremental capital investments. This would help reduce the need for adding recovery capacity in large increments (new units) and operating new recovery boilers well under rated capacity for a number of years until total mill production capacity catches up.

Achieving incremental capacity without a great sacrifice on thermal efficiency will require development of new operating and control strategies, some changes in design and retrofitting of old units, and possibly additional processing steps. There is intense interest in this subject, and results are likely in the near future.

Alternate technologies

A number of alternate technologies to replace the kraft recovery boiler have surfaced in recent years. Some of these are discussed briefly below.

Hydropyrolysis Hydropyrolysis is a method of treating black liquor in which the liquor is heated under pressure, and in the absence of oxygen, to a temperature sufficient to decompose the organic substances and to reduce oxidized sulfur back to the desired sulfide form (*51*). Typically hydropyrolysis reactions are carried out at temperatures of 600-650°F (310-350°C) and pressures approaching 3000 psi (20 MPa). Following hydropyrolysis, the products are separated into three process streams:

1. A combustible gas, containing a significant amount of sulfur, which is subsequently burned and the sulfur recovered by absorbing SO_2 into the feed liquor in a liquor pretreatment step.
2. A carbonaceous char substantially free of sodium, which is washed, dried, and burned in a conventional power boiler.
3. An aqueous stream containing most of the sodium and sulfur in the feed liquor as NaHS, $NaHCO_3$, and Na_2CO_3; which is causticized and returned to the pulp mill.

NSP process This is a Swedish development, and involves the use of a cyclone gasifier to process black liquor solids ahead of a power boiler or conventional recovery boiler (*52*). Pyrolysis and gasification of the organic material in the black liquor and recovery of inorganic chemicals are carried out in the cyclone gasifier. The resulting combustible gases are burned in a power boiler or recovery boiler. Concentrated black liquor (or dried black liquor solids) is introduced axially into a cylindrical furnace where it is dried and pyrolyzed by contact with a tangentially supplied oxygen-containing gas. As the products proceed axially along the furnace, gasification of carbon occurs and molten smelt accumulates on the walls where H_2S absorption can occur. Recycled flue gas is supplied tangentially in this sulfidizing zone. The molten smelt is tapped off and the combustible gases proceed onward to the power boiler or conventional recovery boiler.

Dry pyrolysis process This process involves the production of dry liquor solids in a steam-heated evaporator with multiple-effect steam economies. The dried solids are then pyrolyzed and gasified to produce oils, fuel gas, and reduced inorganic salts (*53*). Dry liquor solids have been produced on a semipilot scale using the Carver-Greenfield process. In this approach, black liquor is suspended in a carrier oil and the air/liquor mixture passes through the evaporator.

Fluidized beds Fluidized bed combustion cannot be used by itself for kraft black liquors because the combustion product is Na_2SO_4 and Na_2CO_3, and a separate high temperature reducing step would be needed. However, fluidized bed burning can be used to provide incremental recovery capacity for kraft mills with overloaded recovery boilers. A part of the black liquor is burned in a fluidized bed and the remainder in an ordinary recovery furnace. The Na_2SO_4-Na_2CO_3 pellets from the fluidized bed are introduced into the reducing zone (char bed) of the recovery furnace, where they are reduced to sulfide and smelted out. This approach has been used successfully on a commercial scale at a kraft mill, although the thermal efficiency of the system is low (*54*).

Literature cited

1. "Performance Test Procedure - Sodium Base Recovery Units," Atlanta, TAPPI, CA Report 39, June, 1972.
2. "Recovery Boiler Test Procedure - Short Form" TAPPI TIS 407-22, Issued 1979.
3. Monaghan, M. T., and Siddall, R. G., *Tappi*, 46(2):89-91(1963).
4. Beckwith, W. F., Kasbohm, D. L., and Hassler, J. C. "Analysis of Black Liquor by Thermogravimetry and Gas Chromatography," paper presented at 89th National Meeting, AIChE, Portland, OR, Aug., 1980.
5. Perry, J. H. "Chemical Engineers Handbook," 4th edition, McGraw-Hill Book Co., New York, 1963, section 9, page 2.
6. Strohbeen, D. T. "An Investigation of the Reactions Leading to the Volatilization of Inorganic Sulfur During Pyrolysis." Ph.D. Thesis, The Institute of Paper Chemistry, June, 1981.
7. Baklien, A., *Appita*, 14(1):5-17(1960).
8. Feuerstein, D. L., Thomas, J. F., and Brink, D. L., *Tappi*, 50(6):258-62(1967).
9. Douglass, I. B., and Price, L., *Tappi*, 49(8):335-42(1966).
10. Merriam, R. L., Richardson, D. L., Grace, T. M., and Taylor, M. L., *AIChE Symposium Series*, 200(76):99-112(1980).
11. Stelman, D., Darnell, A. J., Christie, J. R., and Yosim, S. J., "Air Oxidation of Graphite in Molten Salts," in "Molten Salts," The Electrochemical Society, Inc. 1976, p. 299-314.
12. Creighton, D. M., *Pulp & Paper*, 45(4):61-3(1971).

13. Grace, T. M., "Reduction in Kraft Recovery Boilers" presented at Black Liquor Recovery Boiler Advisory Committee Meeting in Atlanta, GA, Oct., 1979.
14. Richardson, D. L., and Merriam, R. L., "A Study of Black Liquor Recovery Furnace Firing Conditions, Char Bed Characteristics and Performance," Phase II, Report prepared for American Paper Institute, Arthur D. Little, Inc., Cambridge, Mass., Dec., 1978.
15. Cameron, J. H., and Grace, T. M., "A kinetic study of sulfate reduction with carbon," presented at the International Conference on Recovery of Pulping Chemicals, Vancouver, B.C., Sept. 22-25, 1981.
16. Bjorkman, A., "Swedish New Soda-Recovery Process Project" presented at Forum on Kraft Recovery Alternatives, The Institute of Paper Chemistry, Appleton, Wisconsin, April, 1976, p. 205-18.
17. Warnqvist, B., *Svensk Papperstidning*, 76(12):463-6(1973).
18. Borg, A., Teder, A., and Warnqvist, B., *Tappi*, 57(1):126-9(1974).
19. Stelling, O., and Vegeby, A., *Pulp & Paper Mag. Can.*, 70(8):51-77(1969).
20. Plumley, A. L., and Roczniak, W. R., *Tappi*, 58(9):118-21(1975).
21. Dickinson, J. A., Murphy, M. E., and Wolfe, W.-C., "Kraft recovery boiler furnace corrosion protection," presented at 1981 TAPPI Engineering Conference, San Francisco, California, Sept., 1981. pp. 607-12.
22. Tegman, R., and Warnqvist, B., *Acta Chem. Scand.*, 26(1):413-14(1972).
23. Shivgulam, N., Barham, D., and Rapson, W. H., *Pulp & Paper Mag. Can.*, 80(9): 89-92 (T282-5) (1979).
24. Krause, H. H., Simon, R., and Levy, A., "Smelt - Water Explosions," Final Report to Fourdrinier Kraft Board Institute, Battelle, Columbus, Ohio, Jan., 1973.
25. Chamberlain, R. E., "An Approach to Establishing and Maintaining Safe Kraft Recovery Furnace Firing Rates," presented at 66th Annual Meeting of the Technical Section, CPPA, Montreal, Que., Jan. 1980.
26. Roos, T., "Design and Operation Experience of Reductive Recovery Boilers," Proceedings of the Symposium on Recovery of Pulping Chemicals, Helsinki, Finland, 1968, p. 473-89.
27. Erickson, S. L., *Tappi*, 63(2):55-6(1980).
28. Freyaldenhoven, R. J., Peacos, M. M., and Henderson, J. S., "The NCE recovery precipitator wet bottom design and operating experience," presented at 1981 TAPPI Engineering Conference, San Francisco, Calif., Sept., 1981, p. 393-414.
29. Henderson, J. S., *Tappi*, 63(12):71-4(1980).
30. Balasic, P. J., *Pulp & Paper*, 53(10):154-6(1979).

31. Henderson, J. S., *Tappi*, 64(1):55-8(1981).
32. Schabas, W., *Pulp & Paper Mag. Can.*, 80(10):51-5(1979).
33. "Recommended Good Practice - Safe Firing of Auxiliary Fuel in Black Liquor Recovery Boilers," published by The Black Liquor Recovery Boiler Advisory Committee, revised April, 1981.
34. Grace, T. M. and Larsen, R. N., *Tappi*, 58(5):117-20(1975).
35. Smith, V. N., and Typpo, P., *Paper Trade J.*, 163(5):38-46(Jan. 30, 1979) and 163(7):28-39(Feb. 15, 1979).
36. Thoen, G. N., and De Haas, G. G., *Tappi*, 60(2):104-6(1977).
37. Richardson, D. L., and Merriam, R. L., "Study of Cooling and Smelt Solidification in Black Liquor Recovery Boilers," Phase I, Report for American Paper Institute by Arthur D. Little, Inc., Cambridge, Mass., Feb., 1977.
38. Carter, T. M., and Mathieson, R., *Pulp & Paper Mag. Can.*, 82(2):T55-9(1981).
39. Chamberlain, R. E., Smith, A. R., McFarlin, G., and Chose, D. W., *Pulp & Paper Mag. Can.*, 75(12):79-84(T408-13)(1974).
40. Chamberlain, R. E., Lofkrantz, J. E., Norres, R. G., Smith, A. R., and Wostradowski, R. A., *Pulp & Paper Mag. Can.*, 79(2):57-61(T44-8)(1978).
41. Pantsar, O., *Pulp & Paper*, **54**(8):142-5 (1979).
42. Black Liquor Recovery Boiler Advisory Committee, Minutes of Spring Meeting. Atlanta, TAPPI, 1980.
43. American Paper Institute, "Recovery Boiler Operator Training Program," - Lesson 14, Operating Problems, 1980.
44. "Recommended Good Practice - Fire Protection in Direct Contact Evaporators and Associated Equipment," Black Liquor Recovery Boiler Advisory Committee, Nov., 1974.
45. "The Recovery Boiler Reference Manual" prepared for American Paper Institute by J. H. Jansen Co., Bellevue, Washington, Oct., 1979.
46. Brauns, O., *Svensk Papperstidning*, 82(1):5-8(1979).
47. Durrant, O., W., and Lansing, E. G., "Furnace Implosions and Explosions presented to American Power Conference, Chicago, Illinois, April 20-22, 1976.
48. Bhatia, S. P., Marentelle, L. P., De Souza, T. L. C., Barclay, H., Wong, A., and Prahacs, S., *Pulp & Paper Mag. Can.*, 76(Convention issue):88-93(T98-103) (1975).
49. Lang, C. J., De Haas, G. G., Gommi, J. V., and Nelson, W., *Tappi*, 56(6):115-19(1973).
50. Bosch, J. C., Pilat, M. J., and Hrutfiord, B. F., *Tappi*, 54(1):1871-5(1971).

51. Watkins, J. J., and Adams, W. S., "The St. Regis Hydropyrolysis Process as an Approach to Energy Conservation in the Pulp and Paper Industry," presented at 1978 Tappi Engineering Conference, San Francisco, CA, Sept., 1978.
52. Magnusson, H., and Warnqvist, B., "The NSP Project: An Alternative to the Conventional Recovery Furnace," presented at 72nd Annual Meeting of AIChE, San Francisco, CA, Nov., 1979.
53. De Haas, G. G., Hurley, P. J., and Root, D. F., "Dry Pyrolysis Approach to Chemical Recovery," presented at Forum on Kraft Recovery Alternatives, The Institute of Paper Chemistry, Appleton, Wisconsin, April, 1976. p. 103-23.
54. Tomlinson, G. H., Behie, L. A., and Cook, W. H., "Incremental Kraft Recovery with a Fluidized Bed System" presented at Forum on Kraft Recovery Alternatives, Appleton, WI, The Institute of Paper Chemistry, April, 1976, p. 124-46.

5
Cooking Liquor Preparation
Causticizing of sodium carbonate in green liquor

N. K. Mehra, C. F. Cornell, G. W. Hough

Introduction

The recausticizing process, though somewhat neglected in the industry, has an important impact on the operations of the entire pulping operation. The foremost is the ability to produce the required amount of white liquor or cooking liquor at the specified strength and quality for the digester. The amount of active alkali is fixed in relation to the weight of the wood and the production of pulp; therefore, any shortcoming in the production rate of active alkali adversely affects the pulp mill production.

In the recausticizing plant, the green liquor from the recovery furnace dissolving tank is clarified to remove insoluble impurities, and then causticized to convert the sodium carbonate to sodium hydroxide by the addition of lime. The calcium carbonate or lime mud is separated from the causticized liquor and the clear liquor, called white liquor, is sent to the digesters for the cooking cycle.

The recausticizing and lime mud reburning areas have to be considered as integrated operations, as the reburned lime and its characteristics determine whether the mill can successfully causticize the green liquor and produce the white liquor to mill specifications. Likewise, the quality of lime mud produced during the recausticizing operation determines the calcining behavior of the lime mud reburning equipment, such as a rotary lime kiln or a fluidized bed calciner.

Chemistry

In the causticizing reaction, lime (CaO) reacts with sodium carbonate (Na_2CO_3) in the green liquor to produce sodium hydroxide ($NaOH$) and calcium carbonate ($CaCO_3$), called lime mud. The causticizing reaction may be written as:

$$Na_2CO_3 + CaO + H_2O = 2\ NaOH + CaCO_3$$

Because of the low solubility of $CaCO_3$, it is precipitated from the solution.

The causticizing of Na_2CO_3 to NaOH does not proceed to completion due to the common ion effect of the hydroxyl ions. Since both sodium carbonate and sodium hydroxide are very soluble, the reaction proceeds to the right depending upon the relative solubilities of calcium carbonate and calcium hydroxide in the solution. While the concentration of sodium hydroxide is increasing during the causticization reaction, the common ion effect of the hydroxyl ions decreases the solubility of calcium hydroxide, until it is no more soluble than the calcium carbonate, when the following equilibrium is established.

$$Na_2CO_3(aq) + Ca(OH)_2\ (s) = 2\ NaOH(aq) + CaCO_3(s),\ H = -2.1\ Kcal$$

The equilibrium constant, K, for the reaction is expressed in terms of the activities, a, of the reactants and product as

$$K = \frac{a^2\ NaOH(aq) \times a\ CaCO_3(s)}{a\ Na_2CO_3(aq) \times a\ Ca(OH)_2(s).}$$

The activities of the solids in a pure state have the value of unity. For a very dilute solution, the activities of aqueous sodium hydroxide and sodium carbonate may be taken to be equal to their respective concentrations. Assuming complete ionization, these concentrations of the hydroxyl ions and carbonate ions are the same as the stoichiometric concentrations of sodium hydroxide and carbonate, respectively. For such a simplified system, the equilibrium constant becomes

$$K = \frac{[2OH-]}{[CO_3=]}$$

where the brackets indicate the concentrations.

The causticizing reaction proceeds by virtue of the smaller solubility of calcium carbonate as compared with calcium hydroxide. When the system is at equilibrium, each of these two components must obey the law of constant solubility product. Again, for dilute solution,

$$K_1 = [Ca++][2OH-],$$

and

$$K_2 = [Ca++][CO_3=].$$

Hence,

$$K = \frac{K_1}{K_2}.$$

At this point we attempt to make an estimate of the value of K at the usual causticizing temperature of about 212°F (100°C). The solubility of calcium hydroxide in water is 10^{-2}, while that of calcium carbonate in water is 2×10^{-4}, both in g-moles/L. The solubility products, K_1 and K_2, are therefore 4×10^6 and 4×10^8, respectively. The value of K is seen to be of the order of 10^2. The values of the equilibrium constant, as reported by Kobe and Wilkinson (*1*), based on the equilibrium concentrations of sodium hydroxide and carbonate in their causticizing experiments at 207°F (97°C), varied from 47 to 151 for solutions containing 0.4-4.8 equivalents of sodium ions per liter. Their results also showed the trend of decreasing values of the apparent K with increasing sodium concentrations. The true equilibrium constant based on activities, however, should be independent of concentration.

The effect of temperature on the equilibrium constant follows the Van't Hoff equation:

$$\frac{d \ln K}{dT} = \frac{H}{RT^2}$$

where H is the heat of reaction at the absolute temperature T, and R is the gas law constant. In the causticizing reaction, the heat of reaction is nearly zero. The temperature has, therefore, only slight influence on the equilibrium. Its effect on the solubilities of the solid phases would be such that the ratio of the solubility products, K_1 and K_2, remain practically constant. The solubility data of calcium hydroxide and carbonate in water from 32 to 212°F (0 to 100°C) indicate the value of K for a dilute solution to be between 100 and 140, the average being 126.

The yield in a causticizing process is usually expressed by percent conversion defined as follows:

$$\% \text{ conversion} = \frac{\text{NaOH, equivalents in the final solution}}{\text{Na}_2\text{CO}_3\text{, equivalents in the initial solution}} \times 100$$

The extent of conversion is related to the equilibrium constant. For an ideal solution the relation can be derived as:

$$\frac{1 - C}{C^2} = \frac{4M}{K}$$

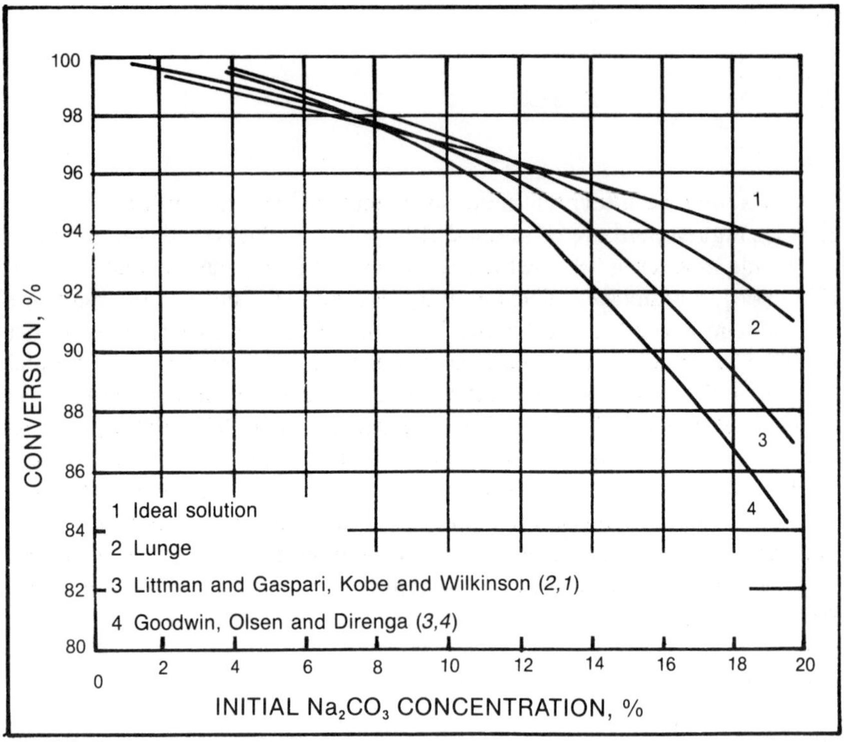

Fig. 5.1. Causticization equilibrium.

where C is the fractional conversion and M the molarity of the initial sodium carbonate solution. Taking K = 126, the conversion may be estimated to be 99.7% for 0.1 M solution, 97.0% for 1.0 M solution, and 94.4% for 2.0 M solution. The experimental data of conversion are shown in Figure 5.1 as a function of percent Na_2CO_3 in original solution.

The agreement between the prediction and the data at low concentrations is reasonable. As the concentration increases, the actual conversion drops much faster than predicted on the basis of the ideal behavior. Among the experimental data, those of Littman and Gaspari (*2*) and Kobe and Wilkinson (*1*) are believed to be more reliable than others (*3,4*).

A green liquor contains, in addition to sodium carbonate, sodium sulfide as a major component, along with other sodium compounds as impurities.

When sodium sulfide is present in a causticizing solution, at least two reactions are possible, causticization and hydrolysis:

$$Na_2S + Ca(OH)_2 = 2NaOH + CaS,$$

and

$$Na_2S + H_2O = NaOH + NaHS$$

Since calcium sulfide has a molar solubility about six times that of calcium hydroxide, causticization of sodium sulfide is unlikely unless the sulfide ion concentration is high. Sodium sulfide, being much more basic than sodium carbonate, hydrolyzes to a considerable extent. The hydroxyl ions so produced would tend to suppress the causticization of sodium carbonate. According to Kobe and Wilkinson (*1*), for concentrations usually encountered in kraft green liquors, the presence of sodium sulfide will lower the conversion by 1% to 2%. The same authors also showed that sodium sulfite is causticized to a large extent (the solubility product of calcium sulfite being of the order of 10^9), while sodium thiosulfate is neither causticized nor hydrolyzed.

Figure 5.2 shows the causticization data for pure sodium carbonate and for commercial green liquors.

Basic process

While the chemistry of the causticizing reaction is simple, the continuous nature of the process involves a number of unit operations: filtration, classification, mixing, sedimentation, calcining, and material handling.

The process may be thought of as a closed circuit operation involving the solids as one cycle and the liquor as another. The two are brought together to produce the cooking liquor, and each subsequently is recycled and reconverted to the chemical form (quick lime and sodium carbonate) necessary to achieve the causticizing to cooking liquor. The manner in which these two cycles function in the recausticizing operation is shown graphically in Figure 5.3.

Liquor circuit

The liquor circuit begins with the green liquor (sodium carbonate plus sodium sulfide), and is completed when white liquor (sodium hydroxide plus sodium sulfide) is prepared for cooking the wood chips. Smelt, primarily a $Na_2CO_3 + Na_2S$ mixture from the recovery furnace, is dissolved in weak liquor from causticizing to form the green liquor, which is reacted with lime from the calciner to produce the sodium hydroxide component of the white liquor.

The close-up of the pulp mill to reduce effluent loadings has increased the quantity of undesirable elements in the liquor cycle. These elements are introduced with the wood, water, and makeup chemicals and tend to build up in the system as the mill is closed up and purge streams are reduced. Some of these elements are potassium, chloride, iron, aluminum, silicon, magnesium, manganese, and calcium. Keitaanniemi and Virkola describe the behavior and detrimental effects of some of these elements in their paper "Undesirable Elements in the Causticizing System"(*5*).

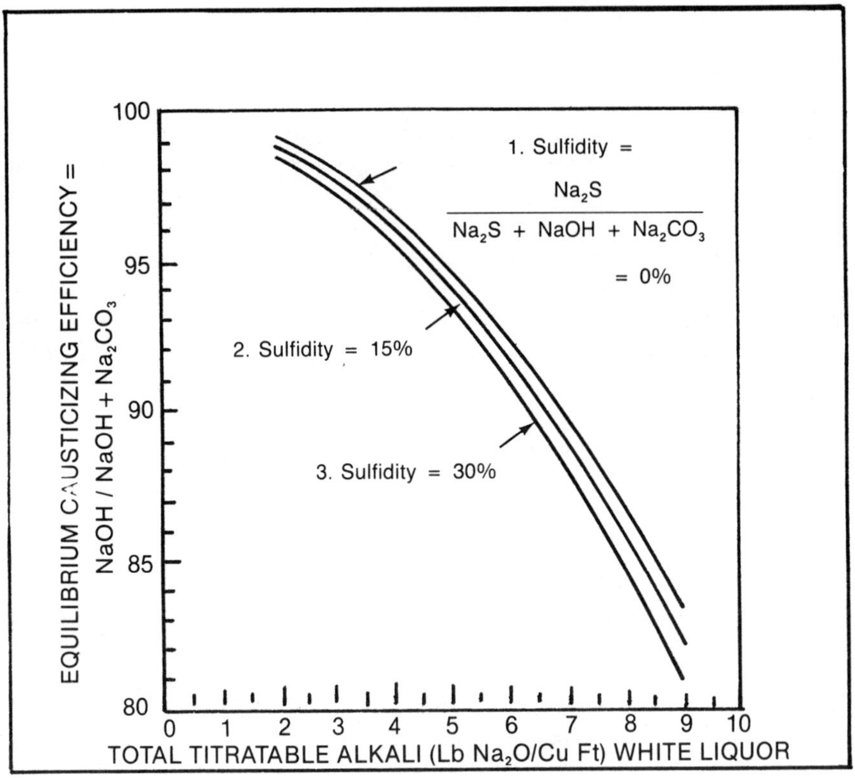

	Sulfidity		
T.T.A.	*0% curve 1*	*15% curve 2*	*30% curve 3*
2.0	99.1	98.9	98.6
3.0	98.1	97.7	97.4
4.0	96.6	96.1	95.6
5.0	94.8	94.2	93.6
6.0	92.5	91.8	91.0
6.5	91.3	90.5	89.7
7.0	89.9	89.0	88.2
7.5	88.4	87.5	86.6
8.0	86.9	85.9	84.9
8.5	85.2	84.2	83.1
9.0	83.5	82.4	81.2

T.T.A. = total titratable alkalinity.

Fig. 5.2. Equilibrium causticizing curves.

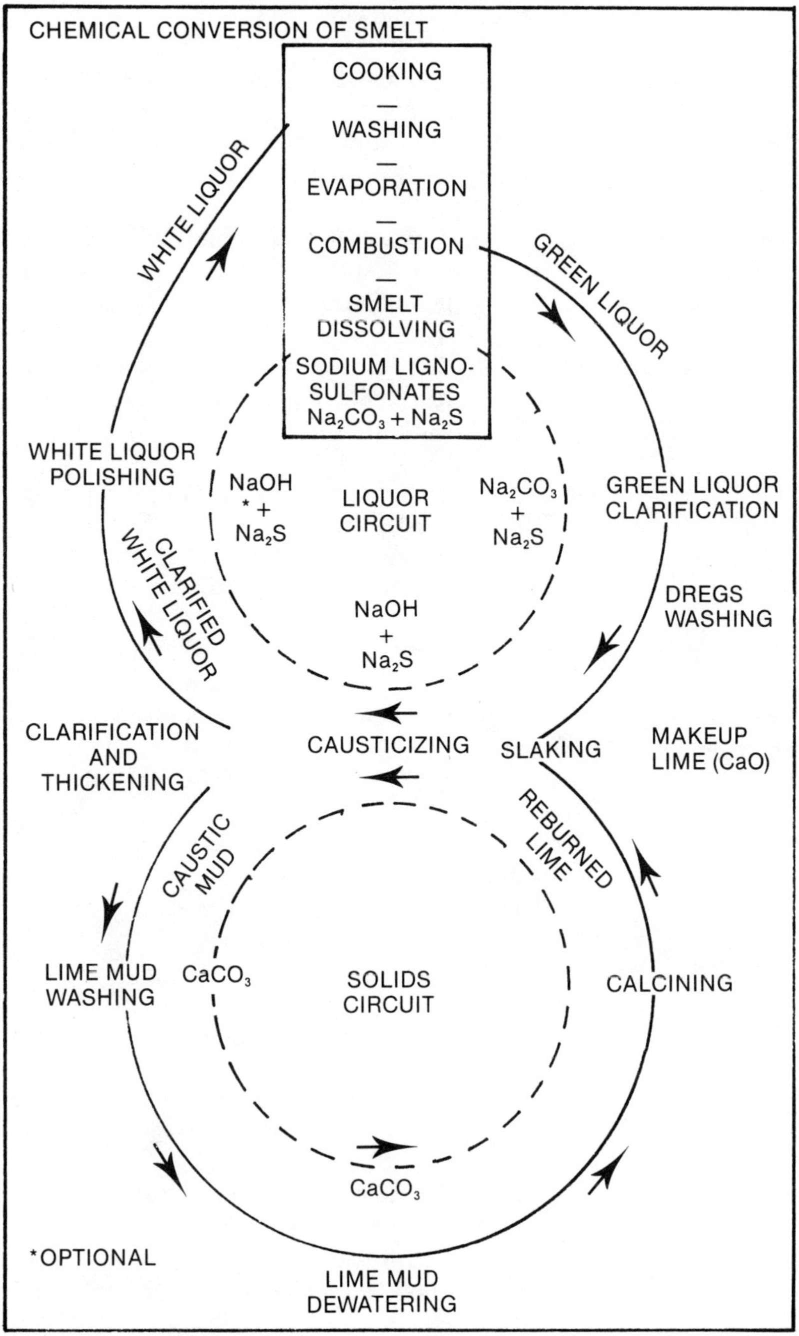

Fig. 5.3. Causticizing process cycle.

The liquor circuit comprises the following steps:

Green liquor clarification	To remove suspended solids (called dregs).
Dregs washing	To remove the residual soda from the settled solids before discharge to waste.
Slaking and causticizing	To slake the quicklime and react it with carbonate in the green liquor for white liquor production.
White liquor clarification	To remove the suspended solids ($CaCO_3$) to provide a cooking liquor of adequate clarity (50-100 ppm).
White liquor polishing (optional)	To remove most of the remaining suspended solids when desired (e.g., for dissolving pulp).

Solids circuit

The solids circuit is comprised of lime and lime mud (calcium oxide and calcium carbonate), although there are solid impurities introduced with the green liquor that must be continuously removed to prevent excessive buildup of inerts in the solids circuit.

Lime added to the green liquor for causticizing results in the formation of calcium carbonate (lime mud). This in turn is washed to remove caustic soda (usually to less than 1%), and, after dewatering, is calcined to produce lime again for reuse in causticizing.

Grit, which is inert material and unreacted lime (65 mesh and larger), is removed after the slaking operation by classification.

The solids cycle comprises the following:

Slaking	To convert quicklime to hydrate and initiate the conversion of sodium carbonate to sodium hydroxide, obtaining calcium carbonate in the process.
Classification	To remove coarse inerts (grit).
Causticizing	To provide sufficient retention time to allow normal degree of completion of reaction.

Thickening	To concentrate the suspended solids (calcium carbonate and inerts).
Washing	To recover the soda for return to the liquor cycle and to reduce the level low enough for calcining.
Dewatering	To concentrate for kiln feed.
Calcining	To produce quicklime from the calcium carbonate.

The unit operations in the combined liquor and solids cycle are carried out in the manner as illustrated in Figure 5.4.

Losses

Lime makeup is required to provide for a small and continuous loss of lime from the system. High grade lime or limestone is used as makeup. Typical composition of purchased lime is given in Table 5.1 and the composition of limestone is given in Table 5.2.

Chemical and physical terminology

The active alkali contained in the pulping liquor (white liquor) is sodium hydroxide and sodium sulfide. The ratio of sodium sulfide to sodium hydroxide varies from mill to mill and also cycles within a single mill.

Table 5.1. Typical chemical analysis — pebble lime

SiO_2 and insolubles	1.20%
Fe_2O_3	0.15%
Al_2O_3	0.17%
CaO (total)	94.4%
MgO (total)	1.20%
Loss on ignition	2.20%
S (sulfur)	0.025%
CaO (available)	92.40%
Phosphorus (P)	0.008-0.010%
Typical Physical Analysis	
Temperature rise in 30 s	28.6°C
Total active slaking time	4.5 min
Total temperature rise	44°C
Slaking residue (on 100 mesh)	1.8%

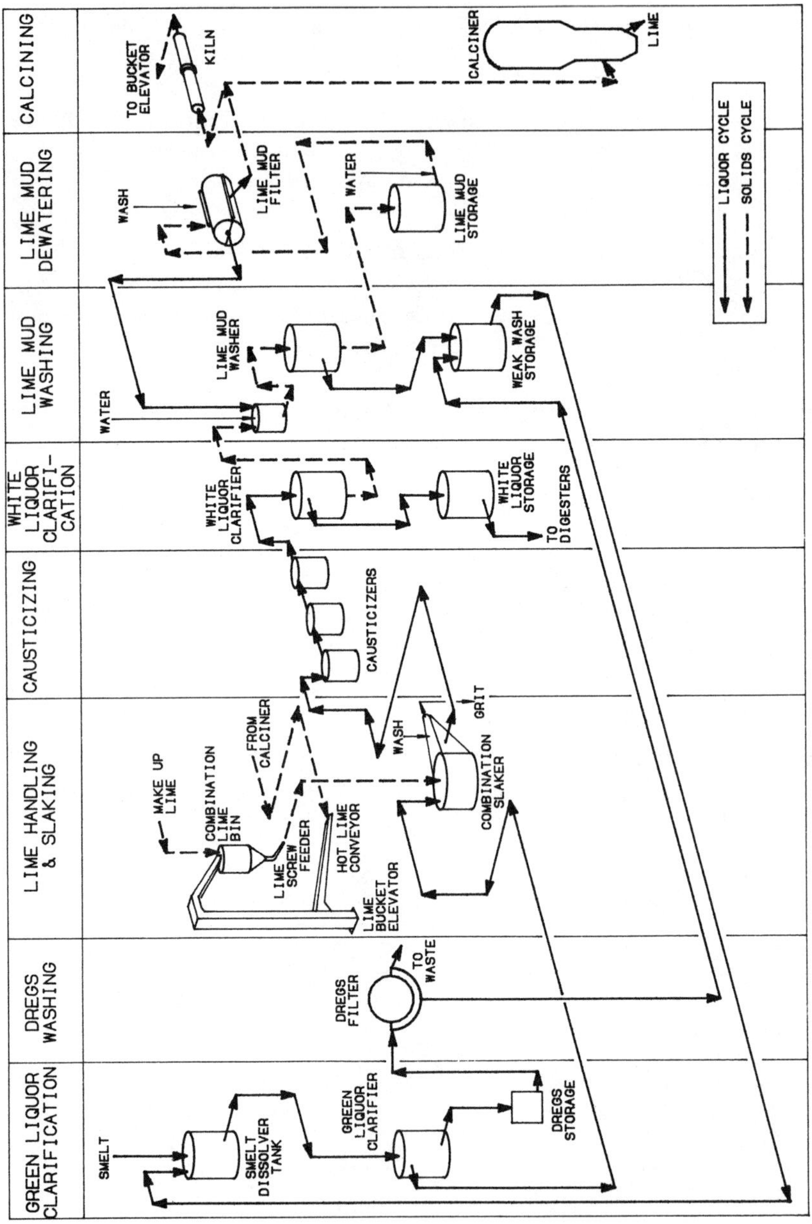

Fig. 5.4. Unit operations flowsheet — conventional kraft mill recausticizing system.

Table 5.2. Typical analysis of high calcium limestone

Moisture — hydroscopic or as received	0.2%
Loss on ignition	43.5%
Calcium oxide CaO	54.0%
Magnesium oxide MgO	1.0%
Silica SiO_2	0.5%
Iron oxide Fe_2O_3	0.1%
Alumina Al_2O_3	0.5%
Miscellaneous & unaccounted for	0.2%
TOTAL	100.0%
Calcium carbonate	96.76%
Magnesium carbonate	1.26%
Insoluble siliceous matter	1.25%
Miscellaneous inerts	0.73%
TOTAL	100.00%

The active cooking chemicals are determined by the types of wood, the degree of cooking and the Kappa number desired, and, to a lesser extent, the manner of operation of the recausticizing and the chemical recovery areas. It is general practice to express concentration in terms of equivalent Na_2O, since sodium is the common element.

The terms used to characterize the cooking liquor are given below. TAPPI useful terms are defined in T 1203 os-61 with all values based on equivalent Na_2O.

Total alkali	$NaOH + Na_2CO_3 + Na_2S + Na_2SO_3$
Active alkali	$NaOH + Na_2S$
Activity	$\frac{\text{Active alkali}}{\text{Total alkali}}$
Effective alkali	$NaOH + \frac{1}{2} Na_2S$
Causticizing efficiency	$\frac{NaOH \text{ (less NaOH in green liquor)}}{NaOH - NaOH \text{ in green liquor} + Na_2CO_3}$
Sulfidity	$\frac{Na_2S}{NaOH + Na_2S}$
Total chemical	All sodium salts

Reduction (in green liquor) = $Na_2S/(Na_2S + Na_2SO_4 +$ Other Soda Sulfur Comp) as Na_2O

Other terms which are are in common use are as follows:

Total titratable alkali (TTA)	$NaOH + Na_2S + Na_2CO_3 + \frac{1}{2} Na_2SO_3$
Sulfidity on total alkali	$\frac{Na_2S}{TTA}$
Causticity	$\frac{NaOH}{TTA}$

The chemical character of the green and white liquors is determined by titrations described in TAPPI Standard Method T 624 os-68, "Analyses of Soda and Sulfate White and Green Liquors." The series of titrations are known as the A, B, and C tests and are also described in "Pulp and Paper Manufacture" Vol. 1, pgs. 563-566. From these three titrations, the amount of the principal chemicals present are readily calculated as follows:

Titration:
Test A = $NaOH + \frac{1}{2}Na_2S$
Test B = $NaOH + Na_2S$
Test C = $NaOH + Na_2S + Na_2CO_3$

Chemical:
Na_2CO_3 = C - B
Na_2S = 2 (B-A)
NaOH = B - 2 (B - A)
All values expressed as Na_2O

Process data

The active alkali requirement for cooking is dependent on the types of pulp produced, type of wood used, and type of digester. The values for active alkali charged, lime required, and lime mud produced are given in Table 5.3 for various grades of kraft pulp. The white liquor characteristics used for the calculations are given as follows:

	Grams per Liter	*Pounds per cf*
TTA @ Na_2O	120	7.5
AA @ Na_2O	100	6.2
Sulfidity on AA	25%	25%

Table 5.3. Comparison of active alkali, lime and lime mud for various grades of kraft pulp

Item	*Unit*	*Linerboard*	*Sack paper*	*Bleachable grades*
Pulp yield	%	52-56	48-50	45 ±
Active alkali on dry wood as Na_2O	%	14-15	15-16	17-19
Active alkali as Na_2O	kg/ODt	250-290	290-330	400 ±
White liquor required	m^3/ODt	2.5-2.9	2.9-3.3	4.0 ±
Lime required at 85% available	kg/ODt	200-230	230-260	320 ±
Lime mud produced	kg/ODt	330-385	385-435	525 ±
Green liquor flow (to clarifier)	m^3/ODt	3.1-3.6	3.6-4.1	4.95 ±

kg = kilogram
m^3 = cubic meter
ODt = oven dry metric ton

The amount of lime required for causticizing is calculated from the active alkali charged to the digester. Thus, makeup lime is added to the calculated amount to allow for losses in the recausticizing/lime reburning operations. The total lime, including makeup is used to calculate lime mud quantities. Water for lime mud washing is, in part, scrubber effluent from the kiln vent stack, which contains lime dust recovered from the kiln flue gas. This amounts to about 10% of the kiln feed, and must be added to the lime mud calculated from lime to determine the load to the lime mud washer and filtration equipment.

Recausticizing plant equipment

General

The recausticizing and lime mud reburning systems in operation today have undergone considerable changes in recent years. The basic flowsheet of Figure 5.5 is still in use, though modifications have been introduced to reduce the emissions and the waste discharges from the system to meet the environmental regulations. Also, there is increased emphasis to produce cleaner white liquor, reduce soda losses, improve fuel economy, and to operate a closed system which is reliable and consistant in operation.

To meet these goals, some equipment has been obsoleted, modified, or improved equipment substituted. Filtration of dregs on a precoat type of dregs filter has been found to produce a dry disposable cake with improved soda recovery efficiency. An oversized lime mud precoat filter has also been installed in some mills to oxidize the reduced sulfur compounds to reduce

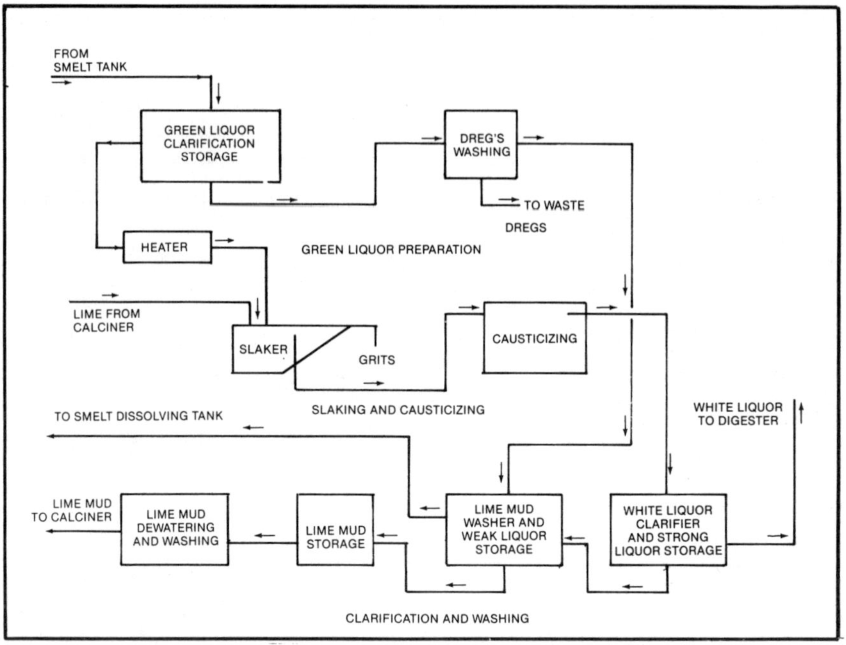

Fig. 5.5. Causticizing flowsheet.

total reduced sulfide (TRS) emissions. Two-stage lime mud washing equipment may also be required if TRS emission limits are reduced to the level first proposed by the Environmental Protection Agency (EPA). Pressure filtration of white liquor achieves superior clarity with attendant low calcium carry-over to the digester. This unit has also been applied as a mud washer to reduce space requirements in housed systems. Many mills have encountered corrosion problems with the carbon steel equipment, and stainless steel has been used for rebuilding or replacing the failed equipment. For reburning of lime mud, fluidized bed type calciners are finding greater industry acceptance in view of its improved fuel economy and high quality reburned product. Computer based control systems are in commercial use for the preparation of white liquor and for the control of the lime kiln.

Green liquor clarification

Impurities

Prior to recausticizing, the green liquor is clarified to remove the bulk of the suspended solids, referred to as dregs. These impurities are continuously introduced with the makeup salt cake, new lime, from the wood chips, from deterioration in the furnace and kiln lining, and from corrosion in the

Table 5.4. Chemical analysis of green liquor dregs

Analysis	*Mill 'A'* *% by wt*	*Mill 'B'* *% by wt*
Ignition loss	56.0	76.0
CaO	16.5	9.0
R_2O_3	4.7	---
Fe	2.2	1.0
Na_2O	2.2	1.0
Total S	3.0	3.0
Acid insolubles	0.32	1.0

digesters, evaporators, and piping. The light, fluffy black material is composed of carbon, iron, silica, calcium, alumina, magnesium, and some sulfides, and is difficult to settle. Analysis of dregs from two different mills is shown in Table 5.4.

Green liquor generally contains from 800 to 1200 ppm of dregs. Recovery furnace improvements through the years, have reduced dregs quantities from over 11 lbs (5 kg) per ton of pulp to as low as 7 lbs (3 kg) per ton. At times, the amount of dregs in green liquor may be found in excess of 2000-3000 ppm, indicating buildup of dregs in the entire recovery system, or carry-over of lime mud solids in the weak liquor. Care should be taken to ensure that sufficient discharge of the dregs from the green liquor clarifier is obtained to avoid buildup. Keitaanniemi and Virkola discussed the undesirable elements in causticizing systems (*5*).

Clarification of green liquor is obtained by settling in a sedimentation type of clarifier. The feedwell dissipates the inlet kinetic energy of the feed stream and also helps in floc formation. An overflow or rise rate of 2.0 ft/h (0.61m/h) has been found to produce an overflow clarity of 100 ppm suspended solids in the clarified green liquor. A detention time of at least two hours is also provided for clarification. Consequently, the sedimentation type clarifiers have a clarification and thickening zone 15 to 18 ft (4.5-5.5m) deep.

The dregs generally settle to about 8-10% solids in the bottom of the clarifier. A rake mechanism moves the solids towards the discharge outlet. Carry-over lime mud in the weak wash introduces calcium carbonate to the green liquor clarifier, and the solids at the bottom of the clarifier may thicken to a higher percent solids. To prevent the buildup of solids in the clarifier, the rate of withdrawal of the dregs is proportioned to the rate of feed to the clarifier.

Centrifugal separators have been tried for green liquor clarification, however, the fine particle size does not lend itself to easy centrifugal separation; as a result, they have not been successful. Direct filtration using a precoat has been tested. Costs have been too high to warrant using this technique to date.

Factors affecting clarification

The suspended solids in the clarified green liquor may be considerably more than 100 ppm and turbidity levels of 500-600 ppm are not uncommon due to incomplete clarification of the green liquor. Factors affecting clarification are as follows:

1. Mechanically, the feed and the overflow connections to the clarifier should conform to the manufacturer's drawings.
2. Speed of the rake arms could affect the operation of the clarifier. Green liquor clarifier rakes rotate slower than other clarifier/thickeners.
3. In an effort to reduce soda losses, the dregs are normally thickened to the maximum achievable percent solids, sometimes at the expense of a large dregs inventory in the clarification equipment. A backup of solids could result in the overflow of dirty liquor from the clarifier.
4. Poor clarification could also result if flocculation of the dregs is disturbed by the use of contaminated condensate in the mud washing circuit. The organic constitutents, such as lignin and tannin, act as a dispersing agent and hinder the natural flocculation and settling of the dregs.
5. To a lesser extent air in the feed steam could also disturb the settling process.

The settling rate of dregs can be increased by use of suitable polymers and the clarification thus improved, particularly in overloaded situations. If the polymer selection and application is handled properly, the overflow rate from the clarifier may be increased to 3.0 to 4.0 ft/h (0.9 to 1.2 m/h). The polymers, however, add to the operating cost, and can contribute to other problems such as poorer filtration of dregs and lime mud.

From time to time, some mills have tried to eliminate green liquor clarification. However, this has generally not proved practical due to the buildup of impurities. One southern mill found that after recycling dregs over extended periods, their lime availability had decreased from the normal levels of 80-85% to as low as 50-60%. In special circumstances where considerable lime mud or lime is diverted to another use, the green liquor dregs cannot build up and clarification may be omitted. Even so, this may not always be the best practice. If the causticizer effluent is to be filtered, the fine dregs can reduce filtration rates and hasten cloth blinding. Also, the clarity of white liquor from sedimentation units will deteriorate.

Iron, silica, and alumina are all slightly soluble in strong hot caustic solutions. Some, therefore, remain in the final cooking liquor and tend to precipitate out when the alkalinity is reduced in the digesters, resulting in digester heater and screen scaling problems. Impurities carried with the black liquor to the evaporators accelerate scale formation.

Reburned lime containing dregs is often unsuitable for calcium hypochlorite bleach liquor preparation, especially if manganese is present. In this case, a mill would require a special lime handling system for bleach liquor preparation.

Burning of lime containing dregs has been found in many cases to reduce dusting and ring or ball formation, particularly with overloaded kilns. The benefits are usually conceded to be outweighed by the additional fuel required to dry and burn a higher circulating load of inert material. From an energy standpoint, the removal of dregs from the system is important, as otherwise, the cake moisture from the lime mud filter is increased considerably, resulting in higher fuel requirements for the lime kiln. The dregs tend to blind the filter medium, resulting in poorly washed and dewatered cake. The capacity of the filter is also seriously affected, resulting in increased frequency for acid cleaning with a consequent loss in the lime kiln production rate.

Clarifiers

Until 1965, multicompartment clarifiers were in common use; however, these machines have been replaced by the unit type design. The unit clarifier with storage is easier to operate, less costly, and can be operated at about 150% of the overflow rate of the earlier shallow tray design machines. A typical unit type clarifier with storage is illustrated in Figure 5.6. The feed slurry is piped directly to a submerged feedwell located at the center of the unit. The dregs settle to the bottom of the clarifier and are raked to the center discharge well by revolving arms mounted on a central vertical shaft. The settled dregs are pumped continuously from the discharge well by a diaphragm pump. The speed and stroke of the pump can be adjusted to obtain any required capacity. Sometimes the settled dregs are withdrawn intermittently by operating a centrifugal pump at preset intervals; however, the continuous withdrawal of the underflow ensures a smoother operation and improves the dregs washing operation as well.

The standard unit clarifier with storage should provide at least 12 hours of green liquor storage. The bottom 14 to 18 ft (4.5 to 5.5 m) of the tank height is used for clarification, and the balance is used for storage. The clarified liquor rises from the clarification section into the storage section and the clarified liquor is drawn off as required through the liquor outlet. Several different arrangements are available for the withdrawal of the clarified

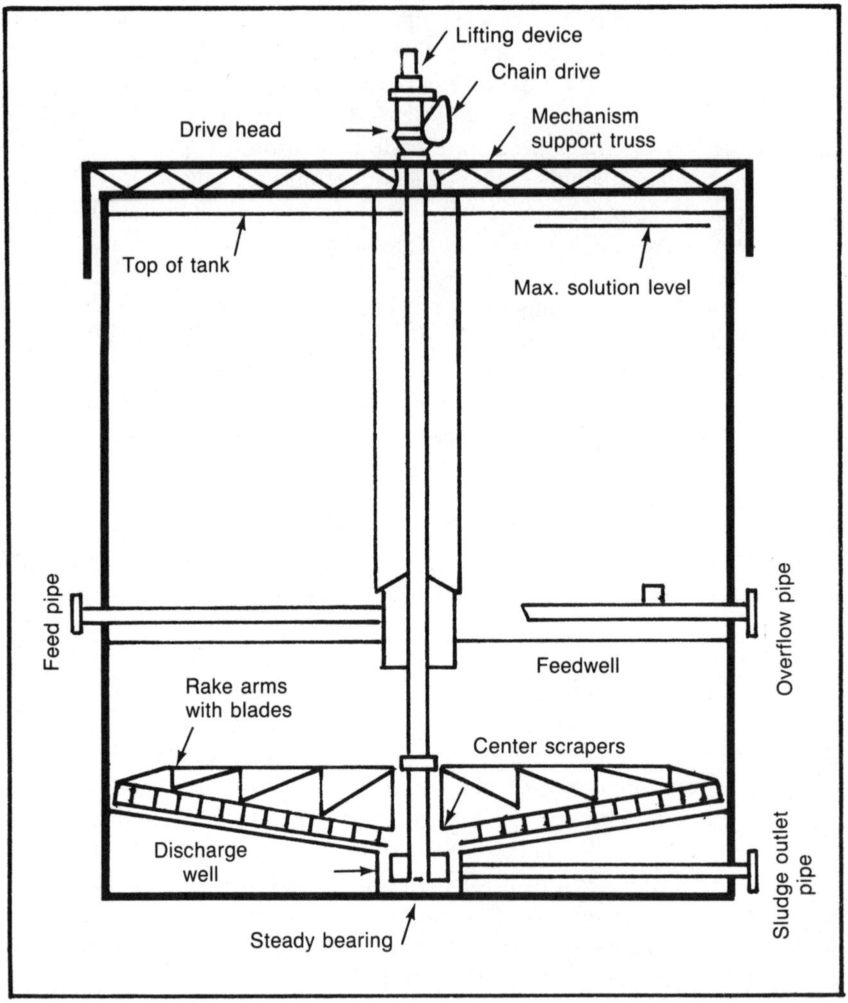

Fig. 5.6. Green liquor clarifier.

liquor, the most common being the internal bustle pipes with several overflow nozzles. Another arrangement includes a floating suction with telescopic pipe; the floating suction rises or falls with the level of the clarified liquor in the tank. Some clarifiers have a single outlet on the side of the tank or several nozzles oriented one above the other. This arrangement, however, is not considered efficient, as short circuiting can occur with entrainment of fine solids in the overflow.

A unit-type clarifier without storage is also used; however, green liquor storage is required, and less capital cost is required to include the storage as the upper part of the clarifier.

The foundation for the clarifier tank is normally a flat slab or a footing around the perimeter of the tank with compacted fill under the tank. In some cases, a tunnel is provided for access to the underflow lines. Underflow line problems can be minimized by providing flush-out lines and a spare underflow line.

The clarifier tank is usually insulated to conserve heat. The temperature of the incoming green liquor from the recovery boiler is generally in the range of 180-190°F (82-88°C). Insulation helps to reduce the heating load of the green liquor heater and the heat savings usually provide a good payback. Polyurethane foam insulation is preferable. The block insulation, which is an alternate method, can create corrosion problems due to moisture penetration between the block and the tank wall.

The green liquor clarifiers have traditionally been fabricated from carbon steel. Antipollution regulations have required a greater degree of mill closure to avoid waste discharges, and there has been a significant change in the process conditions and buildup of the level of corrosive impurities in the recovery system. Some mills now rebuild or redesign the clarifiers in type 304 stainless steel, including the clarifier tank. It is common practice to provide the rake arms, feedwell, overflow launder, and the feedwell rods from type 304 stainless steel. A 304 stainless steel sleeve can also be used on the central shaft. The other parts can be fabricated from a proper grade of carbon steel suitable for caustic service with an adequate corrosion allowance.

The torque requirement for normal raking of the light dregs solids is relatively low. Therefore, it is a very infrequent situation when a torque overload would occur due to an upset condition, and a manual lifting device is normally furnished to allow the operator to raise the rakes to relieve the overloaded condition. In some cases, mills have elected to use the green liquor clarifier as a standby white liquor clarifier. In such cases, the design of the green liquor clarifier must be identical to that used as a white liquor clarifier and stainless protection at the solution level is mandatory. Also, the torque requirements would be higher and an automatic lifting device is employed.

Dregs washing

General

The underflow from the green liquor clarifier contains about 16.5 lbs Na_2O per ton pulp (7.5 kg). Different methods are practiced for the washing of the dregs, utilizing both filtration and sedimentation type equipment. The various dregs washing systems currently in practice are:

1. Drum filter — precoat.
2. Drum filter — scraper discharge.
3. Drum filter — belt discharge.
4. Sedimentation washer — single stage.
5. Sedimentation washer — two stages.

A comparative evaluation, indicating the soda recovery efficiency, in each of the above cases is shown in Table 5.5. The lime mud precoat method recovers the maximum amount of soda and produces a cake of about 45% solids, which is suitable for disposal as landfill. In contrast, both the sedimentation type and the other drum filters have the dregs sluiced to the mill's waste treatment system.

Sedimentation

In a sedimentation type of dregs washer, the underflow from the green liquor clarifier is mixed with wash water in a small agitated tank and settled in a clarifier called the dregs washer. This single compartment thickener is of substantially the same design as the green liquor clarifier, though smaller. Figures 5.7 A and B illustrate the dregs washers. Previously, it was common to find multicompartment units with the compartments arranged in series to obtain two-stage washing.

The operating principle of the dregs washer is to recover the soda by dilution washing. The dilution wash volume is controlled at the point where the weak liquor can be fully utilized for dissolving the smelt. Normally, the overflow from the dregs washer is pumped to the lime mud washer, where it can pick up some additional soda before the weak wash is pumped to the storage tank. The dregs in the dregs washer can normally be thickened to a slightly higher percentage solids (10-12%) than in a green liquor clarifier. A dilution rate of 1:9 to 1:14 by volume between the clarifier underflow and the wash water generally balances out the soda recovery and weak wash requirement. A unit area of 100 sq ft/TPD (10 m^2/tpd) dregs is used to size the dregs washer. The soda recovery efficiency is about 80%. The low efficiency is in part due to the fact that the flow of dregs from the clarifier can

Table. 5.5. Comparison of dregs washing equipment

	Sedimentation		*Filtration*	
	Single stage	*Two stage*	*Drum filter*	*Precoat filter*
Soda removal efficiency	75-80%	90-95%	90-95%	95-97%
Energy usage KWH	4.5	7.5	20	25
Wash water l/t (incl. dilution)	270	200	100	65

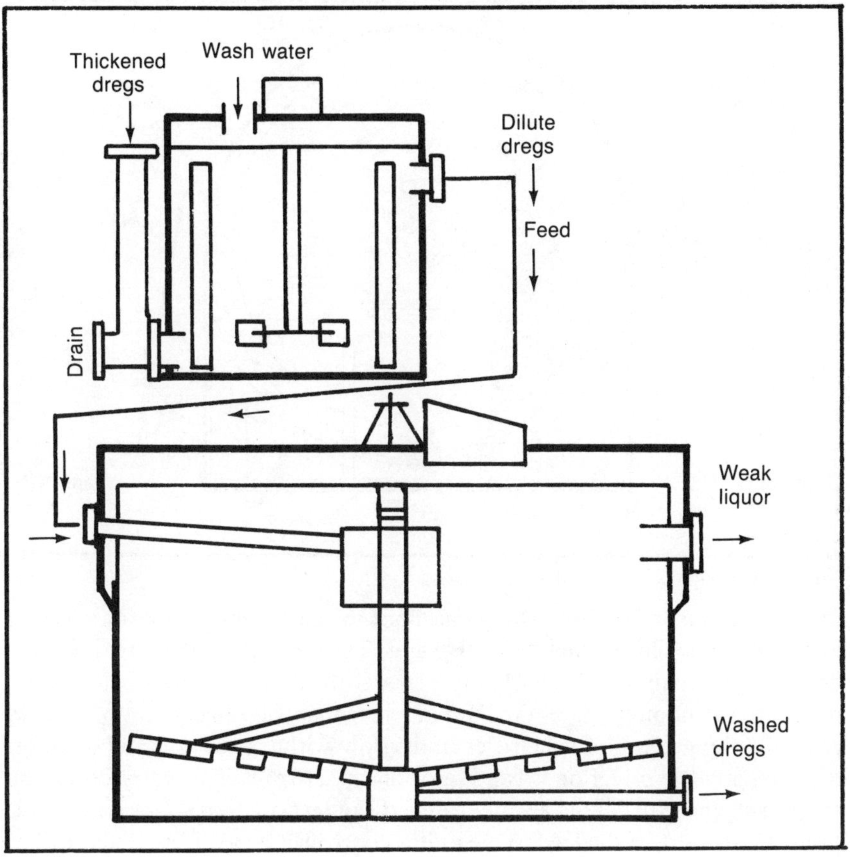

Fig. 5.7A. Dregs washers — sedimentation type.

be intermittent, while the flow of wash water is usually continuous. A separate mix tank with an efficient agitator preceding the washer will assist in maximizing recovery efficiency.

Drum filter

The drum filter with a fixed scraper is the least desirable arrangement, as medium blinding is a serious problem. Without an admixture of lime mud, the filtration rate is very poor, vacuum capacity utilization may be less than 1 cfm/sq ft (0.3 $m^3/min/m^2$) and the cover may be cleaned as often as once every hour. The belt discharge drum filter, on the other hand, operates on the principle of continuous washing of the belt. The thin layer of the cake is sluiced off to the mill waste treatment system. The filtration rate follows a pattern of peaks and valleys depending upon the condition of the belt. The belt requires acid wash with inhibited hydrochloric acid several times per

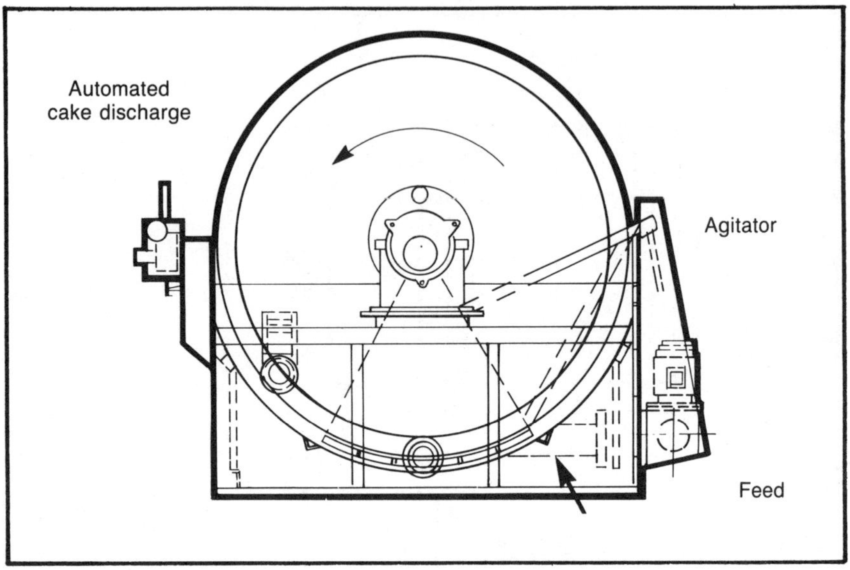

Fig. 5.7B. Dregs washers — filtration type.

month. Because of the wet, slimy nature of the cake, fog type spray washing is required to avoid runback of wash water. The feed rate to the filter is dependent on the capability of the filter to dewater the dregs. Therefore, it is paramount to maintain the filter medium in a reasonably clean condition to avoid reducing the green liquor clarifier underflow withdrawal rate and a consequential adverse effect on green liquor clarity. If this is properly done, the filter can generally handle a wider range of underflow feed solids concentrations, particularly if the feed to the unit is controlled by a differential pressure cell and valve combination. The use of the fog type spray washing provides true displacement of the liquor from the dregs and minimizes dilution of the displaced liquor.

Precoat filter

A lime mud precoat type filter with an automatic scraper advance type discharge is also used for dregs washing. The precoat life generally varies from 8 to 12 hours. A lime mud concentration of 25% is recommended to produce a dense cake; however, mud from the lime mud storage tank can be used directly having up to 40% solids. The lime mud consumption is about 0.2-0.5 kg per kg of dregs. The vacuum requirement is about 1.5 m^3/min per m^2 of filter area. The dregs from the green liquor clarifier underflow are normally pumped to a tank and then to the precoat filter and the operation requires very little operator attention.

Comparative sizing parameters between the different drum filters are summarized in Table 5.6.

Table 5.6. Comparative sizing parameters dregs drum filter

Type discharge	*Precoat*[*]	*Scraper*[**]	*Belt*
Filtration rate kg/hr/m^2	20	17	17
Vacuum m^3/min/m^2	1.5	0.6	1.0
Vacuum (mm Hg)	380-500	380-500	380-500
Lime mud consumed kg/kg dregs	0.2-0.5	1.0	none
Cake thickness, mm	1.6-3 (discharge)	3-15	1-3

[*] Precoat w/lime mud.

[**] Admixed with lime mud @ 1 kg mud/kg dregs.

Steel construction is used for all types of drum filters. Buildup of trace components such as chlorides has resulted in occasional corrosion of the wetted parts, and the use of 304 stainless steel for the piping, division and end strips, and valve may be required. The drum can be fabricated in steel construction with a corrosion allowance. Severe corrosion will occur if the filter is cleaned with uninhibited acid.

An alternate technique when using a dregs filter is to place the filter alongside the green liquor clarifier on a concrete pad. The underflow line can be piped directly to the filter vat feed connection with a valve which is controlled by a differential pressure cell mounted on the filter vat to allow the head in the clarifier to create the driving force, and admit the dregs slurry in direct proportion to the filtration rate. Minor adjustments can be accommodated by adjusting the speed of the filter drum. This eliminates pumping and helps to minimize solids degradation and thereby improves the filtration rate.

Mill data on the operation of dregs filters has been presented in various articles, (*6*, *7*, *8*).

Slaking and causticization

Operations

The reaction between sodium carbonate and lime is carried out in a slaker. The sodium carbonate constitutes about 68% of the total titratable alkali in the green liquor; the balance being sodium sulfide and sodium hydroxide. Small quantities of sodium sulfate and sodium thiosulfate are also present. The percentage of titratable soda (Na_2O) in the green liquor solution is about 10%.

The lime is metered directly to the slaker, and the ratio of green liquor to lime is controlled by manipulating the feed rate of lime at a fixed flow rate of the green liquor. The slaking and the causticizing reactions proceed simultaneously, and about 85-90% of the causticization reaction is completed in the agitator section of the slaker.

During the slaking process, the violent hydration of the quicklime disintegrates the lumps or pebbles, producing more surface area for reaction, and, at the same time, freeing the inerts for removal by the classifying section of the slaker.

Operation of the recausticizing system hinges on what takes place in the slaker. The object is to achieve the desired conversion of the sodium carbonate to sodium hydroxide with a minimum of excess lime. This is essential if the subsequent liquid-solid separations (sedimentation and/or filtration) are to proceed efficiently. Excess lime beyond 1% or so of the equilibrium stoichiometric quantity can cause turbidity in the clarified liquor and lower filtration rates of the lime mud in addition to accelerating blinding of the filter medium. Washing of the soda from the lime mud cake is also impaired, and cake moistures run higher.

The causticized slurry passes through a mechanical classifier, which forms an integral part of the slaking equipment, before overflowing to the first causticizer. The classifying section removes the overburned and unburned lime, sand, and clinkerous impurities down to about 65 mesh. The rejects, known as grit, are usually between 0.5-2.0% of the lime feed during normal operation. The loss of lime at the slaker is principally a function of the impurity content of the lime and the burning conditions in the lime kiln. The loss of lime grits from the slaker is one reason for the makeup, which must be provided on a continuous basis.

Controls

Control of slaking is essential to the production of a uniform cooking liquor of the desired strength and activity. It is customary to maintain the slaking temperature in the range of 210°-220°F (99-104°C). Practically, this is generally left to the operator who must manipulate the green liquor flow, green liquor temperature, and lime addition, while, at the same time, maintaining the liquor strength within the mill's targets.

When insufficient instrumentation is provided to control feed rates and temperatures, liquor uniformity depends upon the operators and their skill. Nearly all mills provide Baume control for the green liquor strength at the dissolving tank, and maintenance of this instrument is paramount to uniformity in alkali strength. Green liquor heaters are also provided to control green liquor temperature. Lime addition, on the other hand, is usually accomplished by means of screw feeders that are quite variable in output. As a consequence, the operators must be continually on the alert to maintain the liquor strength within specifications. It is customary to sample the causticized liquor hourly or every two hours and perform titrations (A, B, and C tests) to determine activity, causticizing conversion efficiency, and sulfidity. Excess liming frequently occurs, especially when the causticizing department is required to attain high causticizing conversion efficiencies that approach the equilibrium point. If conversion efficiencies are held to four to six percentage points below equilibrium, excess liming can be avoided. Equilibrium activity at 0, 15, and 30% sulfidity levels is shown in Figure 5.8A and B.

It is customary for each mill to establish the liquor specifications for their particular cooking practice. The recausticizing department is provided with a target range for the required liquor characteristics. Typical values are shown below:

Item	*Amount*
TTA	7.2-7.5 lb/ft^3 (115-120 g/L)
Active alkali	6.0-6.4 lb/ft^3 (96-102 g/L)
Activity	80-85%
Sulfidity on active alkali	24-26%
Causticizing conversion efficiency	78-83%

The ratio of lime and green liquor is frequently controlled by the temperature in the slaker. The hydration of quicklime is an exothermic reac-

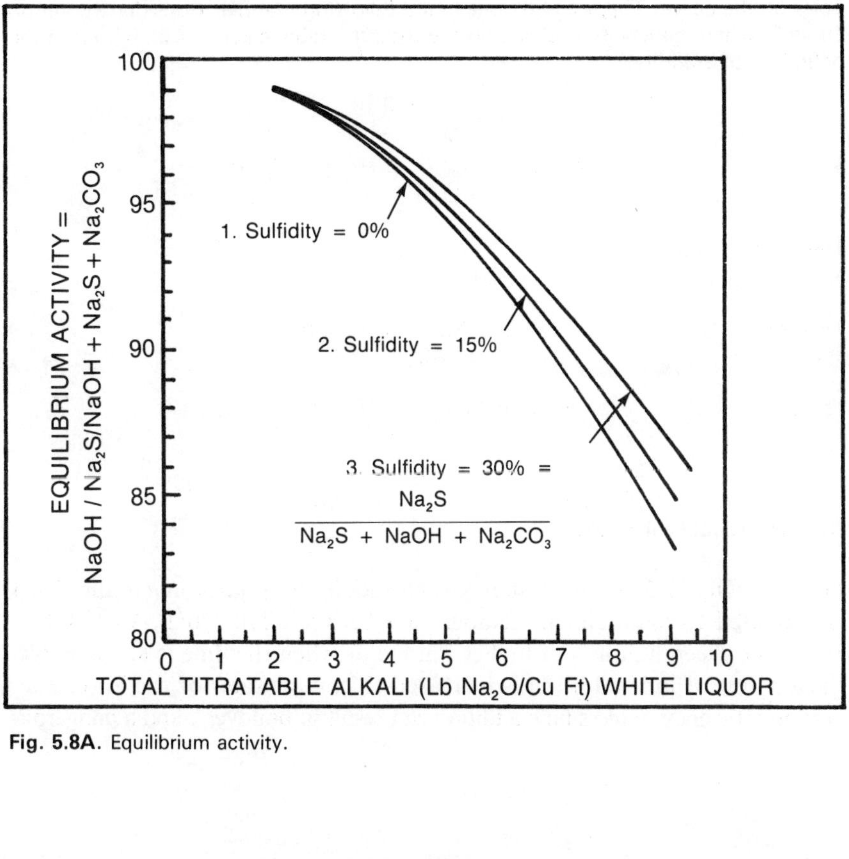

Fig. 5.8A. Equilibrium activity.

	Sulfidity		
T.T.A.	0% curve 1	15% curve 2	30% curve 3
2.0	99.1	99.1	99.1
3.0	98.1	98.1	98.1
4.0	96.6	96.7	97.0
5.0	94.8	95.0	95.6
6.0	92.5	93.0	93.7
6.5	91.3	91.8	92.7
7.0	89.9	90.6	91.7
7.5	88.4	89.3	90.7
8.0	86.9	88.0	89.4
8.5	85.2	86.6	88.1
9.0	83.5	85.0	86.8

Note: If the definition for the percent of sulfidity used is

$$\text{Percent sulfidity} = \frac{Na_2S}{Na_2S + NaOH} \times 100$$

(all expressed as Na_2O) then the value for the sulfidity must be determined for every point on the curve. This may be done in the following manner: Consider a point on curve 2, where causticity is 89% and the total titratable alkali is 7 lb/ft^3. From the definition of sulfidity,

$$\frac{Na_2S}{\text{T.T.A.}} = 0.15$$

where

$$\text{T.T.A.} = NaOH + Na_2S + Na_2CO_3 = 7 \text{ lb/ft}^3$$

Therefore,

$$Na_2S = 1.05 \text{ lb/ft}^3$$

and,

$$NaOH + NaCO_3 = 5.95 \text{ lb/ft}^3$$

Therefore,

$$NaOH = 5.30 \text{ lb/ft}^3$$

and

$$Na_2CO_3 = 0.65 \text{ lb/ft}^3$$

Therefore,

$$\text{New sulfidity} = \frac{1.05}{6.35} \times 100 = 16.5\%$$

Also,

$$\text{Causticizing efficiency} = \frac{NaOH}{NaOH + Na_2CO_3} \times 100 = 89\%$$

Fig. 5.8B. Equilibrium activity.

tion, evolving 27,500 Btu per lb mole. The addition of quicklime to the slaker is regulated to maintain the temperature between 210-220°F (99-104°C) . When the green liquor is of higher Baume or when the lime is less reactive, the amount of lime addition is increased to compensate for the lower conversion efficiency. Excess lime addition can result in boilovers, and a dangerous situation can result if the hot, caustic liquor begins to spill over the sides of the slaker.

Lime mud

To obtain a good settling lime mud, experience has shown that slaking at the highest permissible temperature is generally best. Equipment suppliers have designed their agitators to provide adequate mixing with a minimum of agglomerate degradation. Excess lime is to be avoided, as $Ca(OH)_2$ inhibits the settling of the $CaCO_3$. As a plant operating control, an empirical settling test is carried out once every one or two hours. The sample can be taken from the inlet to the first causticizer. Lime mud is considered to settle poorly when it does not conform to the following typical settling rate in a 500 ml graduated cylinder, while still hot as shown in Table 5.7.

The size distribution of typical lime mud is shown in Figure 5.9.

The possible causes of poor settling lime mud are:

1. Excess lime to slaker.
2. Dregs and other impurities building up in the lime mud. This would be indicative if lime availability was below 80%.
3. High amount of makeup lime. New lime usually does not settle as well as reburned lime.
4. Slaker temperature. Temperatures below 210°F (99°C) cause a coarse, sandy type of lime mud which settles rapidly, but gives a very turbid overflow.
5. Insufficient underflow withdrawal from the white liquor clarifier.
6. Clarifier not large enough to handle tonnage.
7. Overburned lime.

When the lime is fresh from the kiln, the lime mud will generally settle better than when the lime is allowed to cool before slaking, especially if it is stored for some time. High temperatures are considered essential to promote reaction time, and are believed to favor production of a fast-settling lime mud. Good quality reburned lime is very reactive, has a high availabili-

Table 5.7. Typical lime mud settling rate

Start	500cc	10% Solids
5 min.	300cc	16% Solids
15 min.	120cc	35% Solids
20 min.	102cc	40% Solids
60 min.	90cc	44% Solids

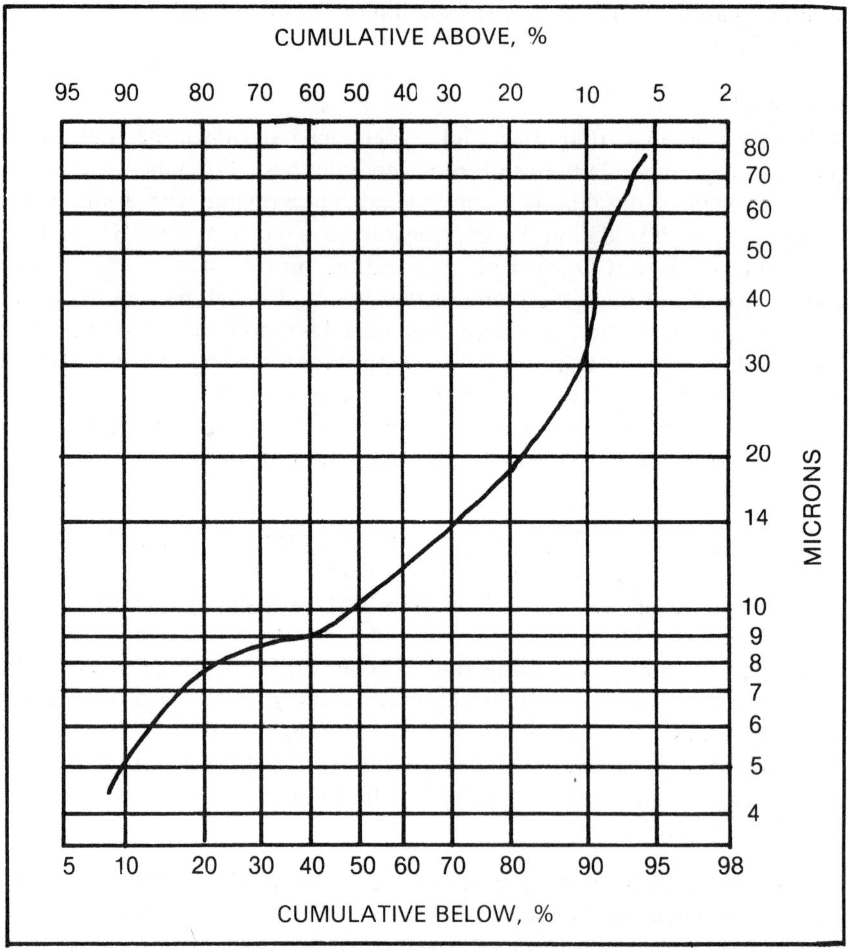

Fig. 5.9. Typical lime mud particle size analysis.

ty, is relatively porous, and slakes rapidly. If the lime is not completely burned, the porosity may not be fully developed so that the reaction with the sodium carbonate is incomplete, resulting in lower causticizing efficiency. Causticizing with incompletely burned lime produces more mud for the same conversion, and the resulting mud is likely to have higher viscosity. Such material is more difficult to pump. On the other hand, lime burned at a higher temperature is likely to have very poor reaction characteristics. The causticizing reaction proceeds very slowly and may produce a lime mud which will settle slowly and is difficult to wash free of alkali. A quick settling lime mud of good filtering and washing properties is obtained if the burning of lime mud is controlled with respect to the calcining temperature, detention time in the burning zone, residual CO_2 in the reburned lime, and the purity of lime.

The relationship of percent solids and specific gravity is shown in Figure 5.10.

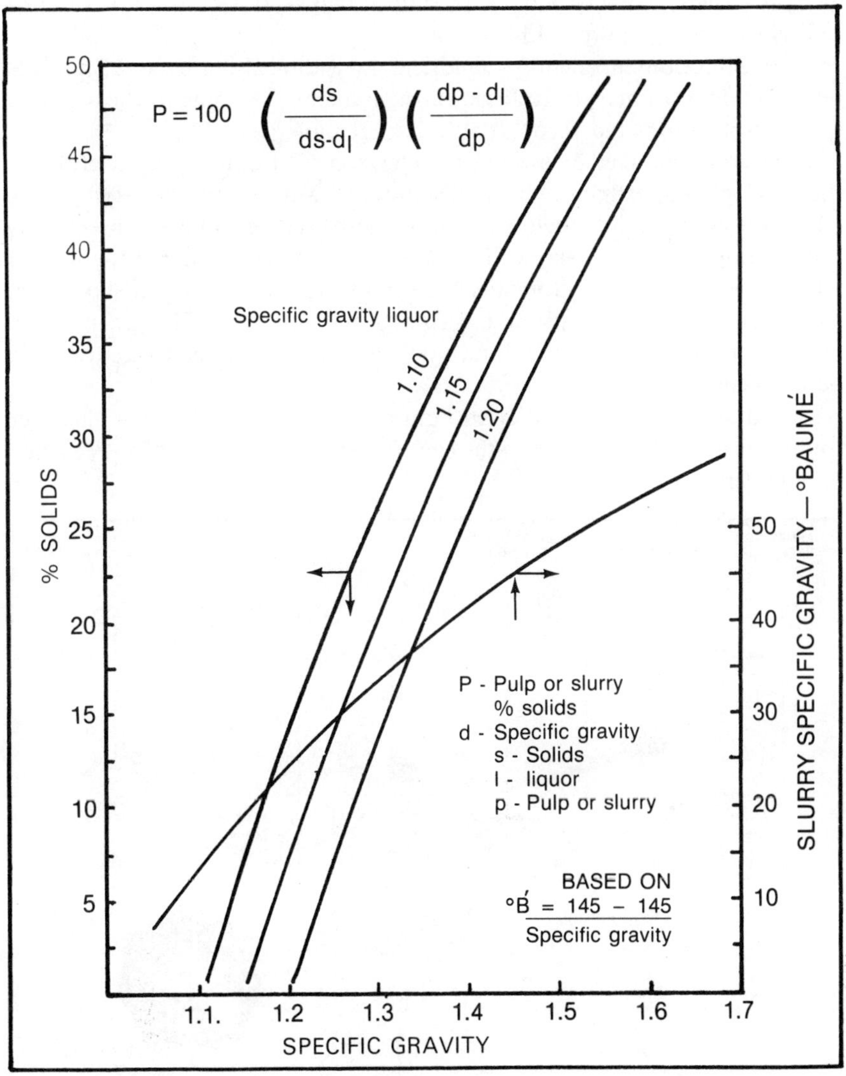

Fig. 5.10. Percent solids vs. slurry specific gravity.

Equipment

Combination slaker-classifier. The combination unit, introduced around 1940, provides for slaking, classification, and grit removal in a single piece of equipment. A slaker incorporating a rake-type classifier is shown in Figure 5.11. A similar type unit employing a screw in place of the rakes, for removing grits, is shown in Figure 5.12. A picture of two slakers at a Southern kraft mill is shown in Figure 5.13.

The slaking compartment is cylindrical, providing about ten to 15 minutes retention time. Lime is introduced through a chute opening in the top, and the green liquor is fed through an adjacent pipe.

The slurry is agitated by means of a pitched blade turbine-type mixer. From the slaking compartment, the slurry flows to a quiescent classifying zone where the coarse solids (below about 65 mesh) settle, and are subsequently removed by the rake or screw. The rake is a set of slow-moving flights that move forward and then automatically rise and return to their original position. The screw is fixed in position and rotates at a low rpm. The grit material is thus transported out of the slurry pool and is spray washed to recover soda before being discharged.

Slaked and partially causticized liquor overflows a weir in the classifier compartment and is directed to the causticizers, where the reaction is carried to completion.

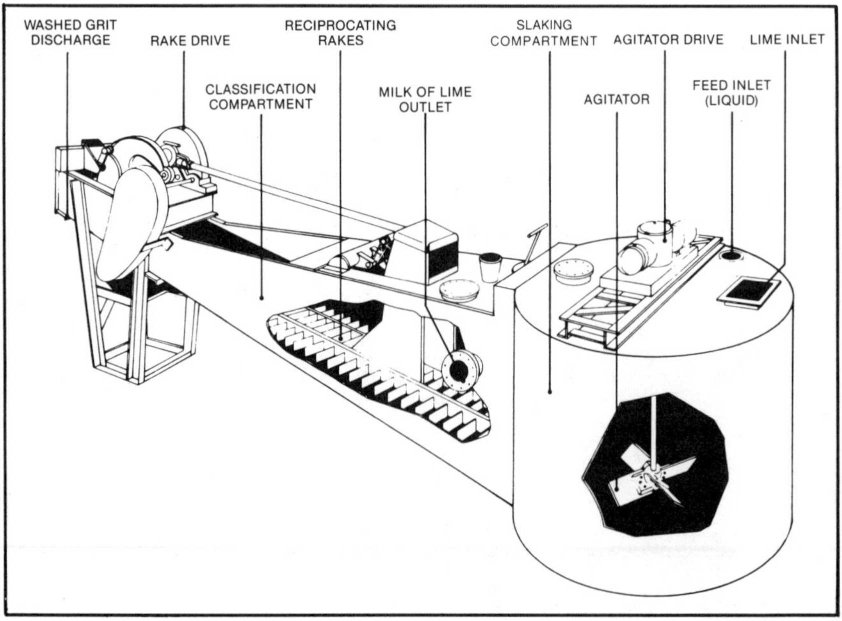

Fig. 5.11. Slaker with rake classifier.

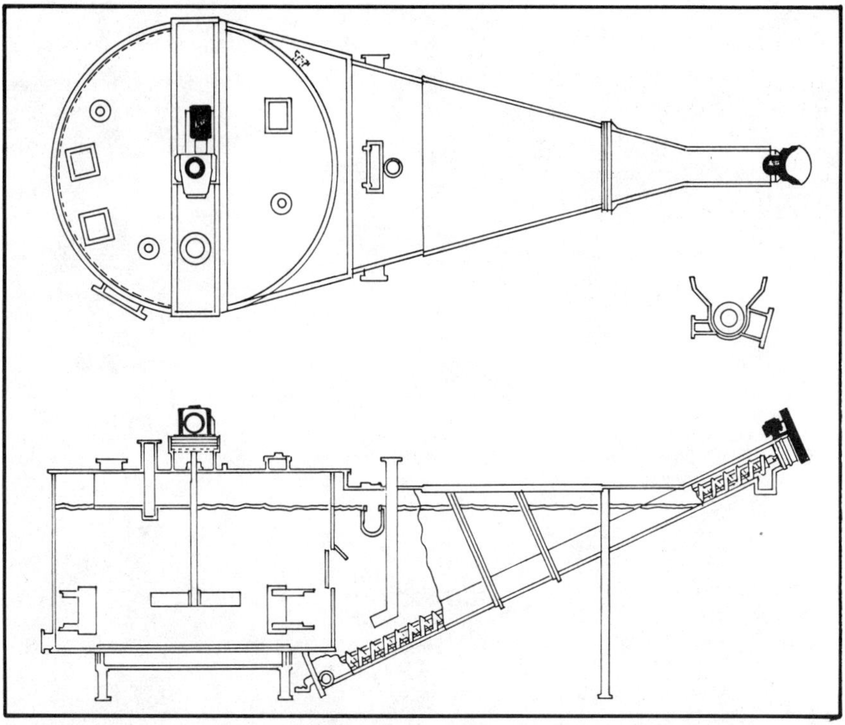

Fig. 5.12. Slaker with screw type classifier.

The round slaking tank is constructed of 1/4 in. type 304 stainless steel, including the internal baffles and the bottom blast plate. The tank cover is usually of mild steel. No tank lining is required. The agitator impeller and shaft are also of type 304 stainless steel. The classifier tank is also made of 1/4 in. type 304 stainless steel, including the rake or screw mechanism. Again, the cover is mild steel.

Rotary slaker. Rotary drum slakers were probably the earliest units to be used. Today, rotaries are still in use but are few in number. They are employed primarily where lime quality is very poor and the grit discharged is high. The rotary slaker is equipped with internal lifting bars to provide mixing, and also for conveying coarse solids to the discharge end. There may be a Trommel screen at this point with 1/2 to 3/4 in. openings to remove the large cores. The remaining inerts are removed by an internal degritter. To prevent excessive dusting, drum seals are required.

Primary slaker. Primary slaking ahead of the conventional slaker was introduced in the early 1950s as a means of avoiding hot lime handling and storage. In this arrangement, the kiln feed was discharged directly into the primary slaker. A secondary slaker with classifier was used for new lime

Fig. 5.13. Two slaker-classifiers with rake discharge.

addition and removal of grits. This unit was normally designed to handle the entire slaking requirement with new lime when a kiln shutdown became necessary.

Problems were encountered in grit removal from the primary slaker as well as the slaker discharge pump. A cyclone classifier was used to remove grit material prior to the secondary slaker. However, these tended to wear rapidly with resulting loss in efficiency. Primary slakers have been abandoned as a result of the operating and maintenance problems encountered.

Lime slaker scrubber. A large amount of steam formed during the slaking reaction must be exhausted, or it will slake the lime in the feed chute causing plugging and spills from the screw feeder. Because of the dust and the spills, housekeeping around the slaker is always a problem.

A scrubber is used in most mills, mounted in the slaker stack. A wet fan mixes the gases, dust, and green liquor as a scrubbing liquid; the latter is introduced into the eye of the fan which flushes out the internal fan surfaces. The liquid and the dust discharge tangentially into the cyclonic section of the scrubber, where the gases are separated from the liquid stream and vented through the outlet, and the scrubbing liquid and captured dust drop into the slaker.

Figure 5.14 illustrates the mechanism of operation of such a scrubber.

Another type of scrubber using a venturi principle is shown in Figure 5.15.

Because of the corrosive nature of the scrubbing liquid, the ducts and the fan are fabricated from type 304 stainless steel.

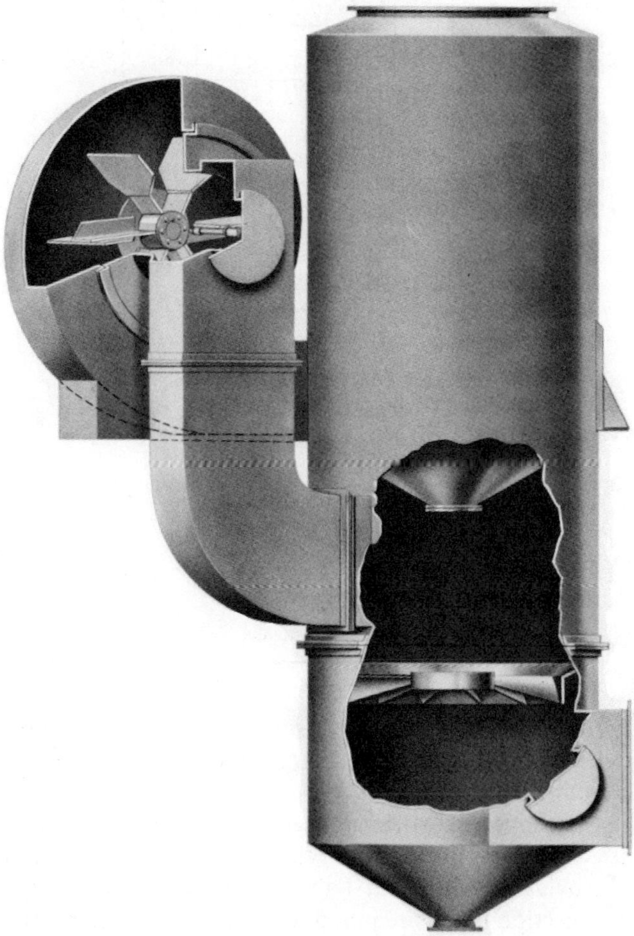

Fig. 5.14. Triple action scrubbing fan.

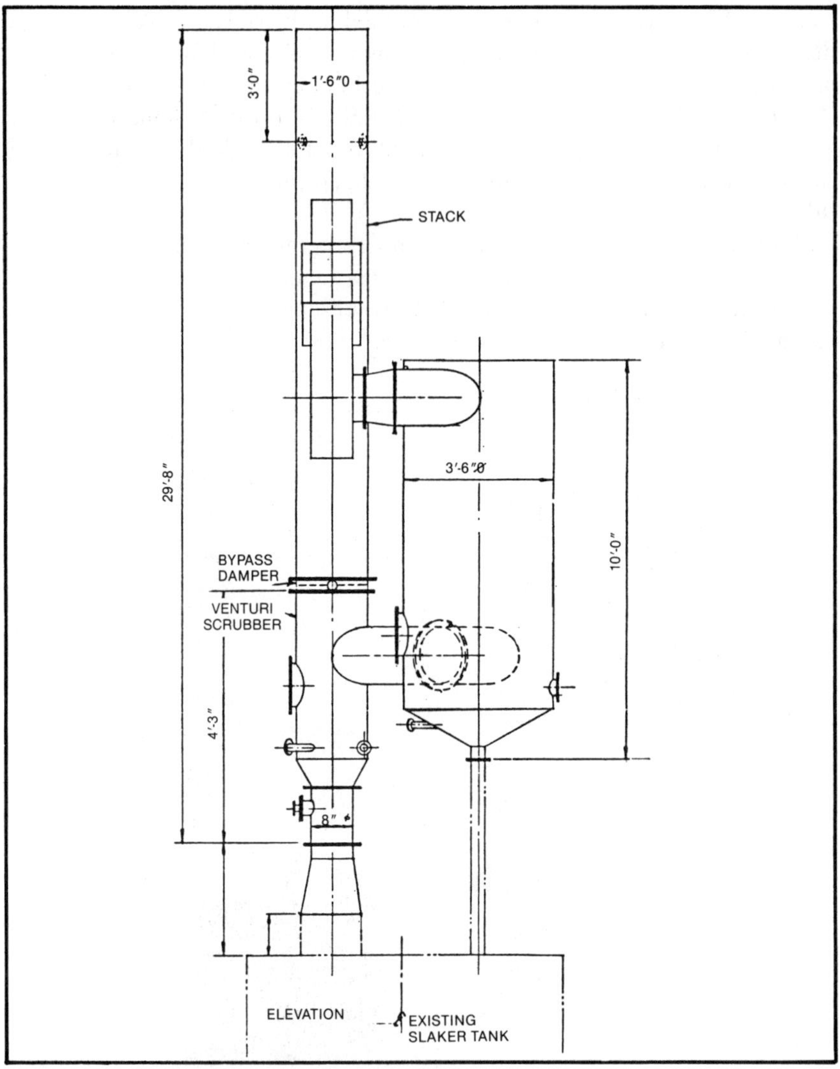

Fig. 5.15. Scrubber using venturi principle.

The causticizer vents can be ducted to the slaker vent. The hot lime bin and bucket elevator vents can be discharged into the slaker to recover lime discharging from those points.

Causticizer. Although causticizing begins in the slaker, the reaction does not reach its equilibrium point in the relatively short detention time (ten to 15 minutes) provided in this unit. Laboratory studies have shown that 60 to 90 minutes reaction time is required to attain 95-98% completion of the equilibrium level as governed by the solution strength. Due to agitation inefficiencies caused by short-circuiting, full-scale units are sized at 90-150 minutes detention time dependent on the clarifier design that follows. The reaction is permitted to continue in a series of tanks equipped with agitators to maintain the solids in suspension, yet not to cause dispersion of agglomerates. The approach used is based on the theory of short circuiting reactors in series, assuming that 90% of the reaction is to be completed. Three causticizers are normally arranged in series to provide the volume required to complete the causticizing reaction. The same conversion can be achieved in two causticizers if the volume can be increased by 25% to 120 minutes total. To achieve the same conversion in one causticizer, the increase in volume would be theoretically 260%.

A typical arrangement of three causticizers is shown in Figure 5.16. Each unit provides 30 minutes of retention time and is arranged for bypassing when maintenance is required.

A typical causticizer is a covered, vertical steel tank equipped with a pitched blade turbine mixer operating at 68 to 84 rpm, depending on tank diameter. The shell is lined with either concrete brick, or sprayed with gunnite, or lined with stainless steel to provide corrosion and abrasion resistance. The recent trend is to provide a type 304 stainless steel tank instead of lining the carbon steel tank. The tanks are vented from the top and a gradient of 1 to 12 is provided so that the slurry can cascade by gravity from one tank to the next.

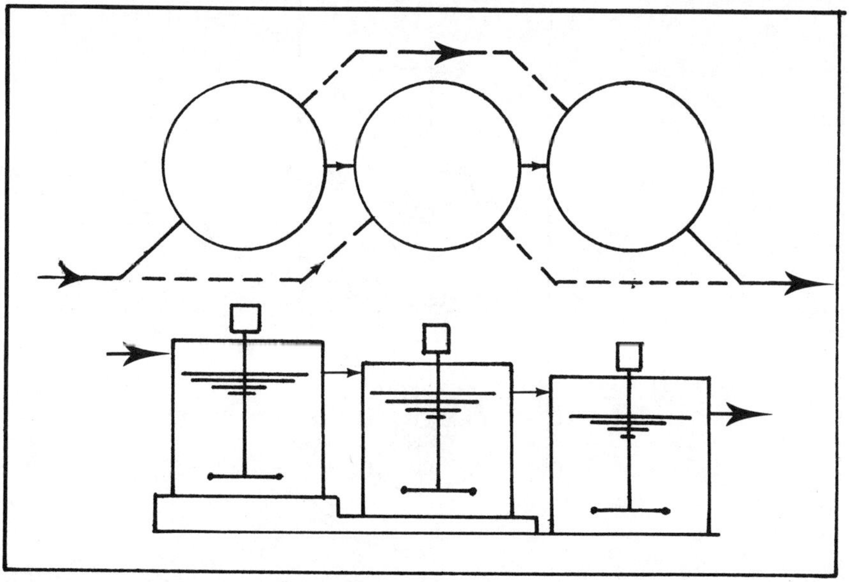

Fig. 5.16. Typical arrangement of three causticizers.

The direction of rotation of the agitator is set up to minimize short-circuiting between the feed and the overflow.

A longer reaction time in the causticizer is provided when the sedimentation type of white liquor clarifier is replaced by a pressure filtration system. The filter is more sensitive to the remaining excess lime in the causticized slurry, and thus requires a longer reaction time in the causticizers. A three-compartment causticizer stacked tank, or a series of these tanks is used with the feed entering the upper compartment, and the slurry passing out through an internal pipe. The volumetric efficiency of such a tank, as shown in Figure 5.17, is relatively high, as the risk of short-circuiting is eliminated. The units are not as flexible as individual tanks, as a failure of the agitator would cause a shutdown of the system.

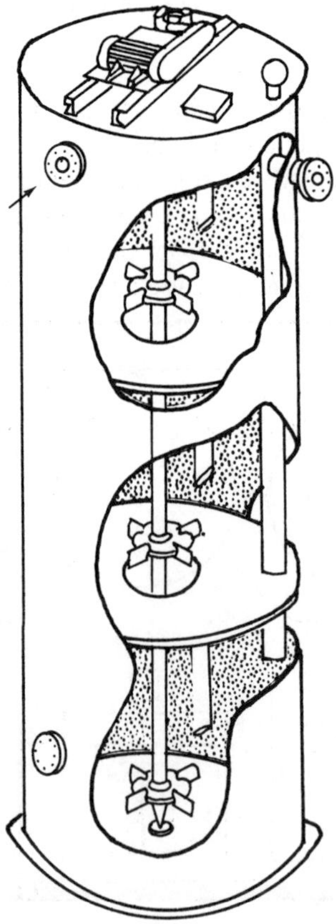

Fig. 5.17. Internal arrangement of a deep (stacked) causticizer.

White liquor clarification

Sedimentation

The slurry from the causticizers contains the calcium carbonate (lime mud) which must be removed from the liquor before it can be used for cooking. In most mills, this is accomplished in gravity sedimentation units called clarifiers. The clarifiers are operated to produce an underflow of 35-40% mud solids to minimize the amount of soda that must be recovered in the subsequent washing step. The primary object of clarification is to attain a clear white liquor, and sedimentation type clarifiers usually achieve 100 ppm turbidity or less when operated at design conditions.

The settling properties of the white liquor solids depend on several factors:

1. Lime burning conditions.
2. Nature and extent of impurities.
3. Slaking and causticizing conditions.
4. Liquor composition.
5. Characteristics of the makeup lime.

As several of these factors are related to operation, changes in operating practices can alter the settling characteristics. The use of polymers can improve the settling characteristics; however, they may impair the operation of a filter used to dewater the mud prior to the calciner.

The sedimentation type unit clarifier is sized on the basis of lime mud 10 sq. ft/TPD (0.93 m^2/tpd). For multicompartment tray machines, the sizing basis may be as high as 16 sq ft/TPD (1.48 m^2/tpd), depending upon the height of the tray compartment. The controlling sizing parameter is the amount of mud to be handled per day and not the overflow rate from the machine; for this reason, a white liquor clarifier is essentially a thickener and not a clarifier.

The underflow density of white liquor clarifier is monitored by density gauges and torque sensing instrument mounted on the clarifier drive head. The underflow rate is controlled by a variable speed pump, thus eliminating the flow control valve. In some mills a control valve may be still in use; however, the present trend is towards variable speed pumps because of the high maintenance and relatively short life of the control valves.

Clarifiers

Unit type. The design of the sedimentation unit is of importance from the standpoint of performance and ease of operation. The simplest of various thickener designs is called the "unit" type. A typical unit design is shown in Figure 5.18A. It contains a suitable feedwell in the center for introducing the feed uniformly to the sedimentation zone, slow-speed rakes for moving the settled sludge to the discharge point, and a collecting launder for the clarified liquor.

Balanced tray type. In order to conserve space, however, multi-compartment clarifiers are used. These are simply several unit-type clarifiers stacked on top of each other. For many years, the preferred type was the four-compartment "balanced tray" thickener as shown in Figure 5.18B.

A feed box and overflow boxes equipped with adjustable overflow pipes provide for a relatively uniform parallel operation of liquors and hydraulic balancing of the sludge on each tray. The settled mud passes down through the central compartment, so that all mud leaves through the same pipe at the bottom. Most tray machines today feature two compartments to simplify the balancing problems associated with the old four compartment units.

Combination clarifier and storage. In the last few years, the use of unit-type thickeners combined with clarified liquor storage has become the accepted standard. Such combinations conserve space, and most importantly, reduce capital costs. Normally, 12 to 20 hours of storage is provided in the section above the clarifier. A typical design is also shown in Figure 5.18C.

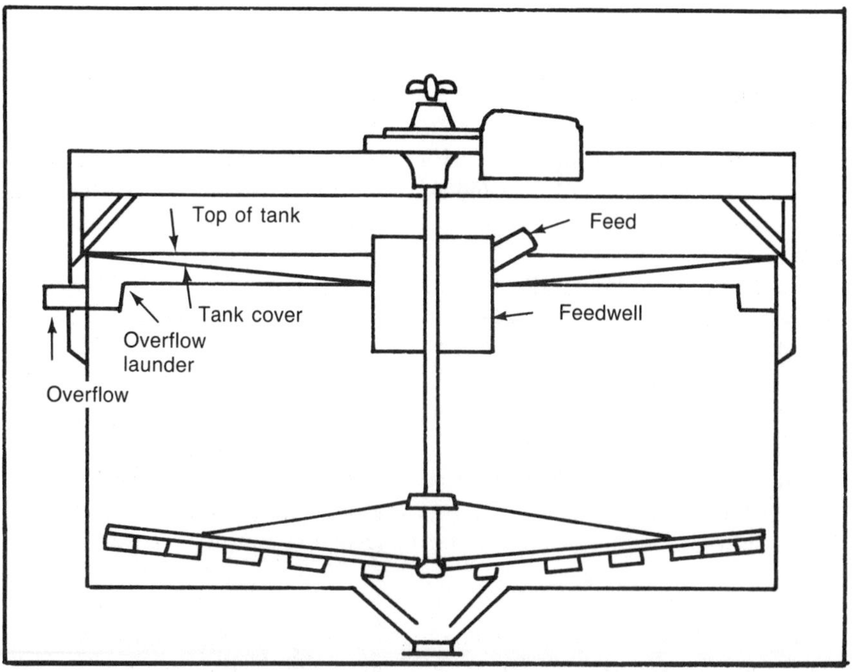

Fig. 5.18A. Unit thickener.

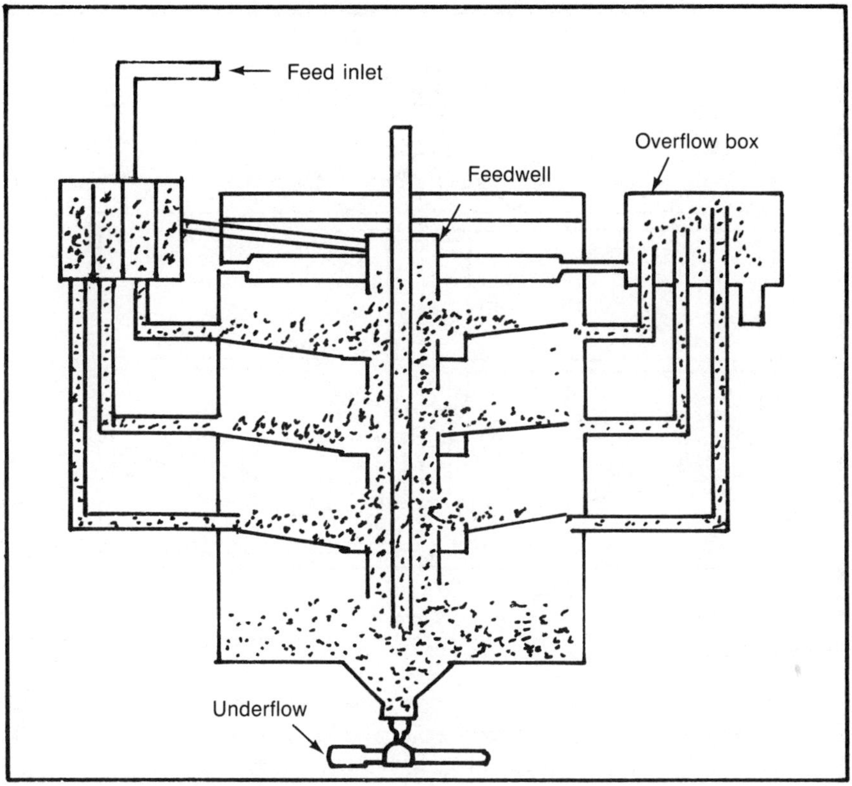

Fig. 5.18B. Balanced tray thickener.

Filtration

Besides the sedimentation type of clarifier, two other types of equipment employing the filtration separation principle for clarification of white liquor are in use in the industry: a vacuum drum filter with belt discharge and a pressure filtration system. The filtration type of equipment started with a premise to produce white liquor of better clarity. Unfortunately, the vacuum drum filter with belt discharge was found to have considerable operating problems, especially when fed slurries containing excess lime, and it uses a higher amount of energy due to the use of a vacuum pump. The pressure filter is commonly used in Europe, principally in the Scandinavian countries. The pressure filter produces white liquor with a turbidity level of about 10 ppm.

Vacuum belt filter. This filter combines white liquor clarification with washing and dewatering of lime mud. Filtered white liquor is pumped to the storage tanks; the filtrate from the washing section is pumped to a weak liquor storage tank.

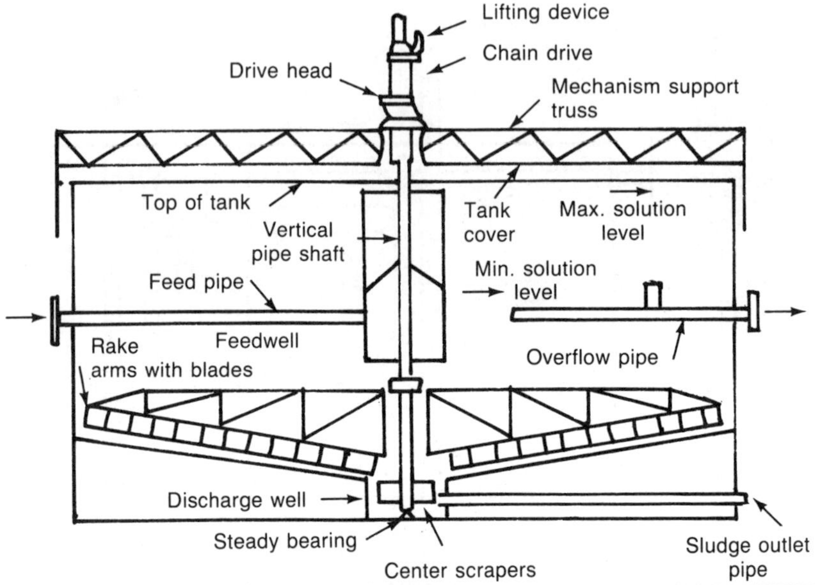

Fig. 5.18C. Combination clarifier and storage.

After a short drying zone on the drum, the endless cloth carrying the mud cake leaves the drum and passes over a cake discharge roll. The cloth then passes through a wash trough for cleaning before meeting the drum again via a guide roll. Some type of belt tracking or guiding system is used, as well as devices for removing wrinkles and keeping the seams straight.

The advantages of a vacuum belt filter are: lower first cost as a separate lime mud washer is not required, less space requirement, and better white liquor clarity. The disadvantages are: increased operating costs due to periodic acid washing and cloth replacement, production rate limited to design level, sensitivity to overliming resulting in cloth binding, higher energy requirement for the vacuum pump, and cooling of white liquor due to flashing in a vacuum operation.

Pressure filter thickener unit. This equipment is a continuously operating pressure filter. It consists of a vertical pressure vessel with conical bottom equipped with a paddle agitator. Between the shell and the upper dished head lies a tube plate supporting a number of perforated stainless steel tubes equipped with filter socks. The top is sectioned by radial walls, which divide the unit into several sections.

Slurry from the last causticizer is pumped into the filter, whereby white liquor passes through the socks and out through the top of the unit. The lime mud forms a cake on the outside of the socks. Periodically the feed to the filter is recycled back to the surge feed tank. This reduces the pressure in the filtration part of the vessel and allows clarified white liquor in the upper portion to flow back down through the filtration tubes, causing the cake to discharge. The lime mud settles to the thickening zone in the bottom from where it is discharged at 40 to 45% solids. The operating principle of the two types in operation are illustrated in Figures 5.19 and 5.20.

The operation of the Eco filter was explained by Juha Lumikko of Enso-Gutseit, Finland at the International Conference on Recovery of Pulping Chemicals, 1981 (*9*).

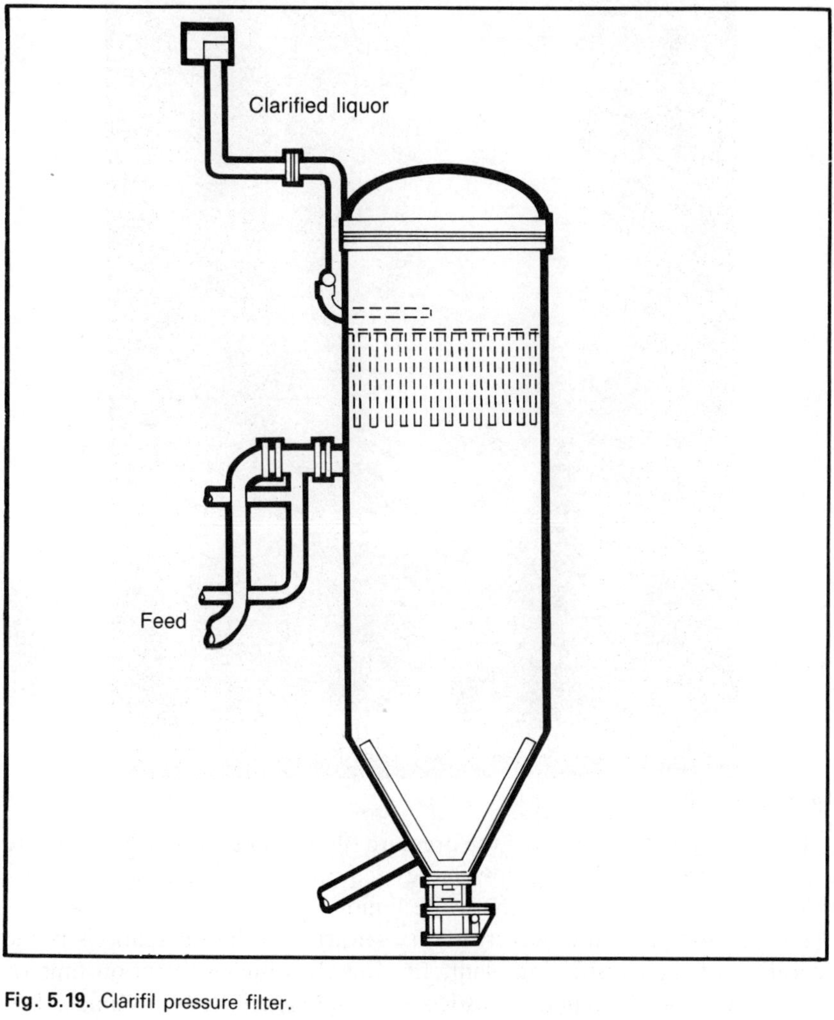

Fig. 5.19. Clarifil pressure filter.

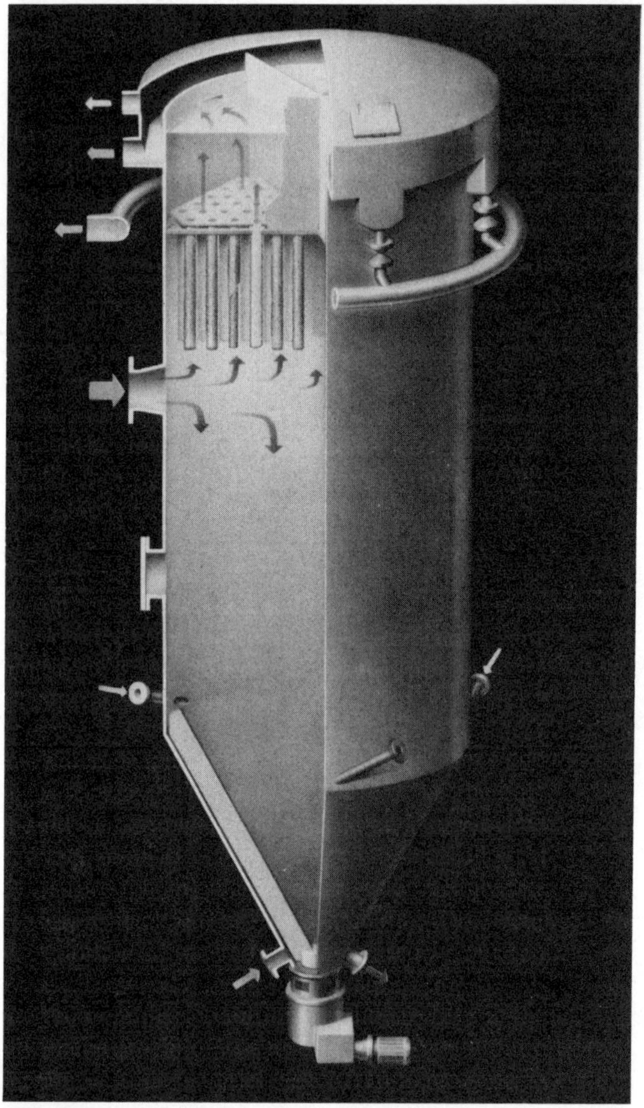

Fig. 5.20. ECO filter.

The principal advantages of the pressure filter thickener are cleaner white liquor, compact installation, low energy requirements, and a closed system which minimizes cooling of the white liquor.

The pressure system however, is very sensitive to the fluctuations in the operation of the causticizing plant. In general, a higher retention time of 2 1/2 hours is provided in the causticizers to fully react the excess lime. Also,

Table 5.8. Typical pressure filter capacities

Unit size		*Clarified W.L. capacity*	
m	*ft*	*m^3/h*	*gpm*
1.8	5′-11″	1,600-1,900	190-350
2.05	6′-9″	2,000-2,500	365-460
2.5	8′-2″	3,200-3,700	585-680
2.8	9′-2″	4,000-4,500	735-825
3.0	9′-10″	5,000-5,400	920-990

due to progressive blinding of the filter cloth, the filter unit must be emptied at periodic intervals to acid wash the filter socks. The interval between acid washes may be one to eight weeks, depending upon the process conditions and unit design.

Pressure filter units are rated on volume through-put basis as shown in Table 5.8.

Lime mud washing and thickening

Sedimentation washers

The lime mud from the WL clarifier is washed and dewatered before the calcining operation. The soda content in the washed lime mud is controlled to ensure satisfactory operation of the calcining equipment, as well as to control the TRS emission in the off gases from the kiln. The washing operation is carried out in two or more steps. The mud from the WL clarifier, having about 16-20% soda (as Na_2O) is normally washed in a sedimentation tank, followed by washing and dewatering on a precoat type of rotary vacuum filter. The combined operation lowers the soda to less than 1% of the weight of dry mud.

Two-stage lime mud (LM) washers, either as unit thickeners connected in series or the tray type machines, were in wide use in the industry before the advent of the precoat filter. The single stage mud washers were combined with liquor storage in a unit type design and were adopted in the late 1960s. The unit machine is basically a thickener, wherein the mud underflow from the WL clarifier is diluted with wash water and settled to produce a clarified weak liquor and washed mud underflow having about 3% soda (Na_2O). The unit design LM washers are sized at the rate of 8 sq. ft/TPD (0.74 m^2/tpd) of lime mud to produce 40-50% mud solids in the underflow. The weak liquor overflow from the unit generally has less than 100 ppm suspended solids. High recycle of mud solids fines in kiln scrubber water to the mud mix tank can result in a higher turbidity level in the overflow.

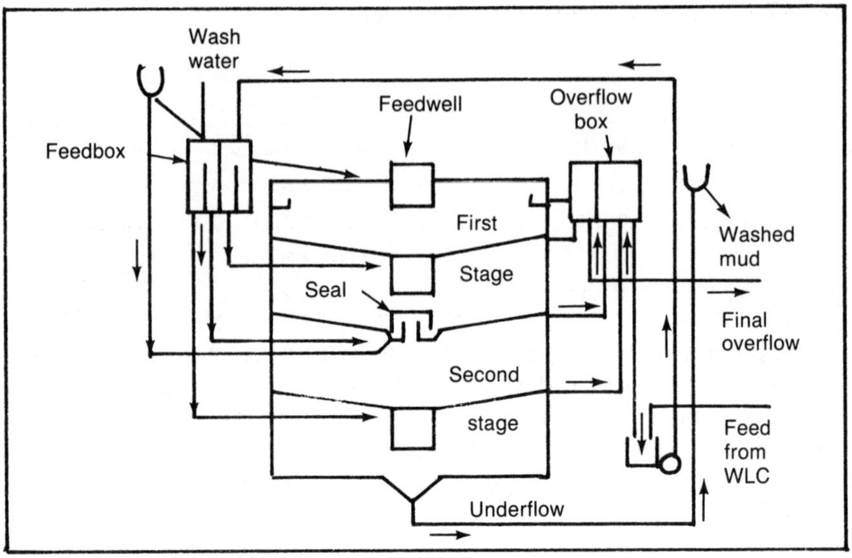

Fig. 5.21. Two-stage balanced tray mud washer.

The most widely used two-stage tray design machine has four compartments; the top pair and bottom pair operate in parallel using balanced trays. A sketch of a two-stage, four-compartment mud washer is shown in Figure 5.21. The second and third compartments of this machine are separated by a labyrinth seal allowing separate mud withdrawals from the middle and bottom compartments of the thickener. The top pair of compartments constitute the first washing stage, where the underflow from the WL clarifier and the weak wash overflow from the second stage are introduced after mixing in an outside feed box. Fresh water is introduced in the second stage after mixing it with the mud from the first stage. A typical two-stage countercurrent washing system is shown in Figure 5.22.

A four-compartment tray machine usually has shallow compartments; the height between the trays may be between 6–7 ft. Such machines have lower overall efficiency and are difficult to balance. Consequently, some compartments are overloaded while others are under-utilized. There is also a higher carry-over of mud solids with the weak wash overflow. For sizing, a unit area of 12-14 sq ft/tpd is considered.

A unit single stage LM washer, with or without the liquor storage, has now been adopted in the industry in view of its ease of operation and relatively low capital cost. Unit machines with storage for weak wash eliminate the separate storage tank and permit a more compact causticizing plant. These machines are identical to the unit white liquor clarifier.

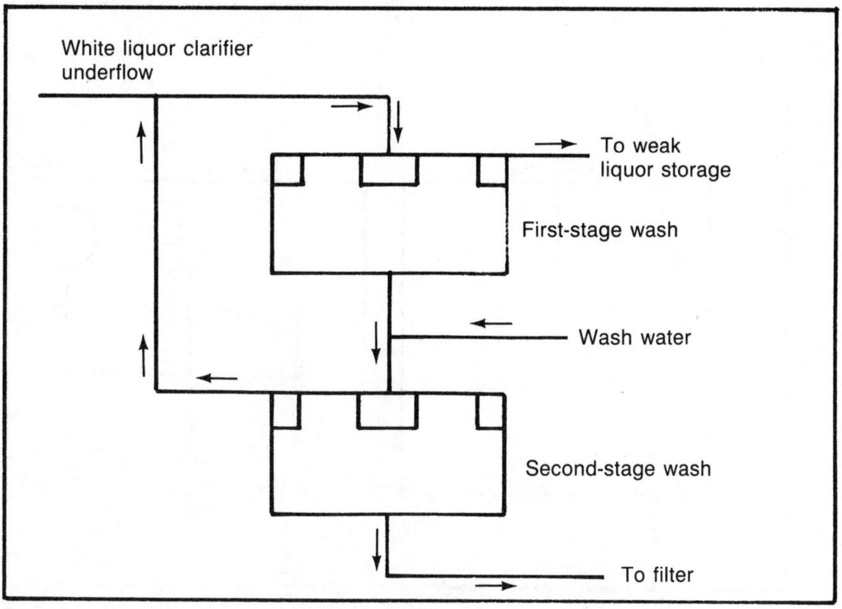

Fig. 5.22. Two-stage lime mud washing system.

For proper operation of the lime mud washer, the underflow mud at a uniform density is withdrawn by controlling the speed of the centrifugal pump. One control logic is based on the premise that the underflow density is proportional to the speed of the centrifugal pump. This type of control, however, is true only within a narrow operating range. In actual practice, the relationship between the variable speed of the pump and the density as measured by the nuclear density device is not constant. The control problem is further compounded by the density gradient in the mud itself, and to the drift in the radioactive detection device after it is calibrated. The general technique for controlling the mud density is through some kind of semi-automatic control, where the operator tries to maintain a uniform condition by manually adjusting the speed of the pump.

A more stable density control is obtained by manipulating the speed of the underflow pump by the torque of the rake mechanism. The torque trend gives the operator a more reliable indication of the density conditions in the clarifier, so that there is less overpumping or underpumping of the mud from the clarifier. A flow diagram of the torque control instrumentation in its very simple form is shown in Figure 5.23. The control loop can also incorporate snap acting relays to quickly open a control valve to recirculate the mud back to the LM washer.

Flow control valves have been found to be a high maintenance item, resulting in their replacement by the variable speed drives.

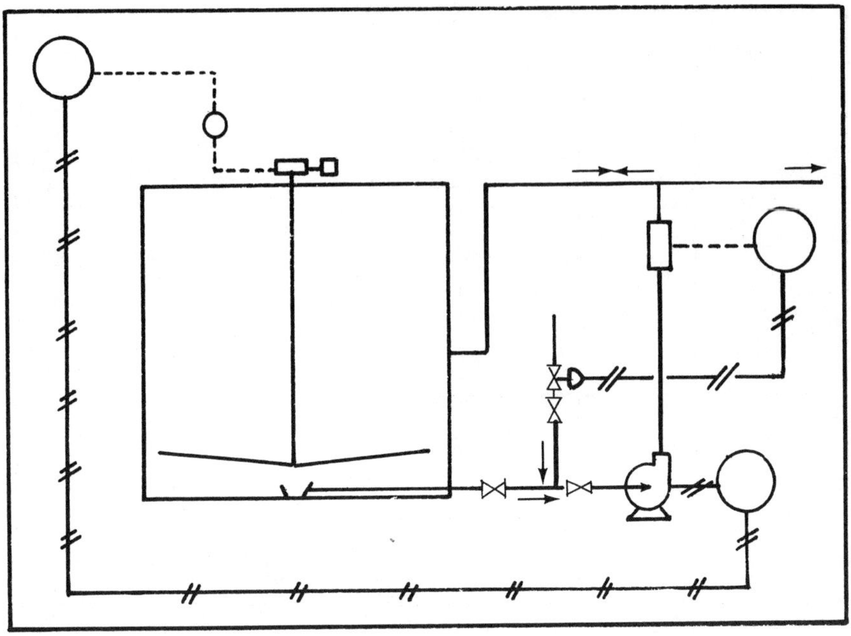

Fig. 5.23. White liquor clarifier or lime mud washer underflow control.

Lime mud filter

After one or two stages of washing in sedimentation type washer, the lime mud is further washed and dewatered on a rotary drum vacuum filter before it is burned in a lime kiln or a calciner. From the storage tank, the mud is diluted to 25-30% before feeding to the filter. A radioactive type density measuring device controls the dilution water. The quantity of mud filtered is controlled as this is essential for the satisfactory operation of the lime kiln. Normally, a magnetic flow meter is used to control the flow with a variable speed driven pump.

The precoat lime mud filter operates at 15 to 20 in. (380 to 510mm) Hg vacuum and the full surface area of the filter drum is under vacuum. The drum is covered with a 24 by 110 mesh face wire of a stainless steel or a polypropylene filter cloth. The scraper is set at a fixed distance of about 3/8 to 5/8 in. (9.5 to 16mm) from the filter drum surface, so that the cake in thickness of about 1/8 to 1/4 in. (3.2 to 6.4mm) above this surface is removed by the sharp edge of the scraper blade, while the heel is left on the drum surface to act as a precoat for the next filtration cycle. Before development of the precoat approach, the entire cake was discharged in each revolution by air blowback. The filter is illustrated in Figure 5.24 and a picture of a precoat filter in operation is shown in Figure 5.25.

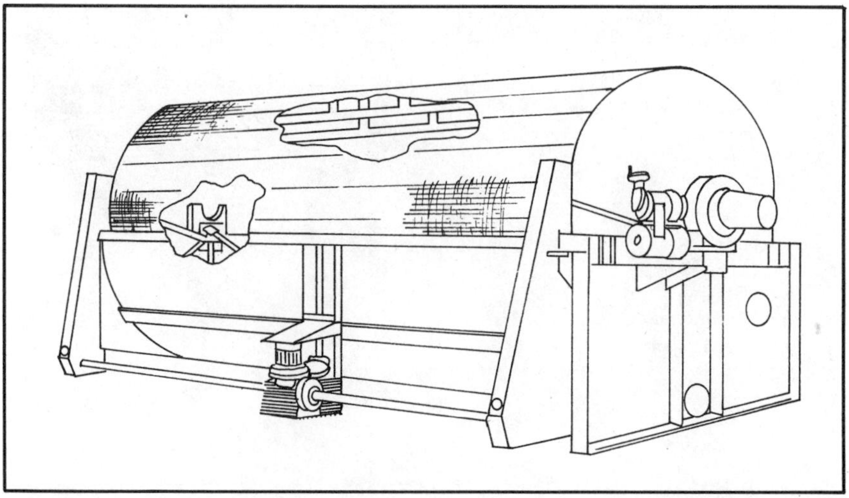

Fig. 5.24. Lime mud precoat filter.

The precoat type of operation is now a standard in the industry due to the improved soda recovery and less moisture going to the kiln, with an associated improved fuel economy. A comparison of the two modes of operation is shown in Table 5.9.

With sufficient amount of dewatering of the mud, the reduction of soda on a precoat type of operation is usually in the range of 50-80%, so that the residual water soluble soda in the filter cake is about 0.5-1.0% (as Na_2O). Although both extraction and displacement washing operations are involved on the filter, the latter provides best results if the shower water is above 160°F (71°C) Operating results from various mills indicate that the filter cake solids range from 62-75% with a typical value of about 70%. Periodic cleaning of the filter face wire with 10% solution of sulfamic acid (or inhibited

Table 5.9. Comparison of lime mud filter operation

	Precoat	*Complete discharge*
Feed solids (recommended), %	25-35	30-35
Vacuum, in. Hg	15-22	10-20
Drum speed, sec. per rev.	20-30	60
Cake thickness, in.	1/8-1/4*	1-1½
Filtrate solids, %	0.2-0.5	1 or more
Cake solids, %	65-75	50-58
Capacity, ton/ft^2/day	1.0-1.2	1.25-1.75

*Actual cake thickness about ½ to ¾ in. (13 to 19mm)

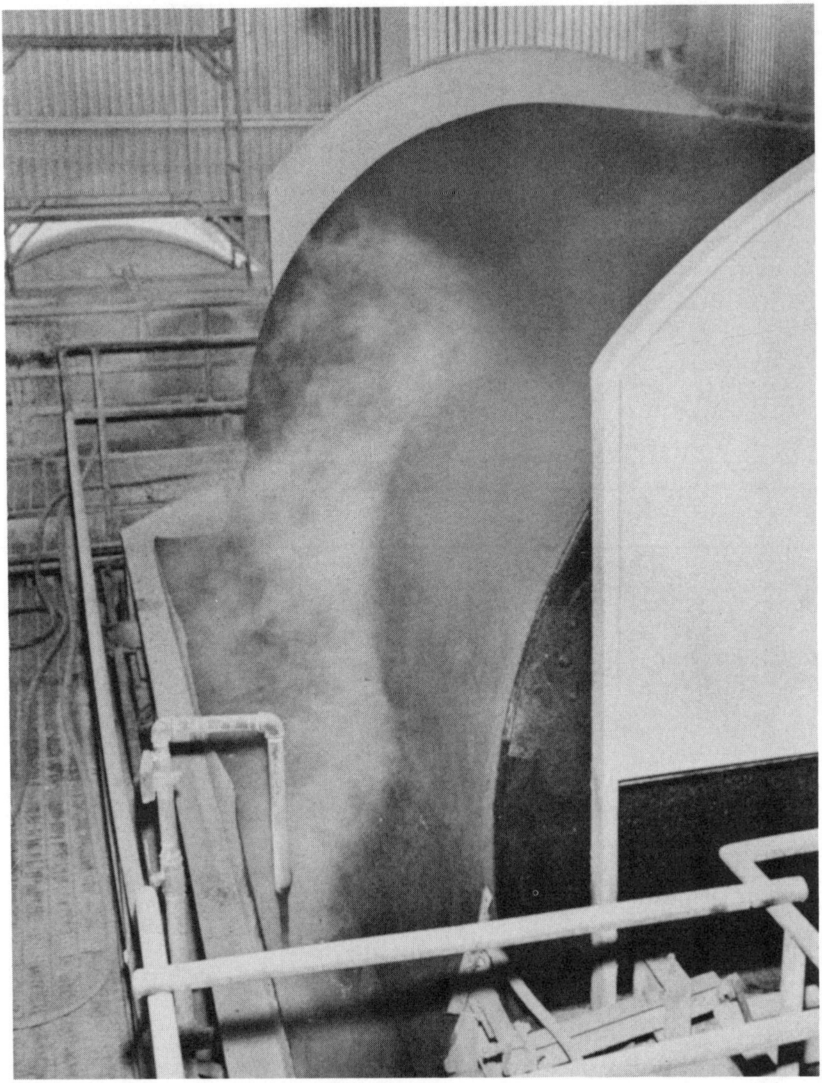

Fig. 5.25. Precoat filter.

hydrochloric acid) is required particularly when irregular cake formation is indicated. This is followed with a water wash using a high pressure shower (100 psig). Many mills have automated the face wire wash cycle with remote control from the control room.

Centrifuge

There are still a number of centrifuges in operation. Very little washing is accomplished in the centrifuge, only thickening. The centrifuge operates on the principle that as the lime mud solids are deposited around the periphery of a rotating horizontal bowl, a spiral screw, operating at a slower speed than the bowl, conveys a layer of material to the end opposite the feed. From the discharge end, lime mud from between 60 to 65% solids drops into a screw conveyor, which discharges into the lime kiln.

The centrate contains a higher solids content than the filtrate from precoat filters. Since the centrate is returned to the lime mud washer, settling area requirements are increased.

Although washing and cleaning of the filter wire is eliminated with the centrifuge, it is claimed that maintenance is higher. The screw flights have to be built-up at periodic intervals and most mills keep a spare rotor on hand.

The largest centrifuge has a capacity of about 300 tons/day of lime mud solids equivalent to a lime kiln capacity of about 180 tons/day of lime. For lime kilns with greater capacity than this, two centrifuges would be operated in parallel.

With the increased cost of fuel, it has become paramount to feed the dryest mud possible to the kiln. Since the precoat filter can produce mud at 5 to 10% higher dryness, centrifuges are no longer installed for mill improvement or expansion projects and are not considered for new mills.

Water balance for mud washing

The makeup water requirements for the entire liquor system establishes the amount of water available for washing the lime mud. A water balance in the recausticizing system must be achieved, otherwise an excess of the weak wash water above that required for the smelt dissolving tank would be sewered.

The makeup water to the causticizing plant is generally supplied at four points: calciner flue gas scrubbing; dilution of lime mud; shower water at the filter; and dregs washing.

Of these, the fresh water passed through the kiln scrubber is about 60-70% of the total makeup water. A typical water balance is shown in Table 5.10.

The operation of the mud washer improves with hot water; therefore, the makeup stream is passed through the kiln scrubber, where it picks up the heat from the hot gases. The use of contaminated condensate in the lime kiln flue gas scrubber will result in stripping of reduced sulfur compound from the condensate by the hot gases, resulting in odor problems (due to the release of TRS) from the kiln stack.

Table 5.10. Water balance — kraft recausticizing (based on 1 ODT kraft pulp)

Assumed conditions:

Pulp yield - 50%; AA on OD wood - 15.5%
AA in WL - 100 g/l (6.25 lbs/cf)
GL dregs 4 kg/ODT
Dregs washing on precoat type filter, discharge at 40% dry
Dryness of lime mud filter cake - 68%

IN	*Liters*	*Gals*	*OUT*	*Liters*	*Gals*
Dregs filter wash	11	3	White liquor	3,100	825
Lime mud filter wash	400	106	Lime mud cake	200	53
Lime mud dilution	640	170	Dregs cake	7.5	2
Kiln scrubber	2,100	555	Evaporation	19	5
Misc & unaccounted for	183	53	Slaker grit	7.5	2
Total	3,334	887	Total	3,334	887

The kiln gases also carry a large amount of particulate matter which is washed out in the wet scrubbers and returned to the lime mud washer. The total mud load handled in the LM washer and the precoat filter is therefore increased by the amount of the circulating load of the carry-over solids. This circulating load has been found to set up an equilibrium at 10 to 20% of the total mud feed rate to the kiln. In a fluidized bed type of calciner, because of the separation of the particulate matter in dry cyclones ahead of the wet scrubber, the carry-over load of the circulating dust is about one half of the circulating load of dust in a rotary lime kiln.

A single stage LM washing followed by filtration can reduce the soda in the mud to less than 1.0%; however, with proper control of the wash ratio during the displacement washing stage, the residual soda in the mud cake can be lowered to about 0.5%. Wash ratio is defined as the volume of wash water per unit volume of mother liquor in the unwashed cake. A two-stage countercurrent washing system, in conjunction with a precoat type of filter, will reduce the soda to 0.2%. A comparison of mud washing shows that a single stage will achieve about 80% reduction of soda whereas two stage will achieve about 95% reduction of soda.

The washing efficiency at the filter is of the order of 60-70%, with a wash ratio of about 1.5.

Equipment loading factors

Average equipment loading factors are summarized in Table 5.11. It should be pointed out that conditions vary from mill to mill and loading factors

Table 5.11. Causticizing equipment loading factors

Equipment	*English units*		*Metric units*	
Slaker	min	10-15	min	10-15
Causticizers	min	90/150	min	90/150
WL clarifier				
Unit tray	ft^2/TPD	9-10	m^2/t/d	0.92-1.0
Multiple tray	ft^2/TPD	14-16	m^2/t/d	1.43-1.64
LM washer				
Unit tray	ft^2/TPD	7-8	m^2/t/d	0.72-0.82
Multiple tray	ft^2/TPD	10-12	m^2/t/d	1.23-1.43
Lime mud filter				
Thickening (precoat)	TPD/ft^2	1.0-1.25	t/d/m^2	10-12
Mud oxidation	TPD/ft^2	0.75	t/d/m^2	7.5
GL clarifier	ft/h	1.7-2.0	m/h	0.52-0.61
Dregs washer	ft^2/TPD	100	m^2/t/d	10.2
Dregs filter	lb/h/ft^2	3.5-4.2	kg/h/m^2	17-20.5

are affected by lime quality, impurities in the liquor cycle, temperature, and the operation of the department. Therefore, a thorough analysis of the operation is required to assess the actual loading factors a particular mill can expect to find acceptable.

Total reduced sulfur (TRS) emissions

The residual soda in the lime mud is partly water soluble and partly tied-up with silica, alumina, and other complex compounds. Part of the washable soda is in the form of Na_2S, which can be the source of H_2S emissions. The release of H_2S is attributed to the pH reduction in the liquid phase by the CO_2 produced in the calcining reaction. To maintain emissions below 20ppm as H_2S, the soluble sulfide concentration leaving the lime mud precoat filter must be maintained below 0.18% Na_2S. The most recent EPA guidelines for TRS emissions recommend 20ppm from the lime kiln.

Recently, there has been a trend to use the conventional precoat type of vacuum filter for oxidation of the residual sulfide, while employing single stage washing of the lime mud. The filter is oversized by using a loading rate of 0.75 tons/day of lime mud per square foot of filter area, in comparison to 1 ton/day when specified for dewatering requirement only. Another method of oxidizing the residual sulfides in the lime mud is air sparging in a tank (*10*). Oxygen can also be used with injection and mixing prior to the filter.

The air sparging system employs a completely mixed tank with air dispersed from a high pressure blower. Two existing units in the northwestern United

States are designed for two and seven hours retention time (*11*). Roughly 10 SCF per gallon of mud slurry are supplied at 10 psig and 17 psig, respectively. Investigations indicated that after two to three hours of air sparging, the sulfide was oxidized as follows (*12*):

Total sulfide conversion 70%
74% to $Na_2S_2O_3$
21% to Na_2SO_4
5% to Na_2SO_3.

For reliable operations, the recausticizing and lime mud reburning operations must function smoothly before the oxidation on the filter can have any chance of success. Precise results on the extent of oxidation on the filter are not available; however, some references available indicate that a monthly average TRS emissions from a lime kiln can be limited to 20 ppm with lime mud oxidation, single stage mud washing, and combustion gases having excess oxygen of about 4% and the kiln cold end temperature between 450-500°F (230-260°C).

The soluble sulfides in the mud are converted to SO_2 during the calcination operation, provided the excess O_2 in the combustion gases is maintained over 4% and the kiln cold end temperature at about 500°F (260°C).

Material and energy balance

The concentration of the pulping liquor varies from mill to mill, depending upon the cooking conditions and the type of pulp produced. All of the water entering the liquor cycle must be evaporated from the black liquor prior to burning in the recovery furnace. Therefore, careful control of mud and dregs washing is required to arrive at a balance and have optimum liquor concentration. As stated above, the makeup water enters the system at four points. These are shown diagrammatically with reference to the complete material balance in Figure 5.26. The following conditions were assumed:

48% pulp yield
16.5% AA on OD wood
25.0% sulfidity on AA
85.0% activity
85.0% lime availability.

The water, soda and heat balances are summarized in Tables 5.12A, B, and C based upon the conditions shown in Figure 5.26. Calculations for the values shown in Figure 5.26 are found at the end of the chapter.

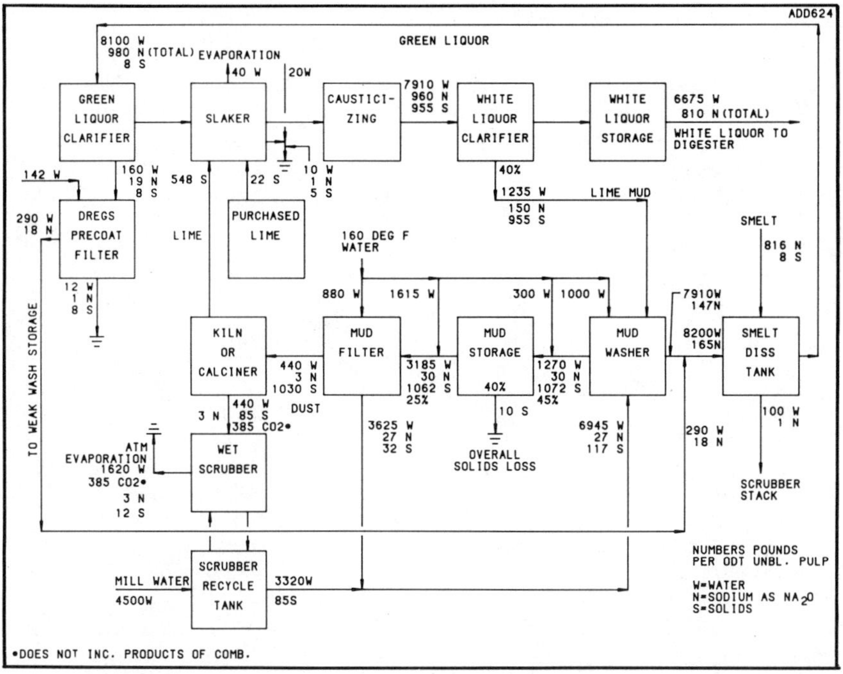

Fig. 5.26. Example of a material balance for a kraft pulp mill recausticizing and lime reburning system.

Process control

Slaker

Process control for the recausticizing plant begins with the metering of lime and green liquor into the slaker. Lime feed is normally controlled with a variable speed drive conveyor, most often a screw conveyor from the bottom of the hot lime bin. The green liquor temperature is recorded and controlled. Direct steam is usually used, because of scaling problems with indirect heaters. The flow of green liquor is controlled using a magnetic flow meter and a control valve. Orifice plates are used; however, scaling of the opening can cause errors in the reading.

The slaker temperature is recorded and used as a guide to the causticizing reaction. Continuous carbonate analyzers are being used in Scandinavia, however, their use is still limited in the United States. Continuous titration analyzers are also being used to control white liquor strengths.

Causticizers

The motor load of the slaker and causticizer agitators is indicated by ammeters on the recausticizing control panel.

Table 5.12A. Water balance

	lb/ODT*	kg/ODt**
IN		
Makeup kiln scrubber	4,500	2,250
Mud dilution	2,915	1,458
Filter showers	880	440
Dregs filter	142	71
Wash slaker grit	20	10
TOTAL	8,457	4,229
	lb/ODT	kg/ODt
OUT		
White liquor	6,675	3,338
Evaporation	40	20
Slaker grit	10	5
With mud cake to kiln	440	220
Kiln scrubber evaporation	1,180	590
Dregs filter cake	12	6
Smelt dissolving tank vent	100	50
TOTAL	8,457	4,229

Table 5.12B. Soda balance

	lb/ODT*	kg/ODt**
IN		
Green liquor from smelt dissolving tank	980	490.5
	lb/ODT	kg/ODt
OUT		
White liquor	810	405
To smelt tank	165	83
Lime kiln stack	3	1.5
Slaker grits	1	0.5
Dregs cake	1	0.5
TOTAL	980	490.5

The flow of the white liquor slurry from the last causticizer is by control of level in a stand pipe. Normally, a variable speed drive on the transfer pump is used; however, type 316 stainless steel ball type control valves are also used.

Storage tanks

Unit clarifiers with clarified liquor storage in the upper section, require recording of the level of the tank. Levels in any white liquor, green liquor, and

Table 5.12C. Heat balance — base temperature 60°F (16°C), makeup water at 160°F (72°C)

	Btu/ODT*	M Joule/ODt**
IN		
Makeup water	860,000	1,000
Steam, green liquor heater	160,000	180
Heat of reaction	215,000	250
Lime, sensible heat	45,000	50
Green liquor, sensible heat	1,110,000	1,290
TOTAL	2,390,000	2,770
	Btu/ODT*	M Joule/ODt**
OUT		
White liquor	990,000	1,150
Weak liquor	730,000	850
Dregs filtrate	40,000	45
Dregs cake	2,000	2
Mud cake to kiln	80,000	93
Slaker evaporation	68,000	79
Precoat filter evaporation	50,000	58
Radiation and convection	330,000	380
Sensible heat, scrubber water evaporated	110,000	130
TOTAL	2,390,000	2,790 (rounded)

*ODT = oven dry short ton (2000 lb)
**ODt = oven dry metric tonne (1000 kg)

weak wash tank are also recorded. Normally, flush mounted type 316 stainless steel dP cells are used to transmit a differential pressure signal. Bubble pipes using air pressure differential are still being used in older mills.

Clarifiers

Torque on the clarifier rake drives is recorded and is often used to control the underflow pump. Underflow from the clarifier is by a variable speed drive pump and normally a spare pump is installed. The lime mud slurry density is recorded using a nuclear density gauge. This signal can be used to control the flow of dilution water going into the bottom well (boot) of the clarifier. Hot water is used to assist or improve the movement of lime mud in the underflow pipe.

Motor operated rake lifts are provided for the lime mud washer and white liquor clarifier. These are remotely controlled from the control room, usually as a manual control. The rake lift can be connected to automatically raise the rakes, if the torque on the rake drive reaches a preset level. Audible alarms are required to signal the operator of a high torque.

Lime mud washing

Water for lime mud washing comes from the lime kiln scrubber sump and from recovered warm water. The flow of makeup water to the scrubber is controlled and the flow to the mud mix tank is on level control of the scrubber sump. Additional lime mud wash water requirement is controlled by level of the weak wash storage tank. Transfer of weak wash from the lime mud washer to the storage tank is by control of level in a standpipe.

Dissolving tank

The flow of weak wash to the recovery boiler dissolving tank is from a green liquor density controller at the dissolving tank. Level of the dissolving tank is normally controlled by overflow at a fixed level in the tank. This eliminates the need for maintenance of a control valve and ensures a fixed solution level in the tank.

The green liquor transfer pump and piping are subject to scaling, and, within a matter of months, scale can build up to the point that the green liquor flow becomes obstructed. Some mills install a spare green liquor pipe line from the dissolving tank to the green liquor clarifier. One line can be being cleaned while the other is in service. Evaporator condensate is often used to wash the spare green liquor line; and in severe cases inhibited muriatic acid is used.

Another method for keeping the green liquor line clean is to alternate the weak wash and green liquor flows between the two lines. Valves are provided at the dissolving tanks and at the green liquor clarifier to allow switching the two lines between green liquor and weak wash. The lines are normally switched once per week on a regular schedule. In this way, the weak wash redissolves the scale which was deposited from green liquor.

Green liquor dregs

Underflow from the green liquor clarifier is pumped to a dregs mix tank. Small centrifugal pumps and diaphragm slurry pumps are used. The underflow ranges from 5 to 10% solids and the flow is relatively small. For instance, flow from a green liquor clarifier sized for 500 tons/day of pulp could be from 3 to 8 gpm. Timers are normally used to start and stop the underflow pump to allow a larger flow to be extracted for a controlled time.

Recovered warm water is normally used for dregs washing. The flow is recorded and may be manual or automatic control. The flow should be sequenced with the dregs underflow.

Dregs may be washed in a decanting type continuous clarifier and flow of diluted dregs from the mix tank would be from level in the mix tank. Weak wash from the clarifier is pumped to weak wash storage. Underflow from the dregs washer can be by a timer-controlled slurry pump, or can be by an automatic valve which directs the dregs to a sewer. The dregs would be removed from the effluent by the effluent primary clarifier. If the underflow sludge from the effluent clarifier is burned, the dregs should not be directed to the sewer.

Disposal of green liquor dregs to landfill is possible by using a filter (with lime mud addition). The flow of dregs slurry to the filter is controlled and the filter may be operated on an intermittent basis. In fact, many mills store the dregs and only operate the dregs filter on day shift. Provisions for acid cleaning of the filter cloth must be made.

Some mills mix a small amount of lime mud with the dregs to help form a filterable mat on the filter. This also serves as a purge for the lime/lime mud circuit.

With a precoat-type dregs filter operation, lime mud is used as a precoat. If the filter is operated continuously, the precoat may have to be changed about once per shift. The filter face cloth is washed and lime mud slurry is pumped into the filter vat. As the filter turns, a layer of mud is deposited. The green liquor dregs flow is then started, and, as the dregs form a layer on top of the lime mud, a scraper removes a thin layer.

Miscellaneous controls

There are a number of miscellaneous recorders and controllers in the causticizing plant which do not necessarily relate directly to the process. For example, the steam flow to the green liquor heater and slaker may be recorded, as well as the temperature and flow of makeup water to the department. The information thus recorded allows energy audits to be made. Warm or hot water tank levels are controlled by makeup addition.

Many mills have clarifier rake position indicators located in the control panel. This is important where a central control room concept is used and one operator controls the lime kiln and causticizing plant operation.

Ammeters for the dregs and lime mud mix tank agitator drives are often supplied.

Computers

Programmable logic controllers are now being used extensively in new mills to simplify wiring and reduce costs. They can control all the electrical drives and are programmed for interlocks, startup and shutdown sequences, and alarming of overload, stoppage, and the like.

Computer control is now being engineered into new mills and recent expansion projects (*13*). The white liquor production is programmed from alkali use in the digester. The slaking reaction is controlled by lime addition to the slaker using an on-line carbonate titrator (*14*), or a conductivity sensor to measure causticizing efficiency.

Better control of the causticizing reaction is claimed with minimum white liquor strength variations. This also maximizes the causticizing efficiency and controls overliming as free lime.

Design considerations

Equipment

There are two suppliers of recausticizing plant equipment in North America: Dorr-Oliver and Goslin-Birmingham. Normally, they supply the slaker, causticizer agitators, clarifier rakes and drives, mix tank agitators, mud storage tank agitator, lime mud filter, dregs filter and special pumps. The information given to them usually consists of the type of pulp produced and the daily tonnage, alkali requirements, white liquor test, and lime availability. The supplier will then prepare a basic flow diagram and equipment specifications to fulfill the liquor-making requirements. They will size the slaker, causticizers, clarifiers, and filters and present a bid on an equipment package. From their past experience, they will propose certain metallurgy, and it is up to the buyer to verify the correct metallurgy for their application.

Tanks

Today, the slaker is generally fabricated of type 304 stainless steel with a type 316 stainless steel agitator. The classifier section is also stainless steel with stainless steel rake or a screw conveyor.

The causticizers are normally mild steel tanks with a gunnite or brick lining. The agitators are 316 stainless steel. The white liquor level tank is normally type 304 stainless steel.

The white liquor clarifier is erected with mild steel plate ranging from 5/8 in. plate in the bottom course to 3/8 in. plate in the top course. Some mills, with a history of severe corrosion problems, clad the upper sections where the liquid level varies up and down, thus exposing the surface to air. The rake shaft is clad with 304 stainless steel from below the lowest liquid level to above the highest liquid level. The rake arms are mild steel; however, the rake shoes are normally 316 stainless steel.

The lime mud washer is erected with mild steel plate ranging from 1/2 in. in the bottom course to 1/4 in. in the top course. The rake shaft may be clad with 304 stainless steel. The rake arms and shoes are normally mild steel; however, the rake shoes should be an abrasion resistant material.

The green liquor clarifier is erected with mild steel plate ranging from 1/2 in. in the bottom course to 1/4 in. in the top course. The rake arms and shoes are normally mild steel, and the rake shaft may be clad with 304 stainless steel.

All other tanks in the causticizing plant are normally mild steel with a minimum corrosion allowance.

Pumps

Recausticizing plant pumps operate under corrosive, very erosive conditions. Special metals and rubber-lined pumps are used in the white liquor slurry and lime mud slurry service. Most of the major pump suppliers furnish chemical pumps for the general causticizing plant service. The lime mud slurry pumps are more specialized, and rubber lined pumps with replaceable liners find extensive use. There are some special metals which give good service: Illium PD by Allis-Chalmers; Worthite by Worthington; Durio by Durcon Co. The precipitated hardened stainless steel CD-4 MCu or the Alloy 20 type stainless steel with high nickel and chromium content appear to be very resistant to the white liquor slurry. The general rule should be to operate these pumps at the lowest possible speed to minimize impeller and casing wear, as well as wear on the pump shafts.

Variable speed drives on the clarifier underflow pumps, mud storage to the mud filter, and white liquor slurry from the causticizers to the white liquor clarifier are preferred to control valves. Fluid couplings are used extensively, and in recent years, due to continued improvement in service, variable frequency electric variable speed drives are being used more and more.

DC variable drive use appears to be more limited apparently due to maintenance of the DC motors in the corrosive atmosphere.

Electrical

All motors in the causticizing plant should be totally enclosed fan cooled (TEFC) and specified for pulp mill and chemical duty. This entails having epoxy coating, stainless steel name tag, and a 1.15 service factor. No aluminum conduit or wire should be used in the causticizing plant. Electric wire can be run in PVC coated conduit or in FRP cable trays. High voltage wire for larger motors would be run in FRP cable trays.

All motor control centers (MCC) are air conditioned with filtered air. It is preferred to locate the MCC above the ground floor, and the trend in newer mills is to have a separate building rather than in the lime kiln feed end, or hot end buildings as in the past.

Programmable controllers are being used extensively in the newer mills to eliminate conduit and wire.

Variable frequency variable speed drives are being used extensively in the newer mills. The control unit is located in the MCC in a controlled environment and a standard, TEFC induction motor is the drive. The speed control is more accurate than with the fluid couplings.

Instrumentation

Stainless steel tubing is used extensively for air and process instrument connections. PVC coated copper tubing is being used for instrument air supply.

Stainless steel transmitters are used extensively because of the corrosive nature of the environment. Dust-tight enclosures are required for field mounted instruments.

Steel bodied control valves can be used with type 316 stainless internals. However, many mills prefer to use all stainless steel valves to minimize maintenance.

The control room is normally common for the recausticizing plant and lime kiln. The room should be air conditioned with a filtered air supply.

Computer control is now gaining wide acceptance in the industry, and microprocessors are being used with data logging.

Potential problems and considerations

The recausticizing area in a kraft pulp mill plays an exceedingly important roll as it produces the pulping liquor for the digester plant. By converting the green liquor to white liquor, the recausticizing step completes the recovery cycle. To produce the pulping liquor in quantity and composition as required by the mill, careful control of the causticizing plant operation and maintenance of the equipment is required. Some of the typical mill conditions encountered in the industry are as follows:

1. The pulping liquor is characterized by fluctuations in quantity, strength, and composition resulting in lost production or higher cost due to makeup chemicals.
2. Intermittent and irregular pumping of lime mud causing imbalance in the process.
3. High degree of operator interface and very little automatic process control or automation.
4. Old and obsolete equipment and, in many cases, overloaded equipment.
5. Unsatisfactory quality of reburned lime.
6. Excessive loss of chemicals due to imbalance in the flow streams.
7. Excessive fuel and energy costs.
8. Impact of waste streams on the causticizing area, both in terms of upset in process as well as on the materials of construction.

Study often indicates that the possibilities for cost savings in this area are substantial, besides improving the operation reliability. Causticizing practice can be improved by: upgrading the materials of construction, introducing on-line analyzers and process control instrumentation to obtain the optimum conversion, reviewing the existing technology and equipment available to improve the conversion of the carbonate and the quality of reburned lime, providing technical skills to operate and manage this area, and allocating greater management attention to this essential part of the mill's operations.

Calculations for the recausticizing plant

Material balance (based on 1 ODT unbleached pulp)

Given conditions

Item	*Unit*	*Amount*
Pulp yield	%	48
AA on OD wood	%	16.5
W.L. total alkali	lb/cf	7.5
W.L. activity	%	85
W.L. sulfidity on AA	%	25
Lime availability	%	85
W.L. specific gravity	-	1.16
G.L. specific gravity	-	1.19
Make-up lime	lb/ODT	22

1. Alkali to digester

Active alkali $\frac{2000}{.48} \times .165$ @ Na_2O	lb	690
Total alkali $\frac{690}{.85}$ @ Na_2O	lb	810

2. White liquor to digester

WL volume $\frac{810}{7.5}$	cf	108
WL weight 108 × 62.4 × 1.16	lb	7,800
Weight of chemicals 10 lb/cf × 107	lb	1,070
Water in W.L. 7,800 – 1,070	lb	6,730

3. Lime calculation

NaOH 690 × .75 @ Na_2O	lb	515
CaO 515 × $\frac{56}{62}$	lb	465
Lime $\frac{465}{.85}$	lb	548

Item	*Unit*	*Amount*
Makeup lime	lb	22
Total lime	lb	570
4. Lime mud calculation		
$CaCO_3$ $465 \times \frac{100}{56}$	lb	830
Inerts 548 – 465	lb	83
Makeup $(22 \times .95 \times \frac{100}{56}) + (22 \times .05)$	lb	42
Total solids	lb	955
5. White liquor clarifier underflow		
Water solids	%	40
Total underflow $\frac{955}{.40}$	lb	2,390
Liquid 2,390 – 955	lb	1,435
Sodium in underflow		
$\frac{1435}{62.4 \times 1.16} \times 7.5$ @ Na_2O	lb	150
Weight of chemical in underflow		
$\frac{1435}{62.4 \times 1.16} \times 10$	lb	200
Water in WLC underflow 1,435 – 200	lb	1,235
6. Lime mud washer underflow		
Underflow solids	%	45
Recirculation from filter & scrubber	%	12
Total solids 955×1.12	lb	1,072
Total underflow $\frac{955 \times 1.12}{.45}$	lb	2,380
Liquid 2,380 – 1,072	lb	1,308
Sodium in underflow		
$\frac{1308}{62.4 \times 1.03} \times 1.5$ @ Na_2O	lb	30
Weight of chemical in underflow		
$\frac{1308}{62.4 \times 1.03} \times 2$	lb	38
Water in LMW underflow 1,308 – 38	lb	1,270

Item	*Unit*	*Amount*
7. Green liquor to slaker		
Total alkali to digester @ Na_2O	lb	810
Total alkali with lime mud @ Na_2O	lb	150
Alkali in slaker grits @ Na_2O	lb	1
Total alkali to slaker @ Na_2O	lb	961
8. Green liquor clarifier underflow		
Volume of GL to slaker $\frac{961}{7.5}$	cf	128
GL dregs (assumed)	lb/ODT	8
Underflow solids	%	4
Total underflow $\frac{8}{.04}$	lb	200
Liquid 200 – 8	lb	192
Sodium in underflow		
$\frac{192}{62.4 \times 1.19} \times 7.5$ @ Na_2O	lb	19
Weight of chemical in underflow		
$\frac{192}{62.4 \times 1.19} \times 12$	lb	32
Water in GLC underflow 192 – 32	lb	160
9. Green liquor from dissolving tank		
Total alkali to slaker @ Na_2O	lb	961
Total alkali to GL dregs @ Na_2O	lb	19
Total alkali from dissolving tank @ Na_2O	lb	980
Water in GL		
$(\frac{980}{7.5} \times 62.4 \times 1.19) - (\frac{980}{7.5} \times 12)$	lb	8,100
10. Lime mud filter		
Lime mud solids 955 × 1.12	lb	1,062
Feed solid	%	25
Weight of feed $\frac{1,062}{0.25}$	lb	4,247
Water in feed 4,247 – 1,062	lb	3,185
Solids in filtrate	%	1

Item	Unit	Amount
Solids in filtrate 3,185 × .01	lb	32
Filter discharge	%	70
Filter discharge $\frac{(1{,}062-32)}{0.7}$	lb	1,470
Water with cake 1,470 – 1,030	lb	440
Filter showers 440 × 2	lb	880
Total filtrate 3,185 – 440 + 880	lb	3,625
Sodium with cake	%	0.25
Sodium with cake 1,030 × .0025	lb	3
Sodium in filtrate 30 – 3	lb	27
11. Green liquor dregs filter (precoat)		
Cake discharge	%	40
Weight of discharge $\frac{8}{0.4}$	lb	20
Water with cake 20 – 8	lb	12
Sodium to filter @ Na_2O	lb	19
Filter efficiency	%	95
Sodium with cake @ Na_2O (lost)	lb	1
Sodium in filtrate @ Na_2O	lb	18
Showers on filter 2.5 × 12	lb	30
Dilution of GL dregs (arbitrary)	lb	112
Filtrate volume	lb	290
12. Weak wash		
A. LM washer		
Sodium from WLC @ Na_2O	lb	150
Sodium from LM filter @ Na_2O	lb	27
Sodium in mud from LMW @ Na_2O	lb	30
Total sodium to weak wash @ Na_2O	lb	147
Water from WLC	lb	1,235
Water from LM filter	lb	3,625
Water from LK scrubber	lb	3,320
Lime mud dilution and makeup	lb	1,000
Water in mud from LMW	lb	1,270
Total water to weak wash	lb	7,910
B. GL dregs		
Sodium in GL dregs filtrate @ Na_2O	lb	18
Water in GL dregs filtrate	lb	290

Item	Unit	Amount
C. Totals		
Total water in weak wash to D.T.	lb	8,200
Total sodium in weak wash to D.T.	lb	165
Conc of alkali in weak wash @ Na_2O	lbs/cf	1.25

13. Water balance (lb/ODT)

Out		In	
White liquor	6,675	Lime mud dilution	2,915
Slaker grit	10	Lime mud filter showers	880
GL dregs	12	G.L. dregs filter showers	142
Dissolving tank vent	100	Slaker/classifier wash	20
Lime mud	440	Lime kiln scrubber makeup	4,500
L.K. scrubber evap.	1,180		
Evaporation	40		
Total	8,457		8,457

14. Soda balance (lb/ODT @ Na_2O)

White liquor	810	Smelt	816
Slaker grit	1		
GL dregs	1		
Lime kiln stack	3		
Dissolving tank vent	1		
Total	816		816

Note: Some rounding of numbers has been done to try to keep indicated accuracy reasonable.

Literature cited

1. Kobe, K.A. and Wilkinson, J.A., *Ind. Eng. Chem.*, **45**(2):307 (1953).
2. Littman, F.E. and Gaspari, H.J., *Ind. Eng. Chem.*, **48**(3):408 (1956).
3. Goodwin, L.F., *J. Joc. Chem. Ind.*, **45**:360 (1926).
4. Olsen, J.C. and Direrga, O.G., *Ind. Eng. Chem.* **33**:204 (1941).
5. Keitaanniemi, Osmo & Virkola, Nils-Erik, "Undesirable Elements in Causticizing Systems". *Tappi*, **65** (7):89-92 (1982).
6. McCumsey, K.R. & Deich, W.E., "New Filter System Controls Dregs," *Pulp & Paper*, **56**(2):158-159 (1982).
7. Swanson, H.X., "Dregs Washing Improvements Reduce Soda Loss", *Pulp & Paper*, **53**(4):101-103 (1979).
8. Barnhill, Geo. D., "Dregs Clarification with an Inclined Plate Separator", *Southern Pulp & Paper*, (7):12-13 (1982).
9. Lumikko, Juha, "Applications of Pressurized Filtration in Recausticizing", International Conference on Recovery of Pulping Chemicals, Sept., 1981.

10. Wilhelmsen, L., "Mud Oxidation for Kiln TRS Control", TAPPI-PIMA Spring Conference, May, 1978.
11. Caron, A.L. & Franklin, M.E., *Refinements in Lime Mud TRS Emission Control*, NCSA.
12. Grandmaison, E.W., Marois, J., Senneville, Que. Domtar, Inc. Research Center. (Private Communications).
13. Elsila, M., et al. "Computer Control of Causticizing and Lime Kiln is Possible", *Pulp & Paper*, **53**(11):152-159 (1979).
14. Hultman, B. & Berglund, A., "A Sensor Based Causticizing Control System", Int'l Conf. on Recovery of Pulping Chemicals, Sept., 1981.

Polysulfide Liquor Generation and White Liquor Oxidation

R. P. Green

Introduction

The use of polysulfide (PS) cooking liquor in the kraft pulping process dates back to a U.S. Patent application by Fuller and Woodside in 1943 (*1*). They claimed the substitution of not less than about one half of the sodium sulfide normally present in the cooking liquor by sodium polysulfide.

In 1946, Haegglund reported a considerable increase in pulp yield when elemental sulfur was added to the kraft cooking liquor (*2*).

At the Norwegian Pulp and Paper Research Institutes, a study of corrosion problems in kraft digesters was reported in 1955 (*3*). It was shown that corrosion could be caused by small amounts of polysulfide acting as a cathodic depolarizer. On the other hand, larger amounts of polysulfide actually prohibit corrosion by passivation of the digester walls.

The benefits of increased pulp yield plus a reduction in digester corrosion prompted several kraft mills to practice polysulfide pulping on a commerical scale. Two mills in Norway converted to polysulfide pulping. One mill achieved a 4% yield increase on wood by the addition of an extra 2.2% sulfur, basis o.d. wood, to the cooking liquor (*4*). Other European and North American mills found similar results (*5*).

However, few kraft mills were able to carry out polysulfide pulping because of the lack of a technically feasible and economically sound recovery process for polysulfide sulfur. Thc addition of sulfur to the recovery system results in large increases in gaseous emissions of TRS and SO_2, and this is no longer permissible in most countries.

Methods of regenerating polysulfide pulping liquor

A number of possible polysulfide recovery processes have been suggested and patented. Ferrigan and Coppick received a patent (*6*) on a process in which sulfur would be added to the digester, and the black liquor evaporated and burned in the recovery boiler in the usual manner. A portion of the green liquor would then be carbonated to release hydrogen sulfide for conversion to free sulfur.

A similar process was patented by Venemark (*7*). In this case, green liquor would be treated with a gas containing a high proportion of carbon dioxide and less than 1% oxygen (flue gas). Another modification of the idea of carbonation was patented by Fogman (*8*). Black liquor would be partially evaporated and then treated with a recycle stream of liquor containing sodium carbonate and bicarbonate. This mixture would be carbonated with a gas stream high in carbon dioxide. The carbonated black liquor would be vacuum stripped to release hydrogen sulfide for conversion to sulfur.

Brannland (*9*) proposed to drop the pH of weak black liquor by carbonation, and then remove hydrogen sulfide gas during the evaporation step. The hydrogen sulfide gas was compressed, and sulfur recovered in a Claus reactor.

Venemark initially proposed electrolytic oxidation (*10*), but the investment costs were very high (*11*). Polysulfide also can be prepared by thermal decomposition of thiosulfate, partial oxidation of sulfide in smelt, or by partial reduction of sulfate; but high temperatures and the absence of water are required (*12*).

Air oxidation of the sulfides present in a mixture of kraft white liquor and black liquor produced some polysulfide (*13*) and increases in pulp yield were obtained (*14, 15*). High sulfidity is required and thiosulfate, as well as polysulfide, is formed.

The concept of oxidation of the sulfide in white liquor was pursued by Union Camp Corporation, and several catalysts were discovered. At the TAPPI Alkaline Pulping Conference in 1969, Barker reported on the use of various manganese oxide compounds as catalytic oxidizers for the oxidation of sodium sulfide to sodium polysulfide and caustic, as shown in the following equations (*16, 17*):

$$(1)\quad Na_2S + MnO_2 + H_2O \rightarrow S^0 + 2NaOH + MnO$$

$$(2)\quad (x\text{-}1)S^o + Na_2S \rightarrow Na_2Sx$$

The reduced MnO was separated from the liquor, and reoxidized with air:

$$(3)\quad 2MnO + O_2 \rightarrow 2MnO_2$$

Pilot plant studies on this process were reported at the TAPPI Alkaline Pulping Conference in 1972 (*18*). Two continuous systems, shown in Fig. 5.27 and Fig. 5.28, were operated.

Using the process shown in Fig. 5.28, continuous oxidation of a kraft white liquor containing 32 g/L of Na_2S, as Na_2O produced a product polysulfide

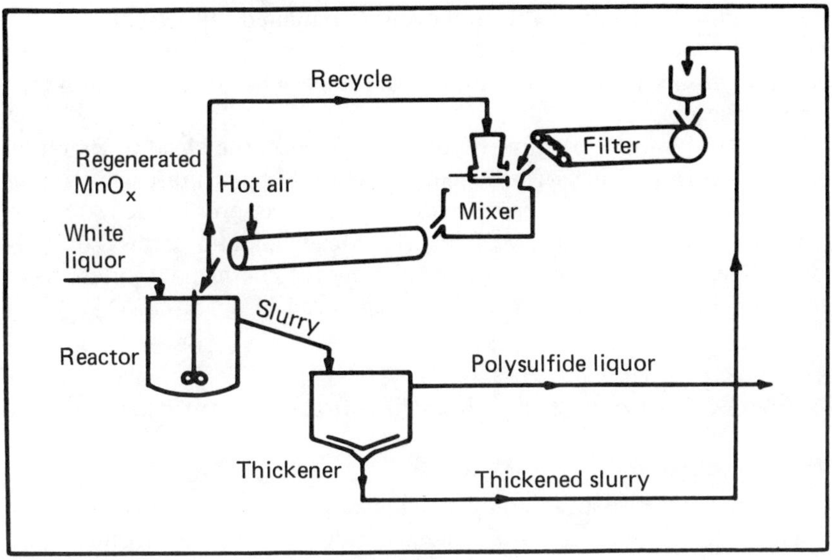

Fig. 5.27. Union Camp filter—solids mixer process.

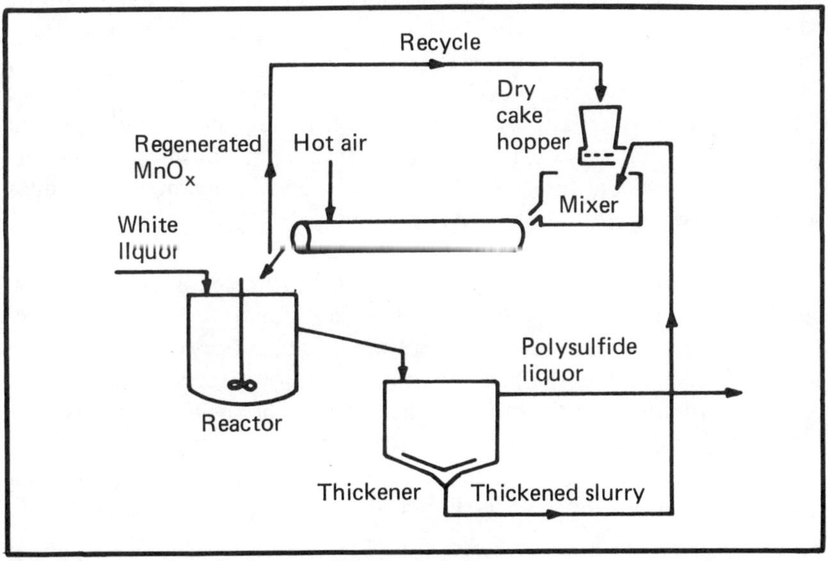

Fig. 5.28. Union Camp thickened slurry—mixer process.

liquor containing 11.5 g/L of Na_2S and 7.0 g/L of polysulfide sulfur. That is, 64% of the sodium sulfide present was oxidized, and of this, 66% formed polysulfide sulfur. The balance of the oxidized sulfide was presented as higher oxides of sulfur, principally thiosulfate.

Other improvements in this process were patented (*19, 20, 21*).

Mueller patented a novel system which oxidizes the smelt from a kraft recovery furnace at high temperatures in the presence of coke and oxygen to form the polysulfide (*22*).

At the 1974 TAPPI Alkaline Pulping Conference, the Mead Corporation reported another means of producing polysulfide from kraft white liquor by oxidation with air in the presence of a partially wet-proofed activated carbon catalyst (*22, 24*). Called MOXY® (for Mead Oxidation), this is the only polysulfide producing process which has achieved commercial status. As of 1981, six commercial MOXY Systems have been in operation.

The MOXY process for polysulfide liquor generation

Process description

The MOXY process is an add-on step to the normal kraft recausticizing system which converts a portion of the sodium sulfide present in kraft white liquor to sodium polysulfide and sodium hydroxide. This conversion is done in a single step oxidation, using compressed air, in a reactor containing a fixed bed of the partially wet-proofed activated carbon catalyst. Figure 5.29 is a schematic flow diagram of the process.

The first part of the process consists of filtering to remove the suspended solids present in white liquor in order to reduce plugging or fouling of the catalyst bed in the reactor. This filtration is done in polishing filters containing anthracite coal filtering media. The filters are operated in parallel, and usually two or three filters are required.

The filtered white liquor is then introduced at the top of the reactors at a controlled flowrate. Compressed air is also introduced at the top of the reactor and its flowrate is ratioed to the white liquor flow to achieve the desired amount of oxidation of the sulfide present. The oxygen in the air and the sulfide react as the liquor and air flow down through the catalyst bed. Retention time is in the order of one minute. The spent air and resulting polysulfide liquor are discharged from the reactor to a sump tank where the two are separated. The air is exhausted to the atmosphere and the polysulfide liquor is pumped to the digester liquor storage tank.

Polishing filters and their operation

The liquor polishing filters are vertical pressure vessels containing six sizes of graded anthracite filter media. This media has proven successful for filtering hot, caustic solutions such as kraft white liquor. Figure 5.30 shows how the various sizes of media are located in the filter. The top layer contains the smallest size media and is the one which does the filtration. The remain-

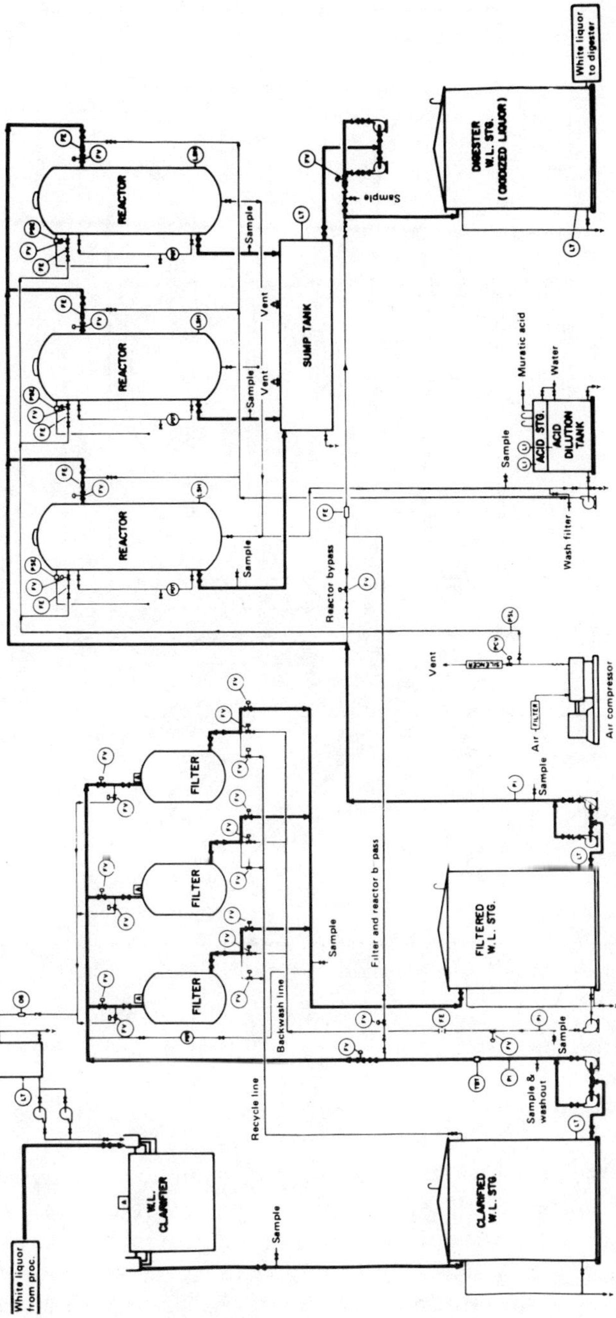

Fig. 5.29. MOXY system 1 flow diagram.

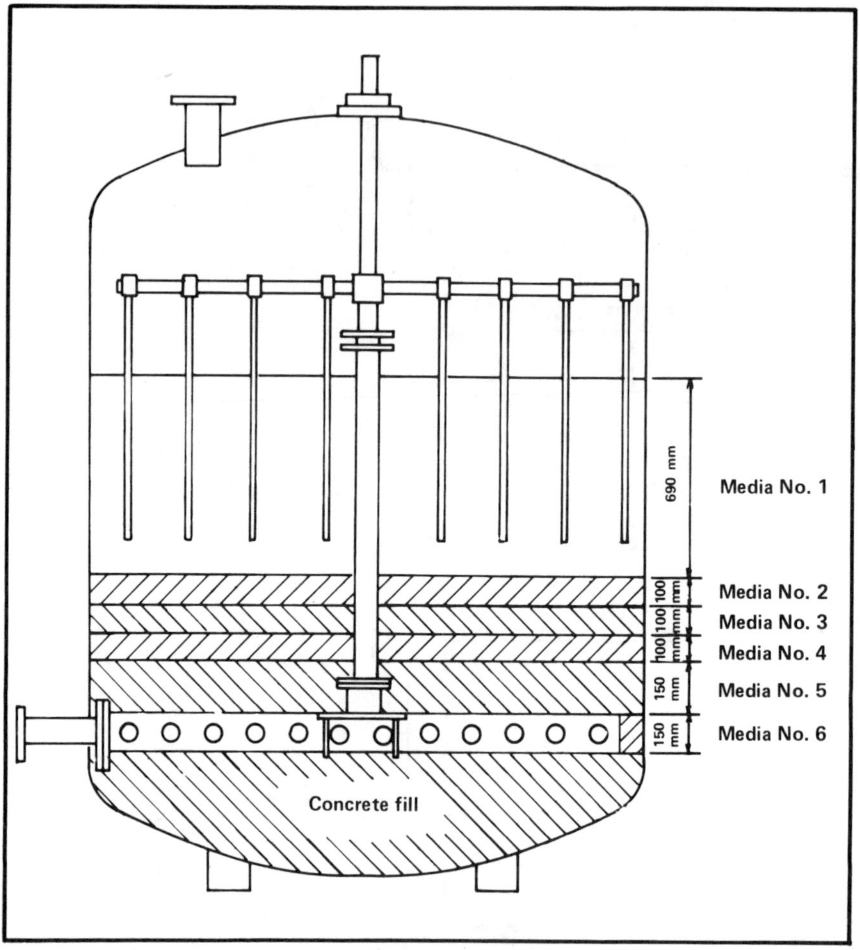

Fig. 5.30. Schematic of anthracite coal polishing filter.

ing five layers are increasingly coarse from top to bottom and serve as a support for the top layer of fine media. Each filter is furnished with a motor driven agitator that rotates at approximately four rpm during the backwashing operation to break up the fine filter media and increase the efficiency of the backwash.

Typically, the incoming white liquor may contain from 50 to 200 mg/L of suspended solids. The filters will remove almost all of these particles, and the average content of the effluent is in the order of 5 mg/L. Normally, the backwash of the filters is done once every six to eight hours. Filtered white liquor is used, and the effluent is returned to the clarifier via a surge tank (see Fig. 5.29). The anthracite coal is usually replaced once per year, but operations in excess of two years have been reported.

Several mills have used this type of polishing filter to remove suspended solids (principally $CaCO_3$) to reduce scaling in digesters, or to remove unburned carbon which sometimes is present in the smelt from the recovery furnace.

In the event of an unusual upset in the white liquor clarifier, large amounts of suspended solids may appear in the white liquor. These filters cannot handle such loads, and will rapidly become plugged. Repeated backwashing with water or filtered liquor may restore the anthracite. If not, then the filter media will have to be replaced.

The MOXY reactor and its operation

The reactor is a vertical, cylindrical vessel which contains the catalyst. The catalyst bed is separated into sections to provide for uniform redistribution of the liquor and air as they flow downward. The reactors are supplied in several diameters (1.1, 1.5, and 2.1 meters) depending upon the liquor requirements of the mill, and the sulfide content of the liquor. As an example, a 2.1 meter reactor has the capacity to oxidize about 2.0 cubic meters per minute (530 gpm) of white liquor with a content of 30 g/liter of Na_2S, as Na_2O. The number and size of reactor will depend upon the pulp mill capacity.

The liquor flow rate to the reactor is measured and controlled by a flow meter and an automatic valve. Air flow is measured by an orifice plate, and controlled by an automatic valve. The pressure drop across the reactor is measured.

The catalyst used is a granular grade of activated carbon which is treated with polytetrafluoroethylene (PTFE) to provide wet-proofing over a portion of the surface of the carbon. Because the carbon is a porous, granular material, it is an excellent filter. It will gradually become contaminated with the small amount of suspended solids still remaining in the filtered white liquor, normally about 5 mg/L. As this occurs, additional air must be supplied to achieve the desired level of sulfide oxidation. As more air is supplied, the pressure drop across the reactor will increase. The normal pressure drop, for clean catalyst, is about 0.3 to 0.7 atmospheres (5 to 10 psig). When the pressure drop approximately doubles, it signals the need for an acid wash to dissolve the entrapped solids and remove them from the reactor.

Acid washing is accomplished by shutting down the reactor, draining the remaining liquor, and then rinsing the reactor with water. Once rinsed, dilute hydrochloric acid (ca. 3 to 5%) is circulated through the reactor for 30 to 45 minutes, the spent acid is drained, and the reactor again rinsed with water.

The reactor is then returned to service. On a fresh charge of catalyst, the need to acid wash is in the order of once every month or two, but as the catalyst ages, the frequency gradually increases to about once every two to three weeks. Normal life of the catalyst has been in the order of one year.

Chemistry

The oxidation of sodium sulfide by air proceeds as follows:

(4) $2Na_2S + O_2 + 2H_2O \rightarrow S^\circ + 4\ NaOH$

(5) $(x\text{-}1)S^\circ + Na_2S \rightarrow Na_2Sx$

(6) $2Na_2S + 2O_2 + H_2O \rightarrow Na_2S_2O_3 + 2NaOH$

Equations 4 and 5 represent the desired reaction, while equation 6 is illustrative of an undesirable side reaction which results in the formation of sodium thiosulfate. Normal operation procedure is to oxidize about 50 to 60% of the sodium sulfide present in the white liquor. Of the sulfide oxidized, about 70% proceeds via equations 4 and 5; 30% by equation 6. Table 5.13 is typical of the results from a commercial reactor.

Table 5.13 represents 60% oxidation of the sulfide in the kraft white liquor, at an efficiency of 70%.

Complete oxidation of kraft pulping liquors

Introduction

Closure of pulp mill systems for pollution or other reasons can upset the desired sodium-sulfur ratio in a mill's recovery system. This problem may

Table 5.13. Typical results obtained upon oxidation of kraft white liquor to produce polysulfide liquor

	White liquor (gpl as Na_2O)	*Polysulfide liquor*
Na_2S	35.0	14.0
NaOH	65.0	82.9
Na_2CO_3	25.0	25.0
$Na_2S_2O_3$	3.0	6.1
Na_2Sx, gpl as sulfur	0	7.6
Active alkali	100.0	96.9
Total Alkali	125.0	121.9

often be alleviated if chemical uses normally supplied with caustic soda are supplied instead with completely oxidized white, green, or orange (polysulfide) liquor.

Oxidized white liquor, free of sodium sulfide, can be used in a number of areas within a pulp mill, including scrubbing of flue gases to remove SO_2 and TRS emissions, and in various parts of the bleaching operation. Oxidized liquors can be used for pH control of oxygen bleaching stages, and is preferred over caustic soda, particularly if the effluent from the oxygen stage is returned to the recovery cycle. The liquor can also be used in caustic extraction stages, or as an anti-chlor in hypochlorite or chlorine dioxide stages. In general, oxidized liquor can be substituted for fresh caustic soda wherever the presence of sodium thiosulfate can be tolerated.

Chemical reaction

The oxidation of sodium sulfide to sodium thiosulfate and caustic soda is, at least in part, a two step reaction as shown in Equations 7, 8, and 9.

$$(7)\quad 2Na_2S + O_2 + 2H_2O \rightarrow 2S^\circ + 4\ NaOH$$

$$(8)\quad 2S^\circ + O_2 + 2NaOH \rightarrow Na_2S_2O_3 + H_2O$$

$$(9)\quad 2Na_2S + 2O_2 + H_2O \rightarrow Na_2S_2O_3 + 2NaOH$$

Equation 9 is the sum of equations 7 and 8. All of the sulfur is oxidized to thiosulfate, while half of the sodium portion of the sodium sulfide appears as caustic soda. Typical results are shown in Table 5.14.

Completely oxidized liquor will be nearly "water-white" in color. Incompletely oxidized liquors will retain a pale yellow color due to the presence of some sodium polysulfide.

Table 5.14. Typical results obtained upon complete oxidation of kraft white liquor

	White liquor (gpl as Na_2O)	*Oxidized liquor*
Na_2S	35.0	0
NaOH	65.0	82.5
Na_2CO_3	25.0	25.0
$Na_2S_2O_3$	3.0	20.5

The MOXY-dizer® system

The principles of the MOXY system described in the section on polysulfide liquor generation can also be applied to complete oxidation of kraft white, green, or orange (polysulfide) liquors (*23, 26*).

Oxidation is accomplished in a single step operation, using compressed air, in a reactor containing a fixed bed of granular catalyst (see Figure 5.31).

The catalyst used is a granular grade of activated carbon which is treated with polytetrafluoroethylene (PTFE) to provide wet-proofing over a portion of the surface of the carbon (*24*). The preferred catalyst is similar to, but differs from, that used for polysulfide liquor generation.

Figure 5.32 illustrates the system for green liquor or white liquor (GW system) oxidation. It is an add-on to the normal causticizing operation. If orange (polysulfide) liquor prepared in a MOXY system for polysulfide liquor generation is the feed liquor, then there is no need for the polishing filter or for the filtered liquor storage tank shown in Figure 5.32.

The operation of the filter portion of the system is the same as was previously described. Liquor substantially free of suspended solids is required to reduce fouling of the catalyst bed within the reactor.

As shown in Figure 5.31, air and the feed liquor are introduced at the top of the reactor at controlled flow rates. With a clean bed of catalyst, the molar

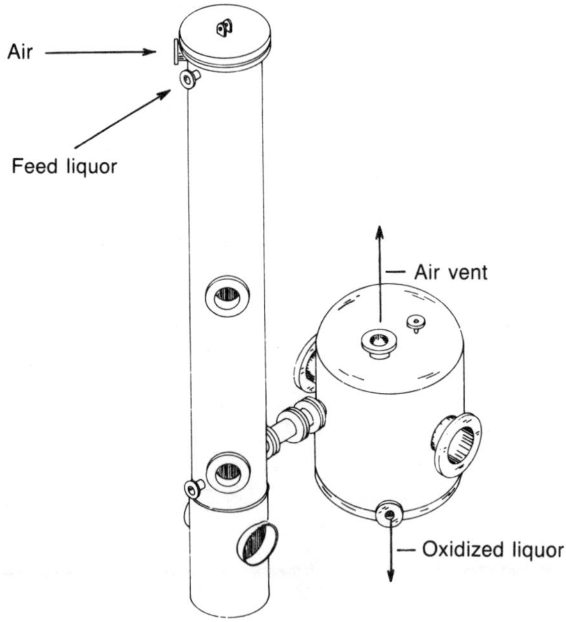

Fig. 5.31. Schematic of MOXY-dizer®.

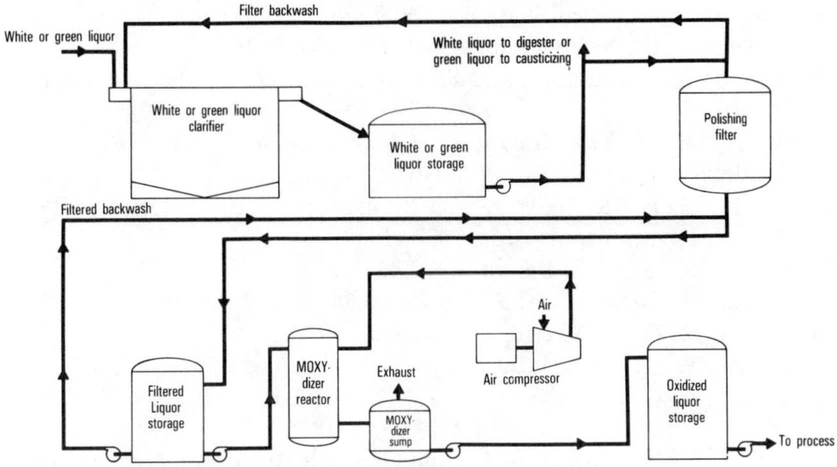

Fig. 5.32. MOXY-dizer® GW system flow diagram.

ratio of the oxygen in the air to the sodium sulfide in the feed liquor should be about 1.2. This is 20% more than the theoretical oxygen requirement for the oxidation of sulfide (see Equation 9). The liquor and air flow concurrently downward through the catalyst bed, and are separated in the sump tank shown in Figure 5.27. Retention time is approximately one minute.

The reactors come in various sizes: 600, 900, and 1100 mm diameter, depending upon the quantity of oxidized liquor required.

Control of the oxidation can be by color. Insufficient oxidation will result in some yellow color remaining in the oxidized liquor (due to trace amounts of polysulfide remaining). Excessive amounts of air can also result in incomplete oxidation due to "blowing" the liquor through the reactor at too fast a rate to permit the reaction to be completed.

Literature cited

1. Fuller, R. R. and Woodside, V.: U.S. Serial No. 515,015 (June 16, 1943), Can. Patent 444,274 (Sept. 23, 1947).
2. Haegglund, E.: *Svensk Papperstidn*. **49**(9):191 (1946).
3. Haegland, B. and Roald, B.: *Norsk Skogind*. **9**(10):351 (1955).
4. Hilmo, P. and Johnsen, K.: *Pulp & Paper* **43**(5):73 (1969).
5. Kleppe, P. J., *Tappi* **58**(8):172 (1975).
6. Ferrigan, J. J. and Coppick, S.: U.S. Patent 3,210,235 (1965).
7. Venemark, E. T., U.S. Patent 3,331,732 (1967).
8. Fogman, C. D., Nelson, H. W., and Tsao, U., U.S. Patent 3,216,887 (1965).
9. Brannland, R., Can. Patent 810,107 (1965).

10. Venemark, E., Swedish Patent 160,736 (1957).
11. Venemark, E., *Svensk Papperstid.* **67**:157 (1964).
12. Ahlgren, P., LeMon, S., and Teder, A., *Acta. Chem. Scan.* **21**(4):1119 (1967).
13. Bilberg, E., and Landmark, P., *Norsk Skogind.* **5**:221 (1961).
14. Landmark, P., U.S. Patent 3,216,887 (1965).
15. Landmark, P. Kleppe, P. J., and Johnsen, K., *Tappi* **48** (5):56A (1965).
16. Barker, R. G., *Tappi* **53**(6): 1087 (1970).
17. Barker, R. G., U.S. Patent 3,470,061 (1969).
18. Barker, R. G., and Ma, J. L., *Tappi* **56**(5): 112 (1973).
19. Barker, R. G., U.S. Patent 3,551,286 (1970).
20. Barker, R. G., and Ma, J. L., U.S. Patent 3,653,824 (1972).
21. Barker, R. G. and Becker, E. S., U.S. Patent 3,860,479 (1975).
22. Mueller, W.A., U.S. Patent 4,153,502 (1972).
23. Smith, G. C., Knowles, S. E., and Green, R. P., *Paper Trade Journal*, Vol. 159 No. 13, p. 38-41 (May 1, 1975).
24. Smith, G. C., and Sanders, F. W., U.S. Patent 4,024,229 (1977).
25. Brunsvik, J. J., "Experience with PS-Pulping in Moss". Paper presented at the Polysulfide Pulping Seminar, Japan TAPPI, Tokyo, Japan (May, 1979).
26. Smith, G. C., and Sanders, F. W., U.S. Patent 4,162,187 (1979).

6
Lime Reburning

The Rotary Lime Kiln

Douglas J. Kramm and Wallace L. Schultz
Edited by Gerald W. Hough

Introduction

The rotary lime kiln has been the prime method of calcining lime mud from the causticizing plant in the kraft pulp industry for many years. The rotary kiln has been used since 1900, and first made its appearance in the paper industry in North America in the mid-1920s. The early kilns had two supports and were relatively small, 6 to 8 ft (1.8-2.4m) diameter and 100 to 140 ft (30-40m) long. Such kilns contained no chains, and, with solids content of the feed as low as 40%, fuel rates as high as 20 million Btu per ton of product were not uncommon. The first installation of chain was made in the 1930s, and improved fuel economy resulted.

The cement industry pioneered in the use of longer, more efficient rotary kilns. Such kilns, which have three or more supports, have been used in the paper industry in the U.S. since the mid-1930s. The improved economy of longer kilns soon became apparent and today with the high cost of fuel only long kilns are considered. Table 6.1 presents a statistical study of long versus short kiln installations in the U.S. and Canada.

Important progress has been made in the mechanical construction of the rotary kiln. A change from riveted to welded design in the early 1930s has

Table 6.1 Lime mud kilns installed in North America

Number of supports	*Prior to 1945*	*1945 to 1964*	*1965 to 1978*
Two supports	98	47	16
Three or more supports	2	93	94

simplified construction, as well as placement of refractory. Advancements in metallurgy have improved shell soundness, roundness and alignment; and have contributed to longer gear, tire, roller, and bearing life. Progress in better design has improved mechanical performance. The use of the computer has been of great assistance in rapid and accurate analysis of each kiln as a further insurance of optimum design.

Lime calcination

A desirable lime is one which has been calcined with as complete an elimination of CO_2 gas as is practically possible, with the resulting characteristics of good slaking qualities, high reactivity, and high availability. Lime calcination is a time- temperature process.

The calcination of lime commences at a temperature of approximately 1500°F (820°C), at which point the charge begins to give off the CO_2 gas which is in chemical combination with calcium oxide. In order to effect this release of CO_2 gas, calcination may proceed rapidly at relatively high temperatures, or alternately, it may proceed slowly at relatively low temperatures.

Control of the calcining process at a higher temperature is more difficult, resulting in possible overburning, poor availability, and poor reactivity. For each specific feed, the proper time - temperature relationship must be determined by adjustment of temperature and kiln speed, or retention time.

A lime sludge kiln can be divided into three distinct zones according to functional purposes. Starting at the cold feed end the chain system comprises the evaporation zone; the preheat zone heats the feed to a calcining temperature; and in the calcining zone final chemical conversion to lime is completed and combustion of the fuel takes place. The design of the chain system, kiln, and refractory in each zone is predicated on efficiently performing a specific function.

In the selection of a rotary kiln, a variety of sizes is available for any given production rate. For example, one selection could be a kiln with a larger diameter and relatively short length, or, for the same production, a smaller diameter, relatively longer length kiln could be used. Both kilns selected would have the same internal volume. The quality of lime produced is dependent upon how each of these kilns is fired and controlled in regard to constant and uniform feed, operating temperature and kiln speed or retention time.

Table 6.2 gives the expected performance from three different proportioned kilns, each with the same nominal rating of 170 tons/day burned product. For purposes of this comparison, the moisture in the feed is taken at 35%,

Table 6.2 Comparative fuel consumptions for equal capacity kilns with different length

Kiln size	*Ratio length/dia.**	*Millions Btu/ton prod.*
11′-6″ × 195′ (3.51m × 59.4m)	18.5:1	9.4 (2611 kgCal/kg)
10′-6″ × 240′ (3.2m × 73.15m)	25:1	8.6 (2389 kgCal/kg)
10′-0″ × 270′ (3.05m × 82.3m)	30:1	8.15 (2264 kgCal/kg)

*Diameter inside brick

Table 6.3 Comparative fuel values for different size and capacity kilns

Kiln size	*TPD production*	*Millions Btu/ton prod.*
8′-0″ × 225′ (2.44m × 68.6m)	85 (77.3 MTPD)	8.5 (2361 KgCal/Kg)
10′-0″ × 270′ (3.05m × 82.3m)	170 (154.5 MTPD)	8.15 (2264 KgCal/Kg)
12′-6″ × 330′-0″ (3.81m × 100.6m)	340 (309 MTPD)	7.8 (2166 KgCal/Kg)

the reactivity or percent CaO in the product is taken at 90%, and Bunker C oil is the fuel. This table shows the improved fuel economy of a kiln with a 30:1 ratio; in this case a 10 ft diameter × 270 ft (3.05m × 82.3m) kiln.

Larger kilns require less fuel per unit of production than small kilns. Comparative fuel values are shown in Table 6.3.

Retention time, slope and speed

Proper retention time in a rotary kiln is of importance if optimum capacity of a highly reactive product is to be expected. Retention time is determined by kiln slope and speed of rotation. Lime sludge kilns are normally operated at slopes from 1/4 to 1/2 in./ft (6-13mm), and at rotational speeds from 0.5 rpm to 1.5 rpm, but an examination of the slope speed relationship should be made to determine the best relationship for each kiln and operation. Experience indicates that proper calcination can be effected if the retention time is somewhere between 120 and 170 min. In order to efficiently transfer more heat to the material, it is necessary to create a rolling or cascading load action rather than a sliding motion. In designing kilns, the rotational speed must be high enough to establish the desirable load action, and slope so as to obtain the necessary retention time to allow complete calcination of the material. A peripheral speed above 35 ft/min (10.5m/min) will usually create a desirable rolling load action.

Retention time is directly proportional to the angle of repose of the material and the kiln length, and is inversely proportional to the kiln inside diameter, rotational speed, and slope. The formula used by the U.S. Bureau of Mines to determine rotational speed at designated retention time is:

$$N = \frac{1.77\sqrt{\varnothing} \times L}{S \times D \times T}$$

N = Rotational speed, rpm
∅ = Angle of repose
L = Length of kiln in feet
S = Slope in degrees
D = Diameter inside brick in feet
T = Retention time in minutes

The angle of repose for lime mud kilns is taken as 35°.

S for different slopes

S = 1.79 at 3/8″/ft
S = 2.39 at 1/2″/ft
S = 2.98 at 5/8″/ft

Heat balances

The basic reaction in a rotary kiln is in accordance with the following equation:

$$\underset{(100)}{CaCO_3} = \underset{(56)}{CaO} + \underset{(44)}{CO_2}$$

There are impurities in the feed and not all the $CaCO_3$ is converted to CaO. If the impurities and unconverted $CaCO_3$ are considered inert, a new equation which more nearly describes the process in a rotary kiln is as follows:

$$\text{Inert} + CaCO_3 = CaO + \text{Inert} + CO_2$$

Dust is discharged in the flue gas from the kiln and the lime mud fed to the kiln contains water, thus, the final equation for a material balance can be taken as follows:

$$\overset{100}{\text{Inert}} + \overset{1710}{CaCO_3} + \overset{920}{\text{Water}} =$$

$$\overset{900}{CaO} + \overset{100}{\text{Inert}} + \overset{710}{CO_2} + \overset{920}{\text{Water Vapor}} + \overset{100}{\text{Dust}}$$

Table 6.4 Heat balance for 10 ft × 270 ft rotary kiln

Conditions:	170 ST/D (154 t/d) product	Bunker "C" fuel
	35% water in feed	5 wind Km/h
	1800°F (980°C) product temp.	90% CaO product
	450°F (230°C) exit gas temp.	6" (150mm) refractory
	10% excess air	Good quality feed
	60°F (16°C) ambient temp.	Uniform feed rate

Heat exchange phase	*Btu/ short ton prod.*	*kgCal/kg*
Evaporate & heat water to 450°F (230°C)	2,410,000	670
Heat into kiln product	820,000	227
Heat of dissociation	2,500,000	695
Heat in CO_2 gas	138,000	38
Heat in dust	20,000	6
Radiation loss*	1,020,000	283
Total net heat	6,908,000	1919
Heat in products of combustion	1,242,000	345
Total gross heat**	8,150,000	2264

* Radiation loss calculated for 5 Km/h wind and shell temperatures as follows: 90 ft (27m) of 400°F (205°C); 27m of 250°F (121°C); 27m of 200°F (94°C).

**If kiln-attached integral tube coolers are used and designed to cool lime to 600°-700°F (300°-370°C), an additional 137 kgCal/kg (500,000 Btu/short ton) of product can be saved. In the above heat balance, the gross heat would become 2127 kgCal/kg (7,650,000 Btu/short ton) product.

The values shown are kilograms for one ton (1000 kg) of kiln product assuming a dust loss of 10% of product and 65% dry mud feed solids.

A heat balance for a 10 ft diameter × 270 ft (3.05m × 82.3m) rotary lime kiln is presented in Table 6.4.

Effect of impurities

The feed, which is never 100% $CaCO_3$, always has impurities. The amount of impurities is variable, depending upon the particular mill. Where these impurities can be held to a uniform minimum, the operation of the rotary kiln will be much more successful. Impurities in excessive quantity have a tendency to build rings or balls within the kiln, which not only restrict the production and efficiency of the kiln, but can necessitate shutdown of the kiln for cleanout purposes. The impurity causing greatest difficulty is sodium which, when present in excess quantity, invariably causes ringing at the discharge end because of its low fusion temperature. Other impurities which have a tendency to promote ring formation include Al_2O_3, Fe_2O_3, and SiO_2. These impurities can be introduced into the circuit in a variety of ways, but a prime suspect is always lime stone when used for lime makeup. Free lime can also cause operating problems and excessive dusting. Therefore, it is imperative that the slaking and causticizing operation be closely controlled to convert the maximum amount of lime to sodium hydroxide and calcium carbonate.

Effect of moisture

In the normal, modern kiln operation, the amount of moisture in the feed to the kiln very rarely exceeds 40% and, in some installations, the moisture content may be 28% or less. Older kilns are in operation with feed moisture contents as high as 50%; it is self-evident that the higher the moisture content, the greater the fuel consumption per unit of production. When considering that the heat of vaporization of water is 970 Btu per pound, with an additional heat loss assigned to each pound of water for heating to the boiling point, plus super-heat in the water vapor form, there is a heat consumption of over 1200 Btu for every pound of water. For this reason it is important from the standpoint of fuel economy to reduce the moisture content in the feed to the lowest practical limit.

A rotary lime kiln is provided with a chain system at the feed end to assist in the evaporation of water from the mud feed. The chain system also acts as a curtain or dust arrestor and dust created in the calcining and drying zone is readsorbed on the lime mud. The design and materials used on the chain system have changed over the years to adjust for changes in kiln size and temperatures and also to resist corrosion from sulfur in the fuel oil and the burning of noncondensible gases from the pulp mill. The chain system must permit a positive feeding of the material through the chain system and must provide sufficient surface area, and be hung in a manner that will permit maximum efficiency of evaporation. It follows that the greater the amount of moisture, the greater the overall quantity of chain required, the greater the portion of the kiln that must be assigned to drying. Table 6.5 shows relative fuel rates at various lime mud dryness.

Flow of gas through kiln

The gas leaving the feed end of the kiln is the combined total of combustion products, CO_2 gas driven off from the material, and water vapor evaporated from the wet feed. Natural gas or oil is burned with air, which is supplied by a separate, dampered, primary air fan. Primary air is usually 10% to 20% of total air requirements. This air is introduced through a tube surrounding

Table 6.5 Relative fuel rates at various feed dryness

Lime mud dryness	60%	65%	70%
Exit gas temperature*	400°F (204°C)	450°F (232°C)	500°F (260°C)
Millions of Btu/ton product	8.6 (2389 kgCal/kg)	8.15 (2264 kgCal/kg)	7.7 (2139 kgCal/kg)

* Note how exit gas temperatures rise as water in feed decreases

the burner pipe, and is intimately mixed with the fuel at the firing end of the burner. Secondary air is induced through the bottom of the firing hood or can in part be provided by a secondary air fan, which also keeps the nose ring segments and nose brick cool. If integral tube coolers are used, preheated secondary air is induced through the cooler tubes cooling the product to about 600°F (315°C).

Fuels

The fuels used in lime sludge kilns are natural or manufactured gas, No. 6 fuel oil, and pitch.

The most widely used fuel, No. 6 fuel oil, is heated and then atomized by using pressure, air, or steam. The oil is heated, strained, and pumped to the burner at pressures up to 1000 psig for pressure atomization, but to only about 350 psig for steam atomization.

Natural gas is used where gas is available from commercial pipelines. When there is a possibility of gas supply interruption, a standby fuel oil system is required.

Pulverized coal is not fired directly in a lime sludge kiln, as is customary in cement and limestone kilns, because of the detrimental effect of silica in the ash. This would remain in the white liquor after slaking. Coal can be fired in a gas producer but this increases capital cost and complicates the operation.

The interest in wood burning is increasing as the cost of fuel oil and natural gas increases. Dried, pulverized wood waste at about 15% moisture is being fired in a kiln in Sweden. It is reported that the wood ash does not contribute any detrimental effects to the lime quality.

Another development is a wet cell burner developed by Lamb-Cargate which burns wet hog fuel in a specially designed cell and the hot gases at 2000-2300°F (1100-1250°C) can be directed to the kiln firing hood. The water vapor from the wet wood increases the gas volume to the kiln and it is estimated that the heat from the wood burning can replace up to 50% of the fuel oil or natural gas requirement. Figure 6.1 illustrates the wet cell burner principle.

Kiln lining

In the hot end, through the calcining zone for a distance equal to 8 to 10 times the internal diameter of the kiln, 6 in. (150mm) and preferably 9 in. (230mm) 70% alumina firebrick for kiln lining is recommended. The balance of the kiln is usually lined with 6 in., or sometimes 9 in., 40-55% alumina firebrick, except that the brick lining may be omitted in the chain section,

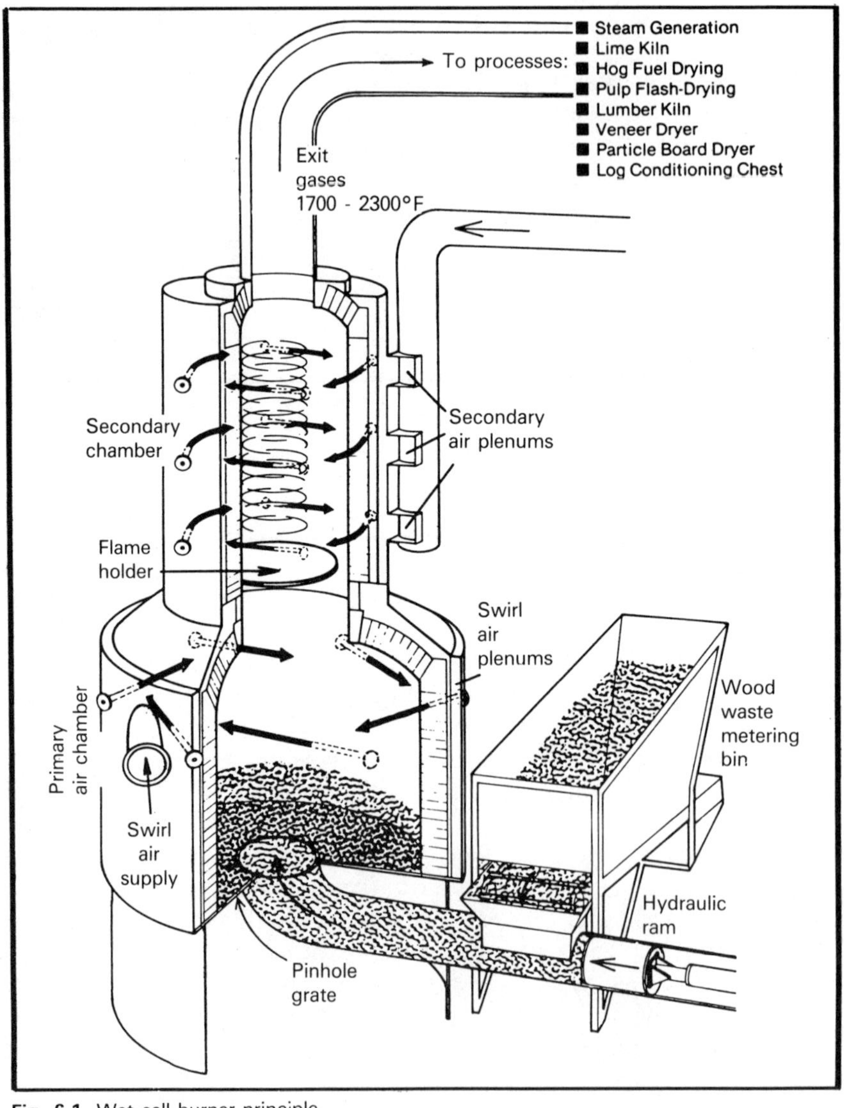

Fig. 6.1. Wet cell burner principle.

where a castable lining may be substituted for chain refractory. The firing hood is lined with first quality firebrick or firebox refractory.

Until rcently, insulating brick between the lining and the shell has rarely been used in the United States. Because of the current great interest, a trend to use various types of insulating materials under the refractory brick has occurred. In Europe, Moler Diotherm insulating brick is sometimes used between the lining and the shell. This is a Danish brick with good crushing strength.

New linings, or repairs to linings, must be thoroughly dried out before the kiln is started. The lining can be dried by keeping a small fire going in the kiln for about 48 hours. The kiln should be rotated 180° every 15 minutes during this period.

Thrust mechanism

Although the normal downhill thrust of the kiln is accommodated by cutting the rollers, it is also necessary to provide a thrust mechanism to hold the kiln in proper position on the supporting rollers. This mechanism consists of two bevel-faced, cast-steel thrust rollers in contact with the bevelled edge of the thrust riding ring.

The thrust riding ring establishes a vertical plane from which thermal expansion takes place. When heated, the kiln shell will expand several inches longitudinally. Provision for this expansion is provided by properly locating the support rollers. The thrust riding ring is always located closer to the cooler, feed end of the kiln adjacent to the main gear.

Great care should be taken in the alignment of the kiln, and after proper alignment, in proper adjustment of the rollers so that the kiln "floats" between the upper and lower thrust rollers. All rollers should be adjusted identically. Improperly adjusted rollers cause abnormally high wear on both rollers and tires. The alignment of a rotary kiln is illustrated in Figure 6.2.

Kiln components

A modern lime sludge kiln with various arrangements is illustrated in Figure 6.3.

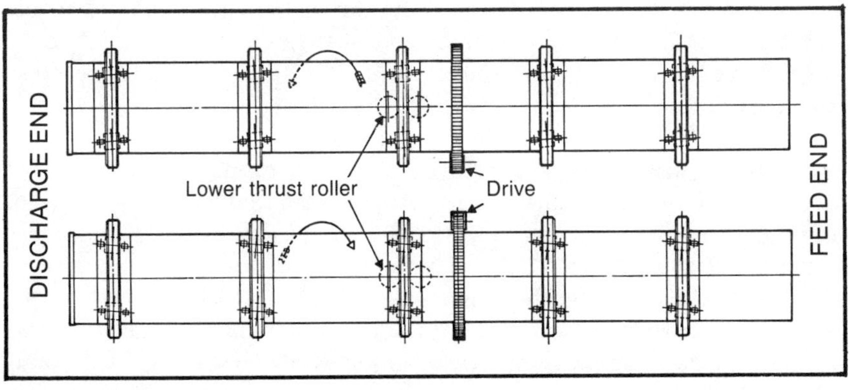

Fig. 6.2. Kiln adjustment.

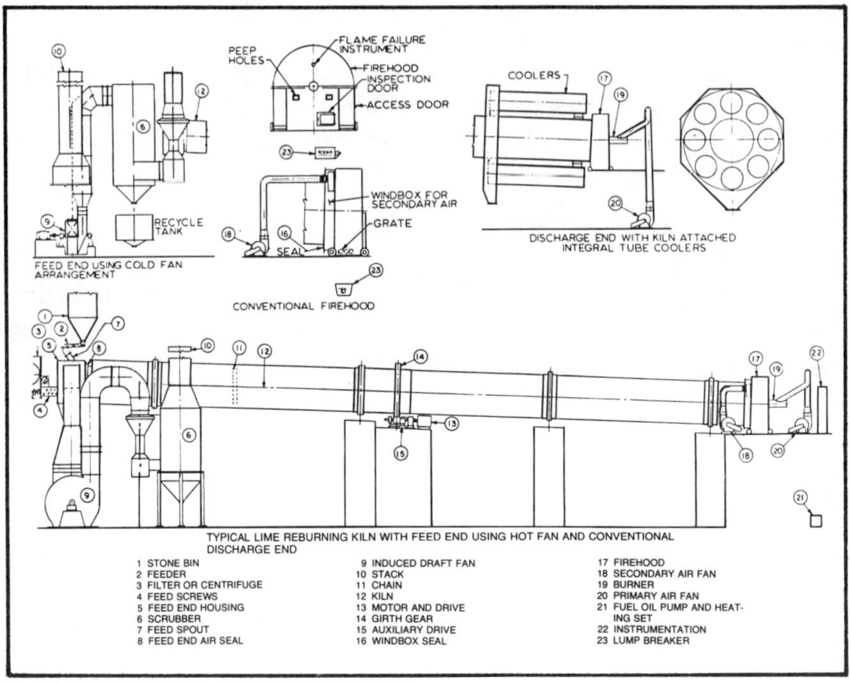

Fig. 6.3. Lime sludge kiln and various arrangements.

Burner

Figure 6.4 shows a typical combination gas and oil burner arrangement. When natural gas is used, oil is usually provided for standby purposes for use in the event that the supply of natural gas is temporarily unavailable. Regardless of the fuel, the burner should be constructed so as to produce a long soft flame such that the release of heat to the material is spread out over a considerable length within the discharge end of the kiln, and this is of major importance in preventing overburning and in promoting a highly reactive, soft burned lime.

The burner should be externally adjustable so as to vary the flame pattern, and both primary and secondary air fans should have generous safety margins in both pressure and volume characteristics, so that the relative proportions of primary and secondary air may be altered as may be necessary to produce the proper flame pattern. Both the burner and fans for the burner must be adequately sized so as not to restrict the capacity of the kiln. If the burner is properly constructed, variation of the flame pattern where gas is used is a relatively easy matter, and almost any length of flame can be readily obtained.

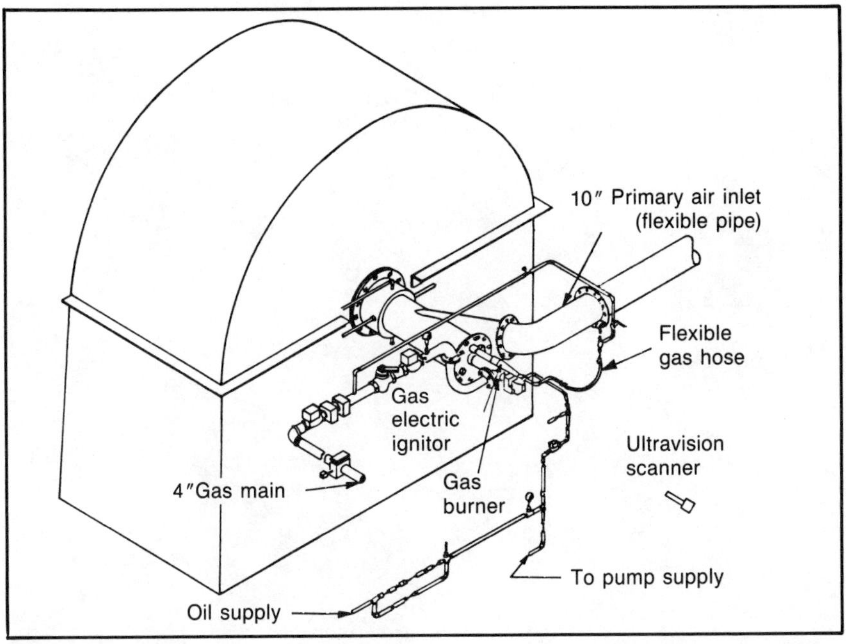

Fig. 6.4. Combination gas and oil burner.

Firing hood

In rotary kilns, a firing hood mounted on wheels is necessary to close the discharge end of the kiln. The firing hood is arranged with various inspection openings, instrument openings, and provision for burner insertion. It should have a cleanout door in the front. It is highly desirable that the hood be deep enough to accommodate a man-sized side door for easy entry and exit. Such a door is of enormous value for access to the kiln, if maintenance inside the kiln is necessary, since it eliminates the necessity of disconnecting the fuel lines and ducting, and then rolling back the entire hood. Figure 6.5 illustrates the firing hood showing openings and windbox arrangement.

Kiln drive

The kiln drive is located next to the thrust ring in order to maintain a proper relationship between the main gear and the driving pinion, regardless of the magnitude of expansion or contraction of the shell. The pinion gear is mounted and keyed to the low-speed output shaft of the enclosed-type gear reducer. The high-speed shaft of the reducer is connected to an electric motor by a V-belt drive, providing flexibility in the event that a change of speed is required.

Fig. 6.5. Firehood with wind box type seal.

An auxiliary gasoline motor equipped with battery, generator, and electric starter is good insurance in the event of power failure to prevent warping of the kiln shell. It is recommended that the auxiliary drive be started every day to ensure that it is available when required.

The kiln should be provided with a variable speed drive so that ideal kiln speed may be easily established. This feature contributes to maximum kiln production at minimum fuel demand.

The proper procedure for cooling the kiln during a shutdown is described in the following paragraph.

A hot kiln should never be allowed to cool down without rotating it. This is to prevent warping the kiln shell and to allow the contraction forces on the carrying rollers to be relieved. The kiln should be rotated continuously until cool, either at low speed of the main motor or on emergency drive, whenever possible. To obtain minimum discharge from kiln during cooling, as desired during a power failure, rotate the kiln intermittently as follows:

a. Rotate kiln on emergency drive for five minutes and then stop kiln.
b. Rotate kiln 180 degrees at five minute intervals for next hour.
c. Rotate kiln 180 degrees at ten minute intervals for the second hour.
d. Rotate kiln 180 degrees at 30 minute intervals until cool.

Chain and plate systems

Since calcination of lime mud involves temperatures in the 1800-2200°F (980°-1200°C) range, some form of heat recovery system is necessary on the exit gases to effect satisfactory heat economy. Early kilns had low thermal efficiencies since the exit gases were around 1200°F (650°C) . In recent years, exit gas temperatures have been reduced to the 300°-450°F (180°-235°C) range through the use of heat-saving devices on the feed end.

The normal type of heat recovery employs chains in the feed (wet end) of the kiln. When the kiln is rotated, the heated chains dip into the wet solids and are then lifted out in a coated condition. This action, being continuous, exposes a large surface area of the feed to the hot gases. When dry, the solids fall free of the chains, which can then absorb heat from the hot gases to repeat the cycle. A straight hanging chain system is shown in Figure 6.6 and a double hanging loop chain system is shown in Figure 6.7.

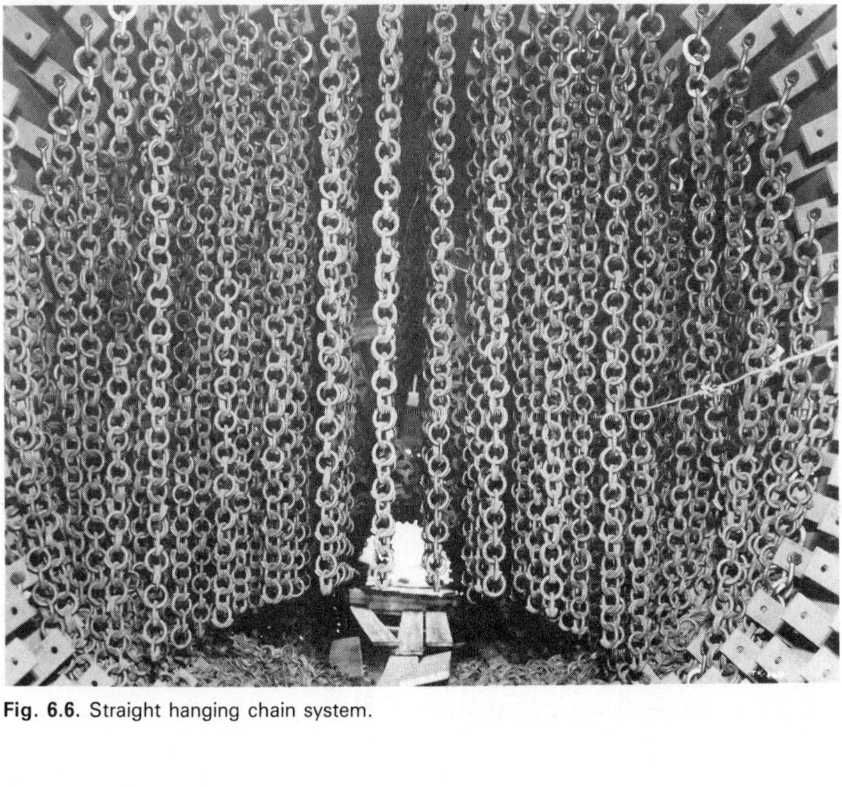

Fig. 6.6. Straight hanging chain system.

Fig. 6.7. Double hanging loop chain system.

The kiln manufacturers recommend installing 5 to 10 ft (1.5 m to 3.0 m) of alloy steel chain in the hot end of the chain section to withstand the higher temperatures and corrosion.

The chain system tends to impede the flow of the material. To offset this action, spirals are sometimes installed to convey the material through the section. The spirals provide additional heat-transfer surfaces.

An efficient heat recovery system offered by Fuller Company uses spirals and steel plates to transfer the heat from the gas to the lime sludge. The sludge is moved to one side of the plate and hot gas to the other. The sludge, when dry, falls off the plate and continues to mix with any wet material while moving down the kiln toward the discharge end. The plate system, when used in conjunction with a chain system, is an efficient heat recovery system.

Feed housing

The feed end housing is a stationary structure providing a connection to the rotating kiln shell for removing exhaust gases and introducing lime mud sludge. A seal is provided to reduce air infiltration at this point. The air seal rides on a reduced diameter section of kiln shell which decreases the area for air induction. The reduced diameter assists in eliminating backspill of raw materials into the exhaust gas system. Feed end spirals are also necessary to further eliminate any tendency of the kiln to backspill. It is important to maintain a close-fitting seal. This reduces air utilization and reduces power demands on the induced draft fan.

Lime mud screw conveyor

Spiral screw conveyors are used to transfer solids from the lime mud dewatering operation to the lime kiln. The conveyor and trough are made of carbon steel, designed to handle the sludges at 60-75% solids by weight. A water cooled jacket is used for the conveyor section extending into the kiln. The rotating elements, when properly designed, do not require water for lubrication. These screw conveyors require high horsepower for proper operation due to the "sticky" nature of the sludge being handled. Conveyors should be arranged so that it is possible to remove an entire unit from the feed hood for inspection and repairs. Edges of screws should be stellited for longer life.

Induced draft fan

There are two possible locations for the induced draft fan. Depending upon the location chosen, the fan may be referred to as a "wet" fan or a "dry" fan.

The "wet" fan is located between the scrubber and the stack and is illustrated in Figure 6.8. It is called "wet" because it handles water-saturated exhaust gases. When located in this manner, even though the gas is almost clean, there will be a buildup on the impeller due to the wet condition of the gas. This buildup must be continuously or periodically removed to prevent vibration. Such fans are often provided with water spray nozzles to wash the rotor during operation. The rotor and casing should be fabricated of stainless steel to minimize corrosion.

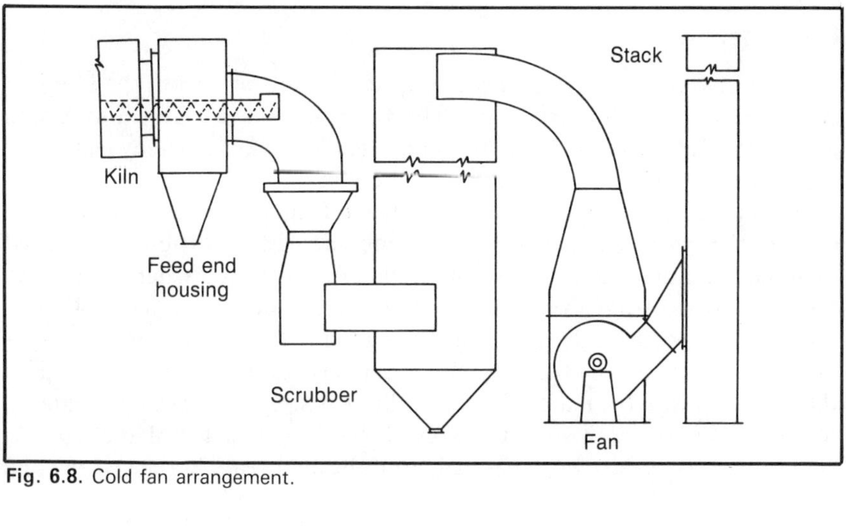

Fig. 6.8. Cold fan arrangement.

The "dry" fan, illustrated in Figure 6.9, is located in the hot gas system ahead of the scrubber and handles the dry gases as they leave the kiln. Here, the fan is not subjected to buildup of wet solids and does not have to be washed during operation. A larger capacity unit is required for the same size calciner but it may be fabricated of carbon steel. The "dry" fan should have water-cooled bearings to obtain long bearing life. A dry fan takes about 20% more power than a cold fan. However, the ease of maintenenace and lower capital cost offset the increased power cost.

All of the recent kiln installations have been provided with a dry or hot fan because of fewer operating problems encountered with the fan in this position.

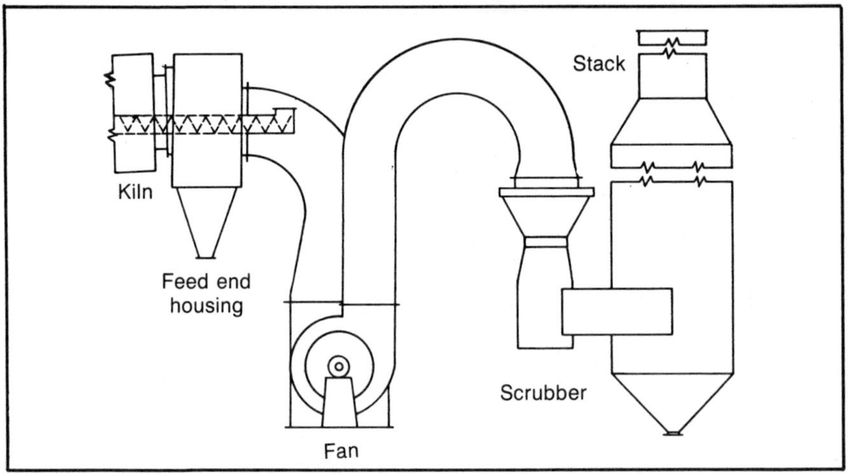

Fig. 6.9. Hot fan arrangement.

Flash drying

Flash drying systems offer another way to recover heat from the kiln exit gases and improve fuel economy. The flash dryer also offers a means to increase kiln capacity, in that dryer mud is fed to the kiln, thereby increasing the preheating and calcining zones in the kiln.

The mud at about 40% moisture is conveyed from the centrifuge or vacuum filter to the system mixer, where it is combined with a portion of the dried material from the cyclone. The conditioned feed is then discharged into the hot gas stream from the kiln. These gases are now at 1000 to 1200°F (540-1680°C), because dry material is being fed to the kiln; the drying operation having been moved from the kiln to the flash dryer. The cage mill mixes the hot gases with the material, and the hot gas stream conveys the material to the cyclone where the vapor flashes off leaving the material at about 1% water. The dry material is separated from the air stream in the cyclone and discharged into the dry divider where a portion is returned to the mixer for conditioning the feed. The remainder is fed to the kiln for calcining.

The latest design eliminates horizontal runs of duct and close couples the equipment handling and material, thereby reducing plugging problems. The supplier claims that up to 40% increase in kiln capacity is possible with a flash dryer. It is also claimed that fuel consumption per ton of lime can be reduced from 20 to 30%.

Although dusting problems are increased, an efficient dust separation and scrubbing system can maintain particulate emission within emission limits. A flowsheet of a typical flash drying system is shown in Figure 6.10. A picture of a kiln with a flash drying system is shown in Figure 6.11.

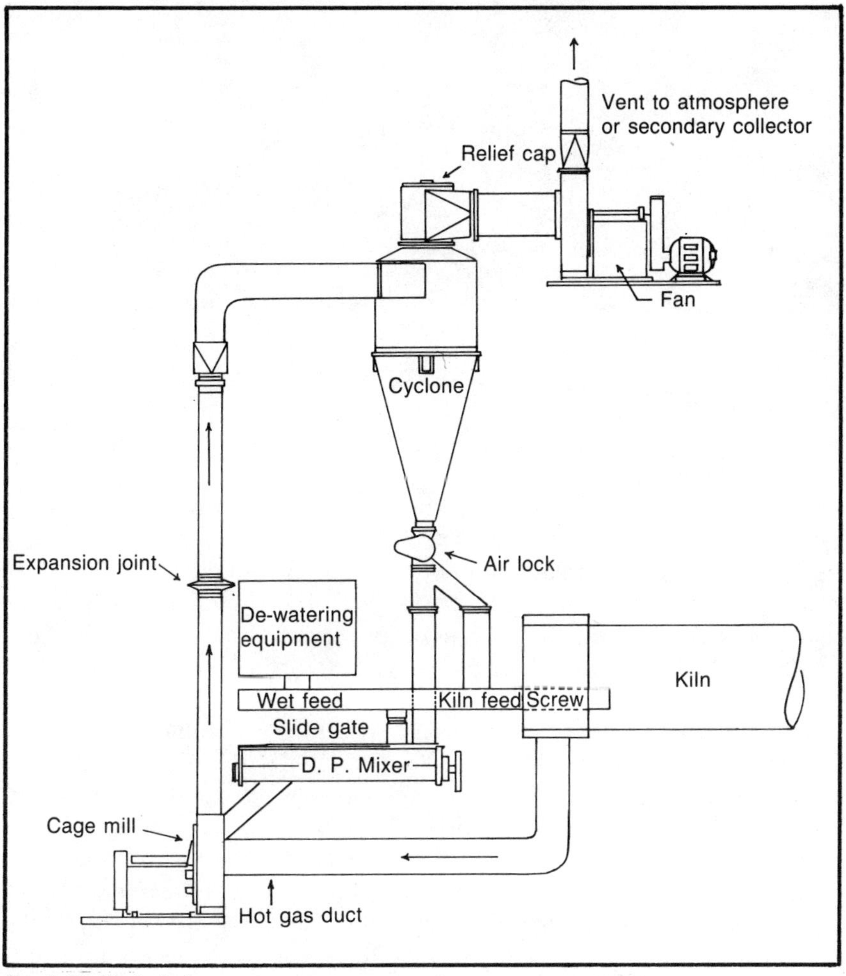

Fig. 6.10. Flash drying system.

Fig. 6.11. Flash drying system.

Dust scrubbers

Particulate emission limits require efficient dust removing equipment. Spray type scrubbers with separators are effective for removing large particulate matter, but are not sufficient for current regulations. Two wet and one dry type scrubbers are used. One wet-type system is an impingement type scrubber which utilizes recirculated water distributed across the top of impingement plates. A stream is continuously bled off to the mud washer, and makeup water is introduced by means of sprays in the final stage of the scrubber. Solids buildup on the plates occurs necessitating periodic shutting down for cleaning.

The most common wet-type scrubber is a venturi type requiring a relatively high pressure drop across the throat to ensure intimate mixing of the flue gas and recirculated water. Several hundred gallons per minute are circulated from a collection sump located either in the lower section of the cyclone separator or in a separate tank under the separator. Normally, a separate pump transfers a portion of the dust sump water to the lime mud washer in the causticizing plant, and makeup water is added at the suction of the recirculating pump. The venturi scrubber arrangement is shown in Figures 6.8 and 6.9.

In the wet-type scrubber, careful consideration should be given to the use of stainless steel for the wetted parts. This is particularly true when the fuel contains a relatively high value of sulfur, which will form sulfurous acid in the scrubber from the sulfur dioxide released in the flue gas.

The electrostatic precipitator may be used as a dry scrubber for kiln flue gas. This is a recent development which was previously not considered because of higher capital cost.

Coolers

A recent development in the U.S. paper industry is the use of kiln-attached integral tube coolers for partial cooling of the lime and for return of this recovered heat to the kiln. If coolers are designed to cool to between 600°-700°F (315°-370°C), the lime will still slake well, and at the same time, there will be a recovery of approximately 500,000 Btu per ton of product. Figure 6.12 shows such coolers on a kiln under construction. Figure 6.13 shows another kiln with integral coolers.

Fig. 6.12. Kiln-attached integral tube coolers.

Fig. 6.13. Kiln with integral coolers.

These coolers are attached to the discharge end of the kiln in such a way that the calcined lime will fall into the coolers at the extreme discharge end of the kiln, and, once in the coolers, will reverse direction of flow and be propelled through heat exchanging arrangement uphill to the opposite end of the cooler, where it will have cooled to the desired temperature. Figure 6.14 shows the grates at the discharge end of the kiln where the hot lime passes into the cooling tubes. A spiral attached tube cooler is illustrated in Figure 6.15. Secondary air required by the burner to complete combustion (about 70% of air requirements) is induced, and enters the coolers through openings at their discharge end. This air, in the heat exchange process, is heated to between 500°-700°F (260-370°C). The heated air then enters the kiln, combining with the fuel and primary air to complete combustion. Such coolers are very efficient and return 70-75% of the heat given up by the lime; the remainder is lost to radiation.

One of the operating problems in lime kilns in the paper industry is plugging of attached tube coolers. Figure 6.16 is a free flowing design recently developed and successfully used.

Cooler tubes will vary in diameter from 3 to 5 ft (1m to 1.5m) and in length from 10 to 20 ft (3m to 6m). Coolers may be designed for either side discharge or bottom discharge. Hand-operated dampers at the discharge end of the

Fig. 6.14. Discharge end of kiln before bricking, showing grates at entry to coolers.

coolers are of assistance in controlling air rates through the coolers. Heat resisting grates are constructed to limit the size of the material passing into the coolers at points of material entry.

A kiln generally has six to ten such cooler tubes. Some of the oversized material breaks up of its own accord, but those lumps which do not break up can be withdrawn through an opening in the kiln fire head or fire hood. A fire head is a circular metal cover lined with refractory, suspended from above, and designed to bear against a wearing plate at the discharge end of the kiln. It has openings similar to that of a firing hood, and forms an air seal. Alternatively, a shallow fire hood may be used with a labyrinth seal. A fire hood is more difficult to seal against tramp air, but can accommodate oversize lumps more readily. Lumps are allowed to accumulate in the fire hood and are withdrawn as desired. Admission of tramp air will decrease the efficiency of coolers and every effort must be made to seal effectively.

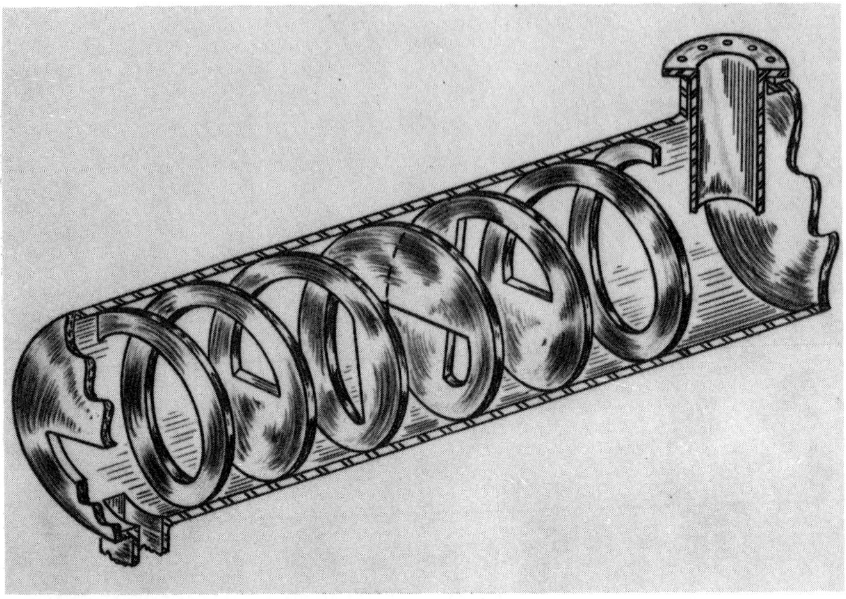

Fig. 6.15. A spiral attached tube cooler.

Crushers

Lime slakers can usually handle 1 to 2 in. (25 to 50 mm) chunks of lime. Protective grizzlies ahead of the hot lime conveyors have been adequate in a great many mills where the quantity of oversize is nominal. Hot lime crushers are often used in modern mills to eliminate the need for removing and manual breaking up the large lumps (Figure 6.17).

These crushers must be rugged units designed to cope occasionally with large pieces of kiln lining as well as chunks of lime. The most widely used have a water-cooled hopper and rotating shaft. Moving cutter blades are rotated by a heavy-duty drive train. The fixed and moving teeth are arranged to pass a maximum 2 in. (50mm) wide piece. A picture of a crusher is presented in Figure 6.18.

Provisions may be made to stop the crusher through a torque-controlled coupling and warning signal in the event of an overload of coarse material. It is advantageous to equip the crusher with means for reversing the direction of the blades to reject excessive accumulations of lime. High-speed crushers are not recommended because of the dust nuisance.

Lime handling

Drag conveyors, oscillating conveyors, pan-type conveyors, and bucket elevators are used to transport hot lime from the kiln discharge to the hot

Fig. 6.16. Feed end internal construction of kiln-attached integral tube cooler.

lime storage bin. Hot lime handling is a severe service and equipment has to be adequately sized and designed. Hot lime temperature ranges from 500°-800°F (260°-430°C) if discharged from a cooler and to 1800°F (980°C) if discharged directly from the kiln. Particle size ranges from dust to lumps the size of a basket ball. The grizzly in the discharge hood of the kiln limits the size of chunks which can reach the conveyor, and crushers are installed to reduce large chunks to approximately 2 in. (50mm) maximum. Hot lime density ranges from 60 to 75 lbs/ft^3 (900-1200 kg/m^3).

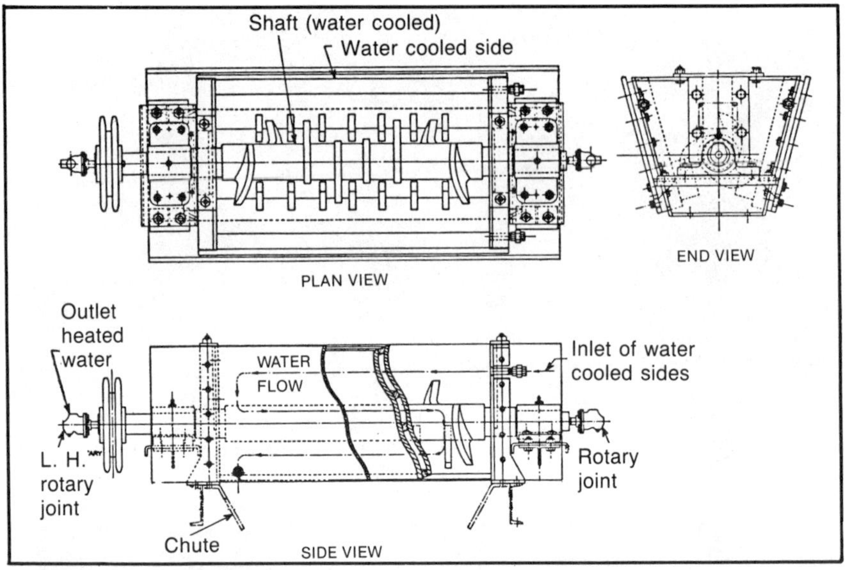

Fig. 6.17. Hot lime crusher.

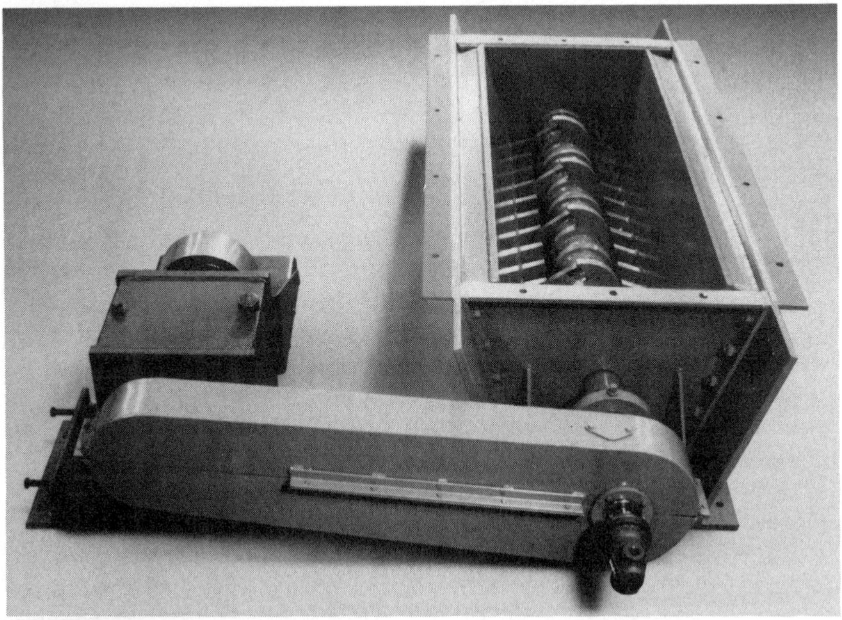

Fig. 6.18. Crusher.

Drag conveyors are built for heavy duty and normally have a water cooled bottom. Care has to be exercised to keep water from leaking into the lime. The conveyor has extra heavy bottom plates and wear strips which are replaceable.

Oscillating conveyors are limited in length of transfer to about 50 ft (15m). They tend to be dusty and require more maintenance than other types.

Pan conveyors are most commonly used as they are rugged, wear well, withstand the hot lime temperature, and, if properly sized, require minimum maintenance.

Hot lime conveyors are normally sized to convey three times the average design production of the kiln to allow for surges which can be caused by hangup in the kiln.

A specially designed lime bin can be used to store both hot lime and fresh lime. A center tube is used to contain the hot lime. When the kiln production rate exceeds the slaking rate, the tube fills up and hot lime spills over to the annular space around the tube. When the kiln production rate is less than the slaking rate, the lime around the periphery of the tube feeds along with the hot lime in the tube. The bin is self-compensating, as the material feeds down the center in preference to sliding down the conical bottom. This cone bottom bin is located above and immediately adjacent to the slaker. Figure 6.19 illustrates the design of the combination hot lime and fresh lime bin.

Instrumentation

Tremendous progress has been made in instrumentation. The early kilns had only one instrument, and this was the operator who used his eyes to see the quality of the product and the temperature of the lime, and who used his muscle to operate valves and dampers. Modern instrumentation can be very sophisticated, and it is now possible to completely automate the kiln system. One of the most significant improvements has been the control of the feed in both rate and moisture content.

Several instruments and control loops are considered essential. They are listed as follows;

1. Mud flow and density control.
2. Temperature measurement at least at feed end housing and at discharge of the kiln. Thermocouples at the discharge of the chain section and in the calcining zone are very helpful.

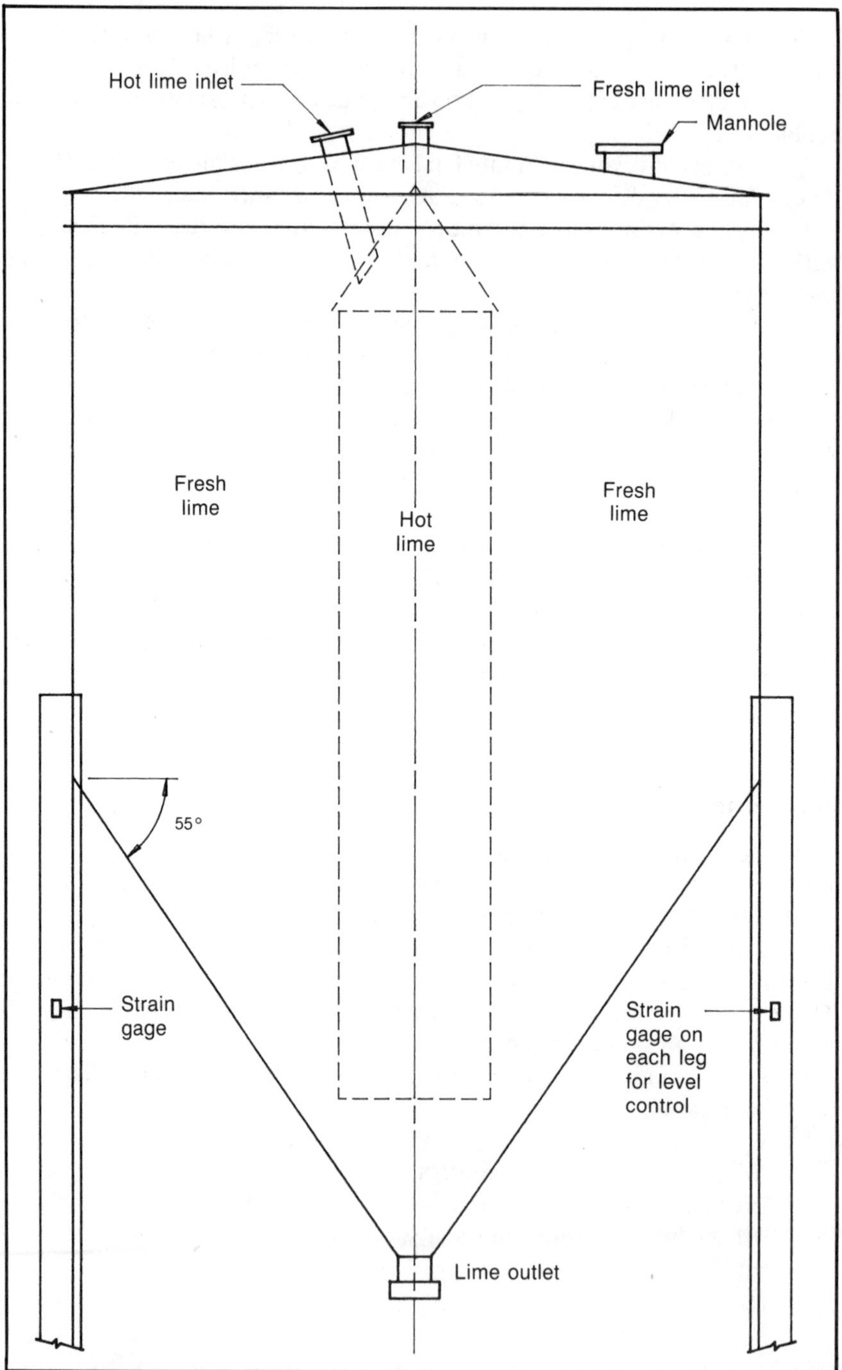

Fig. 6.19. Combination lime bin.

3. Draft measurement in the discharge hood and pressure differential across the scrubber.
4. Fuel flow and pressure control.
5. Primary air and secondary air flow control.
6. I.D. fan speed control.
7. O_2 and combustible recorder for the flue gas.
8. TRS gas emission recorder.

Flame safety controls are provided to shut off the fuel in the event of stoppage of the kiln, air fan or I.D. fan.

Automatic ignition and controls are provided for the fuel system.

The most important operating variables to be watched for deviations from normal or uniform kiln operations are:

- Lime mud density and flow
- Kiln exit gas temperature
- Firing hood draft
- Firing temperature (radiation pyrometer)
- Kiln exit gas O_2 or combustibles content
- Scrubber differential pressure

It must be realized that the reaction time of a rotary kiln system is long and there is a large quantity of lime mud and lime in process within the system at any given time. Therefore, the operator has sufficient time to take corrective action when trends of change in the operating variables are detected from the instrument readings. Changes to compensate for variables should be made in small increments, allowing sufficient time to observe the effect after each change has been made. Overreaction leads to overcompensation, which generally causes cycling and upsets which may take hours to bring back to normal operation.

Testing

It is important that the lime be sampled and tested for reactivity. Tappi Test Methods T617 "Analysis of Lime", and T618 "Analysis of Limestone" outline the procedures for sampling and testing these materials. Many mills check the slaking time of a weighed amount of lime with a measured amount of water. The lime mud dryness and soda content should be checked periodically. Tappi Useful Method 616 presents the Analysis of Kraft Lime Sludge.

Air emission control

The Federal Environmental Protection Agency (EPA) new source standards for emission limits from lime kilns in the pulp and paper industry set a performance standard of 8 parts per million (ppm) by volume of total reduced sulfur (TRS) expressed as sulfur and corrected to 10% oxygen. In May of 1979, the EPA promulgated guidelines for retrofitting emission controls to existing kraft mills and established 20 ppm TRS from existing lime kilns. Most of the states have adopted the 20 ppm as an acceptable limit for TRS from existing lime kiln.

The most critical factor in TRS emissions from the lime kiln is the residual sulfides contained in the mud fed to the kiln. Sulfides in the feed are stripped by the CO_2 in the kiln flue gas and discharged as hydrogen sulfide (TRS), a very odorous gaseous compound which has a smell similar to rotten eggs or cooked cabbage.

The quantity of residual sulfide in the mud is established in the mud washing and thickening operation. A soluble sulfide content of less than 0.1% on the dry mud weight is required to meet the 20 ppm TRS emission limit. This is possible with properly sized and operated mud washers and vacuum thickeners. Oxidation of the sulfides in the mud with air, either through spargers in a tank or on an oversized filter, may be required to achieve the 8 ppm TRS limit for new kilns.

Operation of the kiln can also affect TRS emissions. Excess oxygen in the flue gas is required and the more sulfides introduced in the mud, the higher the excess O_2 must be. An upper limit appears to be 4-5%. Minimum kiln cold-end temperatures appears to be about 450°F (230°C).

The use of untreated contaminated condensate from the digester and evaporator plants in the kiln scrubber, as showers on the mud filter, or part of the lime mud dilution will contribute to TRS emissions and is not recommended.

Good control of the kiln operation through uniform feed and firing conditions will aid in controlling emissions.

Burning pulp mill noncondensible gases

Noncondensible gases containing reduced sulfur compounds from the digester and evaporator plants are often burned in the kiln as a means of controlling odorous emissions from those sources. The warm, moist gas can be introduced into the primary air fan and thus discharged directly into the fuel stream at the burner. Since only 10-20% of the total air is introduced as primary air, care has to be taken that the oxygen supply is not reduced below an acceptable limit for proper burning of the kiln fuel. The limit appears to be 17% by volume. Below this amount the flame becomes hard to control and it becomes difficult to maintain kiln temperatures. Noncondensible gases are often diluted with 20 parts air to 1 part noncondensible gas prior to introduction into the burner in order to prevent explosions. If tube coolers are used, the introduction of diluted noncondensibles with the primary air fan is not practical, because the cooling tubes would be starved for secondary air. In this case, the noncondensible gases are injected directly into the burner flame through a separate opening or nozzle within the burner assembly.

Backup flame safety devices are required when burning noncondensibles, and the noncondensible gas line should be fitted with flame suppression and venting devices. Controls are required to divert the noncondensibles to a backup incinerator or to the atmosphere in the event of a kiln outage. There appears to be no detrimental effect on the kiln operation from injection of noncondensible gases, provided the oxygen required for combustion is not reduced to a critical level.

It is estimated that approximately half of the sulfur is recovered as calcium sulfate in the lime and the rest in the scrubber, thereby returning to the mud washer as wash water.

The noncondensible gases tend to mask the burner flame, and two flame detection devices are used to ensure proper detection of the flame for control of the flame failure system.

The Fluid Bed Calciner

Dorr-Oliver, Inc.

Introduction

The application of the fluid bed principle to calcining is a relatively recent development. The initial work was conducted on water softening sludge in the mid-1950s when the first commercial unit was installed for the city of Lansing, Michigan. Lime mud reburning for the pulp and paper industry followed with the first installation at the S.D. Warren mill in Muskegon, Michigan. Currently, twenty installations at capacities up to 250 tons per day of lime are reported to be in operation or under construction.

Fluid bed calciners are vertical units, in contrast to rotary kilns, which are essentially horizontal. Consequently, substantially less space is required for the fluid bed calciner.

When solids of a uniform size are suspended in a gas stream, their flow properties are similar to liquids. This dynamic mixture of solids and gases is known as a fluid bed. Fluidization requires that the proper gas velocity be maintained. If the velocity is too low, the particles settle and the bed defluidizes, and if too high, the particles will obviously be conveyed upward and into the exhaust stack.

Solids in a fluid bed are in constant turbulent motion of a random nature. The turbulent motion ensures thorough mixing and intimate contact between the solids and gases. As a consequence, heat transfer is rapid and temperatures are quite uniform (fuel combustion, solids, and exit gas are all essentially the same). Thus, the fluid bed is an ideal means for transferring heat from the combustion of fuel to the product that is to be calcined.

In fluid bed calcination, the fine calcium carbonate particles pelletize because of the action of a small amount of alkali present in the feed, or otherwise added in the form of soda ash. Without this pelletization, fluid bed calcination would not be practical due to the small size of the solids (1-30 microns). Pellets form up to 1/8 in. or so in diameter as the molten alkali at calcining temperatures 1600°F (880°C) enables the newly formed CaO particles to adhere to others, gradually building up in size.

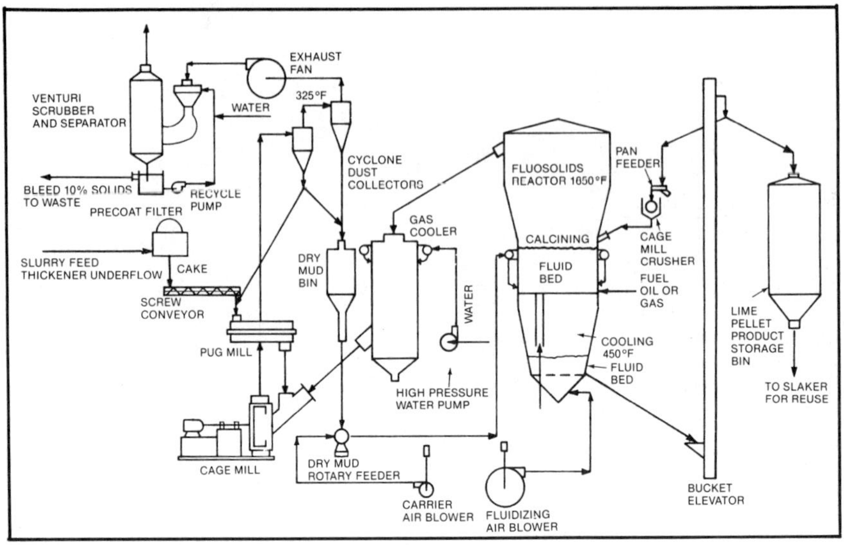

Fig. 6.20. Typical lime mud reburning system.

A typical lime mud reburning system is shown in the flowsheet (Fig. 6.20). The system is divided into two basic parts. The first is the drying system, where the wet lime mud filter cake is dried to form dry feed for the calciner. The second part is the calcination system which produces uniform pellets of the calcined product.

Drying system

Lime mud from the recausticizing system is fed to a rotary vacuum filter where the mud is washed for final reduction of alkali content and dewatered to 60-70% solids. The filter cake is conveyed to a paddle mixer, where it is mixed with dry recycled calcium carbonate along with a small amount of water as required for the control of the drying system temperature. Moisture in the paddle mixer discharge is maintained at 10-15%. The lime mud from the paddle mixer is fed to the cage mill along with the calciner gases where the moist solids are dried and agglomerates dispersed. The gases sweep the fine dry solids out of the cage mill through a duct to two-stage cyclone collectors. Part of the collected dust is recycled to the paddle mixer and part is gravity-fed to the calciner feed bin.

The cyclone stack gases pass through an exhaust fan into a wet scrubber, where they are cleaned before they are discharged to the atmosphere. Scrubber effluent containing fine lime hydrate and carbonate may either be sent to the lime mud washer or to a solid bowl centrifuge for recovery of solids. After centrifuging, the solids may be fed directly to the paddle mixer, avoiding a recycle load to the lime mud filter.

Calcination system

The calcination reactor is a two-compartment vessel: the top bed for calcining the carbonate and pelletizing the calcium oxide, the bottom bed for cooling and heat recovery. A suitable blower is used to supply air to the reactor for fluidization and fuel combustion. The dry carbonate feed is metered by a special rotary valve to an air-swept pipeline, and is then blown into the bottom part of the calcination bed through a series of "feed-guns" located around the periphery of the reactor. Heat for the calciner is obtained by direct bed burning of a fuel which is distributed through "fuel-guns" also located around the periphery of the reactor. The calciner can be equipped to burn either heavy No. 6 oil or natural gas. Complete combustion of the fuel is obtained by maintaining a slight excess of air over stoichiometric requirements. It is important to note that in the use of heavy fuel oil, the oil is not atomized in the conventional manner, nor is there concern with conventional flame propagation resulting from the mixing of fuel and air. Fuel burns in a fluid bed without visible flame; combustion being accomplished as a result of a turbulent mixing of air, fuel, and solids at an elevated temperature.

Simultaneous calcination and pelletization of the carbonate feed occurs in the calcining bed maintained at 1600°F (880°C). The space above the fluid bed, called the freeboard, is expanded in area to decrease the velocity of the uprising gases to allow entrained solids to disengage from the gas stream and fall back into the fluid bed, where they act as nuclei for pellet growth. If insufficient solids are present for pellet nuclei, a portion of calcined pellets is recycled through a roll crusher for re-entry into the reactor.

The flow of calcined lime pellets (-4 mesh to +65 mesh in size) through an internal underflow pipe is regulated at a rate that is automatically controlled to maintain a predetermined bed depth. Within the cooling compartment, the hot pellets are cooled and the incoming air heated to a temperature of 400-600°F (205-315°C). The discharge rate of cooled pellets from this chamber is controlled to constantly maintain the depth of the cooling bed. The product pellets, conveyed to a product bin, are then ready for reuse in the causticizing system. Figure 6.21 is a schematic diagram of a fluid bed calciner.

Instrumentation and control

For proper and efficient operation, calcination temperature, drying system temperature, and fluid bed levels are accurately controlled.

The calcination temperature is maintained at 1600°F ± 10°F (880°C ± 5°C) by a temperature recording controller. The air flow is set for a given capacity, and the fuel rate manually adjusted to optimum fuel air ratio for efficient fuel combustion. The temperature controller automatically adjusts

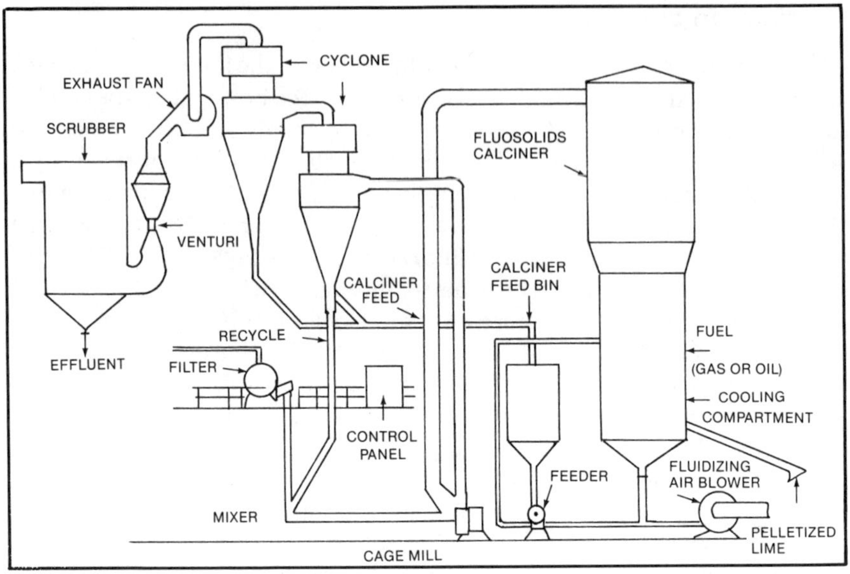

Fig. 6.21. Schematic diagram of a fluid bed calciner.

the carbonate feed rate to maintain calcination temperature at the desired preset level.

The drying system temperature is held between 135° and 175°C (275° and 350°F) by automatic adjustments of temperature control water added to the paddle mixer.

The two fluid beds (calcining and cooling) are maintained at the desired levels by bed level recording controllers. Pressure taps sense the fluid bed depth by measuring a differential pressure across the fluid bed. This pressure signal is transmitted to the level controller that automatically adjusts the transfer valve controlling pellet flow from the bed, thereby maintaining the fluid bed at the desired preset level or depth.

To assist the operator, pressures and temperatures at selected places throughout the system are measured and recorded.

A preheat burner firing into the freeboard region of the cooling chamber of the calciner is used in startup operations to heat the calcination bed to fuel combustion temperature.

A fluid bed calciner can be operated by one person, as controls are centralized and the process is readily controlled. The system can be rapidly shut down in 15 to 30 minutes, since the reactor contents do not have to be removed. Restarting requires about one hour even with shutdowns up to as long as two days. A cold reactor, on the other hand, requires about six to twelve hours in order to reach the calcining temperature.

Performance and operating problems

The fluid bed system can be operated at varying production rates (70-100% of design capacity). Where future expansion is contemplated, oversize reactors can be installed and "bricked in" to handle the current production requirement.

The fluid bed calciner produces a pelletized product of good uniformity. It is reported that the grit volume after slaking is quite low and lime availability is in the 85-88% range. Uniform calcining temperatures primarily are responsible for these characteristics. Overburning is essentially eliminated in the fluid bed unit due to the low temperature operation and the close temperature control throughout the calcined bed.

The heat consumption depends upon the manner of operation. With proper control, the heat requirement per ton of product need not exceed 7.0 million Btu. Power consumption is higher with fluid bed units than ordinary kilns, due to the power required for the fluidizing air flow. The total power requirement, including all auxiliaries, is on the order of 125 kw per ton of reburned lime.

The rotary feeder and paddle mixer blades are subject to severe wear and require rebuilding every six months. Normally, spare rotating elements are kept in stock for replacement of the worn units to minimize downtime. The worn elements can be rebuilt and returned to stock.

A slag formation can build up in the reaction chamber from impurities in the mud feed. This buildup can necessitate shutdown two or three times per year for clean out of the unit. Total time for cooling down, cleanout and startup is two to three days.

Air emissions

The control of particulate emissions from the fluid bed calciner system is with a wet scrubber, usually a venturi type, after the dry cyclone separators. Emission limits are readily controllable below EPA standards.

TRS emissions from the fluid bed calciner system are reported to be lower than from conventional rotary kilns, and thus, meeting EPA standards appears to be no problem.

Index

For information on publications of related interest, or for a catalog of new publications, please write:

TAPPI PRESS
Technology Park/Atlanta
P.O. Box 105113
Atlanta, GA 30348
USA

or call the TAPPI PRESS book sales department at (404)446-1400.